Color Atlas of Pediatric Anatomy, Laparoscopy, and Thoracoscopy

Merrill McHoney • Edward M. Kiely
Imran Mushtaq
Editors

Color Atlas of Pediatric Anatomy, Laparoscopy, and Thoracoscopy

Springer

Editors
Merrill McHoney
Paediatric Surgery
Royal Hospital for Sick Children Edinburgh
Edinburgh, United Kingdom

Edward M. Kiely
Great Ormond Street Hospital for Children
London, United Kingdom

Imran Mushtaq
Great Ormond Street Hospital for Children
London, United Kingdom

ISBN 978-3-662-53083-2 ISBN 978-3-662-53085-6 (eBook)
DOI 10.1007/978-3-662-53085-6

Library of Congress Control Number: 2017930186

Printed on acid-free paper

This Springer imprint is published by Springer Nature
The registered company is Springer-Verlag GmbH Germany
The registered company address is: Heidelberger Platz 3, 14197 Berlin, Germany

Foreword

This beautifully illustrated Atlas will, I believe, prove invaluable to paediatric surgeons at all stages of their career. Those who are already proficient in minimal access surgery will find it useful in planning a new procedure as well as rehearsing more familiar ones. Trainees will benefit from it as they develop their surgical skills. Many paediatric surgeons have been slow to adopt minimal access techniques, continuing to perform "conventional" open surgery for most procedures. This book will hopefully stimulate those yet to be convinced, to ensure that they acquire the appropriate skills and adopt these techniques.

The authorship of this Atlas is international, from the United States, Europe, the Far East, and the United Kingdom, including, I am proud to note, several from Edinburgh, one of the first paediatric centres in the United Kingdom to pioneer these techniques. The operative stages of each procedure are clearly illustrated in a step-by-step sequence to aid understanding and facilitate successful completion of the entire operative procedure endoscopically.

Endoscopic surgery, mainly laparoscopy and thoracoscopy, not only offers benefit to the patient (improved cosmesis, less pain, less post-operative ileus, shorter hospital stay, faster recovery, etc.), but it is also advantageous to the surgeon. The view of the operative site, as seen in this Atlas, is usually far superior to that obtained through an open incision. The light is better and the magnification of the image on the screen gives a much clearer view of the detailed anatomy. I am certain that the next generation of paediatric surgeons will look back and say, "Did they really make those large, unsightly and disfiguring incisions when the same procedure can easily be performed using minimal access techniques?"

"In the 21st century it is unacceptable to perform any surgical procedure on a child by the open route if it can be safely and easily be carried out through minimal access surgery" (MacKinlay GA, BAPS Liverpool 1999). When I made this statement, it was considered heretical, but increasingly in the past two decades, the minimally invasive approach has evolved into routine paediatric surgical practice. This volume will likely become an essential component of every paediatric surgical department.

Edinburgh, UK — Gordon A. MacKinlay

Preface

We have been privileged to have worked with trainers and colleagues who have encouraged the development of minimally invasive surgery in children. The popularity of minimally invasive surgery in children is increasing, as well as the need for an atlas to help with the step-by-step approach to common operations, not only for those who are learning but also more advanced practitioners needing refreshing pictorial tips or reminders. This is not a textbook with details of disease pathology, clinical presentation, or even indications for surgery for each procedure. Instead, this atlas purely focuses on the operative steps once these steps have already been achieved. The impetus to edit this book has come from our desire to help students, trainees, and colleagues in developing minimally invasive surgery in children. We hope that this atlas will be of value to those who are developing those skills.

Contents

1 **Introduction and General Principles** 1
Merrill McHoney, Edward Kiely, and Imran Mushtaq

2 **Equipment and Ergonomics** 9
Alex C. H. Lee and Hugh W. Grant

3 **Thoracoscopic Debridement of Empyema** 27
Fraser D. Munro, Merrill McHoney, and Malcolm Wills

4 **Spontaneous Pneumothorax** 31
Merrill McHoney and Michael Singh

5 **Thoracoscopic Lung Biopsy and Segmentectomy** 39
Kokila Lakhoo

6 **Thoracic Sympathectomy** 45
Giampiero Soccorso and Michael Singh

7 **Thymectomy** 49
Melissa Short and Dakshesh H. Parikh

8 **Mediastinal Cysts** 55
Michael Singh

9 **Thoracoscopic Nuss Procedure** 59
Joanna Stanwell and Robert Wheeler

10 **Esophageal Atresia and Tracheoesophageal Fistula** 65
Merrill McHoney, Fraser Munro, Jimmy Lam, and Gordon MacKinlay

11 **Thoracoscopic Aortopexy** 75
Joanna Stanwell and Edward Kiely

12 **Thoracoscopy for Congenital Lung Malformations** 81
Fraser D. Munro

13 **Thoracic Neuroblastoma** 89
Michael Singh and Giampiero Soccorso

14 **Thoracoscopic Placation for Eventration of the Diaphragm** 93
Ashok Daya Ram and Michael Singh

15 **Laparoscopic Repair of Morgagni Hernia** 97
Michael Singh

16 **Primary Button Gastrostomy** 103
Clarie Clark

17 **Nissen Fundoplication** 109
Merrill McHoney

18 **Laparoscopic Thal Fundoplication** 119
Khaled Ashour, Alex C.H. Lee, and Hugh W. Grant

19 **Pyloric Stenosis** 125
Merrill McHoney

20 **Duodenal Atresia Repair** 133
Aimee Gibson and Nada Sudhakaran

21 **Meckel Diverticulum** 143
Brian MacCormack and Philip Hammond

22 **Laparoscopic Appendectomy** 147
Shabnam Parkar and Simon A. Clarke

23 **Laparoscopy for Intussusception** 161
Kate Cross

24 **Laparoscopic Cholecystectomy** 165
Augusto Zani and Niyi Ade-Ajayi

25 **Laparoscopic Splenectomy** 173
Khalid Elmalik and Sean Marven

26 **Laparoscopic Surgery for Choledochal Cysts** 183
Nguyen Thanh Liem

27 **Laparoscopic Colectomy** 189
Maurizio Pacilli and Hugh W. Grant

28 **Laparoscopic Duhamels Pullthrough for Hirschprung Disease** 201
Merrill McHoney

29 **Laparoscopic-Assisted Soave Pullthrough for Hirschsprung Disease** 211
Amanda J. McCabe

30 **Laparoscopic-Assisted Swenson-Like Transanal Pullthrough for Hirschsprung Disease** 225
Michael Stanton, Bala Eradi, and Marc A. Levitt

31 **Inguinal Hernia Repair** 235
Merrill McHoney

32 **Undescended Testis: Laparoscopic Fowler-Stephens Orchidopexy** 241
Francisca Yankovic and Naima Smeulders

33 **Laparoscopic Varicocelectomy** 253
Vassilis J. Siomos, Cole Wiedel, and Duncan T. Wilcox

34 **Retroperitoneoscopic Adrenalectomy** 259
Imran Mushtaq and Francisca Yankovic

35 **Laparoscopic Nephrectomy** 265
Pankaj Kumar Mishra and Abraham Cherian

36 **Retroperitoneoscopic Nephrectomy/Heminephrectomy** 271
Jimmy Lam

37 **Pyeloplasty** 279
Harish Chandran

Index 291

Contributors

Niyi Ade-Ajayi Department of Paediatric Surgery, King's College Hospital NHS Foundation Trust, London, UK

Khaled Ashour Department of Paediatric Surgery, John Radcliffe Hospital, Oxford University, Oxford, UK

Harish Chandran Department of Paediatric Surgery and Urology, Birmingham Children's Hospital, Birmingham, UK

Abraham Cherian Department of Paediatric Urology, Great Ormond Street Hospital, London, UK

Melanie Claire Clark Royal Hosptial for Sick Children, Edinburgh, UK

Simon A. Clarke Department of Pediatric Surgery, Chelsea and Westminster NHS Foundation Trust, London, UK

Kate Cross Paediatric Surgery Department, Great Ormond Street Hospital for Children NHS Foundation Trust, London, UK

Khalid Elmalik Sheffield Children's Hosptial, Sheffield, UK

Bala Eradi Department of Paediatric Surgery, Leicester Royal Infirmary, Leicester, UK

Aimee Gibson Paediatric Surgery, Gold Coast University Hospital, Queensland, Australia

Hugh W. Grant Department of Paediatric Surgery, John Radcliffe Hospital, Oxford University Hospitals, Oxford, Oxfordshire, UK

Philip Hammond Royal Hosptial for Sick Children, Edinburgh, UK

Kokila Lakhoo Department of Paediatric Surgery, Oxford Children's Hospital, Oxford University Hospitals and University of Oxford, Oxford, UK

Jimmy Lam Department of Paediatric Surgery, Royal Hospital for Sick Children, Edinburgh, UK

Alex C.H. Lee Department of Paediatric Surgery, Oxford University Hospitals NHS Trust, Level 2, Oxford Children's Hospital, John Radcliffe Hospital, Oxford, UK

Marc A. Levitt The Ohio State University, Columbus, OH, USA

Nationwide Children's Hospital, Columbus, OH, USA

Nguyen Thanh Liem Vinmec International Hospital, Hanoi, Vietnam

Brian MacCormack Department of Paediatric Surgery, Royal Hospital for Sick Children, Edinburgh, UK

Gordon A. MacKinlay Royal Hospital for Sick Children, Edinburgh, UK

Sean Marven Sheffield Children's Hostpial, Sheffield, UK

Amanda J. McCabe Department of Paediatric Surgery, Royal Hospital for Sick Children, Edinburgh, UK

Merril McHoney Department of Paediatric Surgery, Royal Hospital for Sick Children, Edinburgh, UK

Pankaj Kumar Mishra Department of Paediatric Urology, Great Ormond Street Hospital, London, UK

Fraser D. Munro Department of Paediatric Surgery, Royal Hospital for Sick Children, Edinburgh, UK

Maurizio Pacilli Department of Paediatric Surgery, Oxford University Hospital, Oxford, UK

Dakshesh H. Parikh Department of Paediatric Surgery, Birmingham Children's Hospital NHS FT, Birmingham, UK

Shabnam Parkar Department of General Paediatric Surgery, Chelsea Children's Hospital, Chelsea and Westminster Hospital, London, UK

Ashok Daya Ram Department of Paediatric Surgery, Birmingham Children's Hospital, Birmingham, UK

Melissa Short Department of Paediatric Surgery, Birmingham Children's Hospital, Birmingham, UK

Michael Singh Department of Paediatric Surgery, Birmingham Children's Hospital, Birmingham, England

Vassilis J. Siomos Department of Surgery-Urology, University of Colorado Hospital, Children's Hospital Colorado, Aurora, CO, USA

Naima Smeulders Great Ormond Street Hospital for Children NHS Foundation Trust, London, UK

Giampiero Soccorso Birmingham Children's Hospital NHS Foundation Trust, Birmingham, UK

Michael Stanton Department of Paediatric Surgery, G Level, Southampton Children's Hospital, Southampton, UK

Joanna Stanwell Department of Paediatric Surgery, University Hospital Southampton NHS Foundation Trust, Southampton, UK

Nada Sudhakaran Paediatric Surgery, Gold Coast University Hospital, Queensland, Australia

Robert Wheeler University Hospital Southampton NHS Foundation Trust, Southampton, Hampshire, UK

Cole Wiedel Department of Surgery-Urology, University of Colorado Denver School of Medicine, University of Colorado Hospital, Children's Hospital Colorado, Aurora, CO, USA

Duncan T. Wilcox Department of Pediatric Urology, Children's Hospital Colorado, Aurora, CO, USA

Malcolm Wills Department of Cardiothoracic Surgery, Royal Infirmary of Edinburgh, Edinburgh, UK

Francisca Yankovic Department of Paediatric Urology, Great Ormond Street Hospital NHS Foundation Trust, London, UK

Augusto Zani Division of General and Thoracic Surgery, The Hospital for Sick Children, Toronto, ON, Canada

1 Introduction and General Principles

Merrill McHoney, Edward Kiely, and Imran Mushtaq

Abstract

Minimally invasive surgery (MIS) has become relatively commonplace in paediatric surgery, and is becoming more popular. Paediatric surgeons perform laparoscopic and thoracoscopic surgery with the commonly held belief that MIS is associated with a dampened stress response, more rapid postoperative recovery, and early discharge from hospital. There are also long-term cosmetic advantages. Depending on the operation in question, some of the potential advantages hold, but others do not, and we need to be conscious of potential disadvantages and difficulties when embarking on MIS.

Keywords

Minimally invasive surgery • Laparoscopy • Thoracoscopy • Retroperitonoscopy • Children

Minimally invasive surgery (MIS) has become relatively commonplace in paediatric surgery, and is becoming more popular. Paediatric surgeons perform laparoscopic and thoracoscopic surgery with the commonly held belief that MIS is associated with a dampened stress response, more rapid postoperative recovery, and early discharge from hospital. There are also long-term cosmetic advantages. Depending on the operation in question, some of the potential advantages hold, but others do not, and we need to be conscious of potential disadvantages and difficulties when embarking on MIS.

As an introduction to the rest of this atlas, this chapter discusses some of these issues (albeit very briefly) in addressing the "Why, When, Who, Where, and How" of MIS in children.

M. McHoney (✉)
Department of Paediatric Surgery, Royal Hospital for Sick Children, Edinburgh, UK

E. Kiely
Department of Paediatric Surgery, Great Ormond Street Hospital for Children NHS Foundation Trust, London, UK

I. Mushtaq
Department of Pediatric Urology, Great Ormond Street Hospital for Children NHS Trust, London, UK

1.1 Why Minimally Invasive Surgery in Children?

We can address the reasons for performing MIS surgery in children by thinking of the advantages or benefits of MIS, but we will also touch on the risks and potential downfalls. There is evidence for some of the perceived benefits of MIS in children, but some aspects lack substantial evidence at the moment. The evidence base is increasingly being accrued and investigated, however, and a few operations have been evaluated in randomised controlled trials [1].

1.1.1 Potential Benefits

1.1.1.1 Postoperative Pain and Recovery

Both thoracoscopy and laparoscopy are associated with a significant reduction in the amount of tissue trauma and thereby a reduction in postoperative pain. Studies have shown varying reductions in postoperative pain after MIS in both adults and children. Clinical evidence in adults shows that laparoscopic surgery reduces postoperative stay, respiratory complications, and postoperative pain when compared with open surgery [2, 3]. The decreased postoperative pain of tissue trauma after laparoscopy must

M. McHoney et al. (eds.), *Color Atlas of Pediatric Anatomy, Laparoscopy, and Thoracoscopy*,
DOI 10.1007/978-3-662-53085-6_1

be balanced with the possibility of shoulder tip pain, perhaps accounting for the fact that decreased postoperative pain is not always proven. Laparoscopic surgery for moderate to severely invasive operations has proven quicker recovery in many studies. Thoracoscopy in children greatly improves postoperative recovery [4, 5] and the minimisation of postoperative pain.

1.1.1.2 Cosmetic Advantages

The improved cosmesis after MIS is one of the hardest advantages to quantify and report. The exchange of large laparotomy and thoracotomy incisions for keyhole incisions is undeniably beneficial. The minimisation of the visible scar associated with the incisions in an important long-term advantage to patients.

Reduction of physical deformity, especially on the chest wall, is also very important. Long-term chest wall deformity is minimised by MIS, and sometimes is eliminated completely. Winging of the scapula, kyphoscoliosis, pectus deformities, and other deformities seen after thoracotomy are reduced by thoracoscopy [6, 7]. Although most often associated with chest wall incisions, such deformities also can be associated with large abdominal wall incisions.

1.1.1.3 Blunting of the Metabolic Response

MIS is associated with minimisation of the degree of tissue trauma (as the incisions into the body wall are smaller than in the comparable open operation) and is of benefit in reducing some of the postoperative complications by blunting of the metabolic and stress response. The cytokine response is reduced after operations of a major magnitude performed by MIS [8–10]. One of the major determinants of the metabolic response to surgery is the magnitude of the operative stress [11, 12]. Operations of greater magnitude are associated with a greater metabolic response [13]. Therefore the benefit is more pronounced when bigger operations are performed by MIS.

1.1.1.4 Thermoregulation and Energy Metabolism

There is an important association between alteration in thermoregulation and the metabolic response. In the 1960s, it was demonstrated that maintaining a 30 °C environmental temperature blunted the metabolic response to trauma and could therefore play an important role in determining the postoperative metabolic response [14]. Morbidity and mortality were also influenced by thermoregulation. Infants and children are more susceptible to alterations in thermoregulation and environmental temperature than adults. Physiological differences in thermoregulation may be partially responsible for differences between neonates, children, and adults in patterns of metabolic response.

Because MIS is not associated with large open wounds, heat loss and evaporative water loss are prevented, in turn altering thermoregulation. Studies have shown maintenance of core temperature and oxygen consumption in children undergoing thoracoscopy [15, 16] and laparoscopy [17], which was more marked in younger and smaller children. Changes in intraoperative thermoregulation may alter postoperative metabolism and changes in energy expenditure.

Luo et al. performed a trial in adults randomised to open or laparoscopic cholecystectomy [18]. Rest energy expenditure (REE), as measured by indirect calorimetry, was elevated on postoperative day 1 in both groups, but the rise in REE was significantly higher in the open group than in the laparoscopic group. Postoperative energy metabolism is also altered by laparoscopy in children, with a preservation of energy metabolism in comparison with open surgery [19]. There are possible effects on postoperative protein metabolism alongside these alterations. It seems, therefore, that MIS is associated with preservation of homeostasis with regard to energy expenditure.

1.1.1.5 Visualisation and Magnification

The visualisation obtained with MIS is often superior to visualisation with open surgery. Access to many deep recesses and folds can be improved with the use of the scope. For instance, access to the oesophageal hiatus, pelvic structures, and apical areas of the lung is greatly facilitated with MIS, compared with open surgery.

A much greater degree of magnification also can be obtained using MIS. Structures that may be difficult to see with the naked eye (e.g., the vagus nerve and its branches during fundoplication and oesophageal atresia repair) are often easily visible on the screen with the optical and digital magnification allowed with MIS.

1.1.2 Potential Hazards of MIS

1.1.2.1 Carbon Dioxide Absorption from the Surgical Cavity

One of the new dimensions introduced by MIS is the creation of a working space. This technique can involve abdominal wall lifting, but the method most commonly used is insufflation of CO_2 to create a capnoperitoneum (or capnothorax). CO_2 absorbed from the body cavity during MIS causes an increase in CO_2 elimination via the lungs. In adults undergoing laparoscopy, there is typically a brief period of increased CO_2 elimination, but after 10–30 min of insufflation, a plateau is usually reached [20]. In children, the CO_2 profile is different: there is a continuous increase in CO_2 elimination throughout intraperitoneal insufflation of CO_2 in children [21]. The increase in CO_2 elimination was more marked in younger and smaller children, suggesting that age modifies the intraoperative handling of CO_2, and the same difference was true for thoracoscopic surgery [15]. The increased CO_2 load has been calculated to be approximately 16% accounted for by absorption from the abdomen in one study [22]. In the case of thoracoscopy, nearly 50% of expired CO_2 is absorbed from the thorax [22].

Neonates are particularly prone to acidosis during thoracoscopic surgery owing to the markedly increased CO_2 load, the decreased respiratory elimination from lung collapse, and exaggerated absorption in smaller children [15, 21]. Patients with congenital diaphragmatic hernia, for instance, are also at risk of significant acidosis and secondary effects [23–25]. Thoracoscopic surgery therefore should not be performed without suitable expertise and monitoring, or if the patient is unstable.

1.1.2.2 Mechanical Effects of Carbon Dioxide Insufflation

Insufflation of CO_2 used during laparoscopy increases intra-abdominal pressure. The optimal intra-abdominal pressure for laparoscopy in children has been established to be between 8 and 12 mm Hg [26], with neonates tolerating lower pressures than older children. The increase in intra-abdominal pressure causes a rise in intrathoracic pressure, which alters respiratory dynamics and leads to impaired respiratory function, including reduced functional residual capacity, increased airway pressure, and decreased lung compliance. Absorption of CO_2 from the abdomen seems to peak about 30 min into surgery, with up to 20% of expired CO_2 derived from absorption; it decreases back to preoperative levels 30 min postoperatively [27]. During laparoscopy in self-ventilating patients, this change translates into an increase in end-tidal CO_2 and arterial CO_2 tensions [28] that can lead to acidosis.

In children undergoing controlled ventilation during laparoscopy, there is generally a good correlation between end-tidal CO_2 and arterial CO_2 pressures ($PaCO_2$) [28, 29]. If ventilation parameters are maintained at pre-insufflation values, both end-tidal CO_2 and $PaCO_2$ increase as intra-abdominal pressure increases. Occasionally, the increase in $PaCO_2$ is out of step with the increase in end-tidal CO_2 [30]. A 20–30% increase in minute ventilation is usually sufficient to compensate for the increased CO_2 load [31–33], thus avoiding an increase in end-tidal CO_2 or acidosis.

Intra-thoracic insufflation of CO_2 has different mechanical effects on respiratory dynamics than intra-abdominal insufflation. Greater impaired respiratory capacity imposed by lung collapse has significant implications for oxygenation and CO_2 excretion [34]. Thoracic insufflation of CO_2 may also have a different absorption profile than abdominal insufflation, as it seems not to reach steady state within 30 min [23]. A greater percentage (up to 30%) of exhaled CO_2 is derived from absorption during thoracoscopy, compared with 20% during laparoscopy. The greater absorption of CO_2 insufflated into the chest, coupled with the impaired ventilation, can lead to a marked increase in arterial CO_2 concentration, which is especially of concern in neonates and smaller children, who have been shown to have greater CO_2 increases than bigger children [15]. Acidosis can be severe and prolonged in neonates undergoing thoracoscopy [23, 24]. The ability to increase CO_2 excretion in the face of the increased load created by its absorption is crucial to safe thoracoscopy in children. To avoid harm, the anaesthetist must anticipate, monitor, and expertly manage this requirement. Therefore thoracoscopic surgery in these circumstances should be performed only in experienced centres and with good prospective monitoring and management of CO_2 load.

1.1.2.3 Learning Curve

The impact of learning new tasks needed for MIS must be taken account in embarking on such a venture. Many skills are of course transferable between operations, but not always between open and laparoscopic surgery. Intracorporeal suturing is a part of some MIS procedures and must be learned before embarking on operations requiring this technique. There is a role for learning these basic skills first on a form of trainer (of which there are several types available), before or while simultaneously attending a basic course. More advanced courses teaching the combined steps and skills for specific and advanced operations also can be used. Many training models have been developed for specific operations.

Whereas the learning curve can be measured in terms of operative time and hospital stay, better measures are patient safety outcomes such as complications and recurrence rates. Many MIS operations can take significantly longer than the corresponding open operation, especially during the learning curve. This difference must be appreciated by the surgeon, anaesthetist, and theatre staff (as well as patients and family, of course), for good teamwork and success. For most surgeons with advancing skills, however, this difference in time taken lessens and becomes clinically (and occasionally actually) insignificant.

There are various estimates of the number of MIS procedures required to reach the peak of the learning curve. For example, the number of procedures needed for laparoscopic hernia repair is estimated to be between 10 and 30 cases [35, 36]. It must be remembered, however, that the learning curve is both surgeon-specific and procedure-specific.

Being mentored at the outset of the MIS venture is one means of quickly and safely negotiating the learning curve. Inviting experienced operators to mentor surgeons at the beginning of their venture should facilitate quick and safe advancement up the learning curve.

1.2 When Should MIS Be Used in Children?

1.2.1 Indications and Contraindications

More and more operations are being performed by MIS in children. Indications for each specific operation are beyond the scope of this book. Some general indications and contraindications can be given.

1.2.2 Specific Operations

Some operations lend themselves nicely to MIS. Operations that are particularly suitable for MIS may have the following characteristics:

- A small, focused area of interest that would otherwise require a large incision for access (e.g., the oesophageal junction for myotomy or fundoplication)
- Access to areas that are relatively difficult to reach (e.g., deep recesses) but are suitable for access with a scope (e.g., operations around the oesophageal hiatus or pelvis)
- Operations that have incisions associated with poor cosmetic outcome (e.g., chest wall deformity) but that can be improved with MIS
- Operations in which diagnostic uncertainty exists or when MIS offers opportunity for diagnostic benefit not easily available with open surgery (e.g., assessment of contralateral inguinal ring and pelvic organs in hernia surgery, and investigation of impalpable testis)

Some operations may pose a relative or absolute contraindication to MIS, but absolute contraindications are becoming fewer with advancing experience, instrumentation, and innovation. Contraindications are suggested by the following considerations:

- If the MIS approach is associated with higher complication rates, it is contraindicated.
- The MIS approach can be sanctioned for cancer surgery only if the cancer surgery principles can be adhered to (e.g., nodal sampling or clearance, wide tumour margins, and intact tumour retrieval without rupture).
- If MIS ports do not allow safe organ or specimen retrieval, open surgery may be needed. Often hybrid techniques are possible, however (e.g., see Splenectomy chapter), using alternative innovations or techniques or a more appropriate abdominal incision.
- The need to alter the steps of the "classic" open operation is often cited as a contraindication for laparoscopic surgery—an idea both correct and incorrect. Often various innovations in instrumentation and technique allow the MIS operation to be performed using the classic steps, and this should be the first intention. But in other operations, the outcome is equivalent even though the classic steps of an open operation are not performed laparoscopically. The key consideration is whether the efficacy and outcome of the laparoscopic approach have been shown to be equal to those of the open approach.

1.2.3 Clinical Status

Patients being considered for MIS should be specifically clinically evaluated for the potential physiological changes discussed previously. In general, they should have achieved some physiological stability, if not normality. Emergency operations in unstable patients are associated with higher rates of complications. Active bleeding is a relative contraindication for MIS, as bleeding will severely obscure visualisation in the cavity being explored. Furthermore, blood itself causes difficulty by light absorption, thereby further decreasing visibility.

There is no age or weight limit for the application of MIS in children. Even preterm neonates can be candidates for diagnostic and therapeutic interventions, and even operations requiring advanced MIS skills are being performed in younger and smaller children.

There are some contraindications:

- Inability to tolerate the additional challenges of MIS surgery and the CO_2 load required, as shown by evaluation
- Active bleeding (relative)
- Physiological instability—a relative contraindication but an important parameter that may prevent MIS

1.3 Who Should Perform MIS in Children?

1.3.1 Training and Competence

Not all surgeons may suit advanced MIS surgery and the skill sets that are required, but with adequate training, nearly every surgeon can perform simple MIS procedures or operations. Many models of training exist and a combination of some or all is usually employed in a stepwise fashion.

- Simple box trainers allow the novice MIS surgeon to test, evaluate, and develop the skills required. Simple box trainers may also employ a manual or electronic scoring system to allow documentation (and audit) of the developing psychomotor skills. An example of a simple box trainer used in one of our centres is shown in Fig. 1.1. Box training has been shown to successfully contribute to laparoscopic competence [37].
- Complex box trainers allow the trainee to develop the sometimes more complex mix of the many different skills and techniques required during MIS. Acquisition of data regarding developing skills is again possible.
- Specific training models for specific operations also exist, simulating the steps required for the completion of an operation from start to finish. For example, box trainers exist for operations such as repair of inguinal hernia, pyloric stenosis, diaphragmatic hernia, or oesophageal atresia and tracheo-oesophageal fistula. These models can use a combination of simulated reconstructions and realistic body cavities with simulated tissue.
- Training courses are a very good means of gaining exposure to MIS surgery. Courses are available at a variety of levels, from those targeted at the novice and most junior trainee to advanced courses for established MIS surgeons. Some of the advanced courses allow realistic exposure to animal tissue or models (including live operating) with expert tutorship and teaching.
- Clinical exposure is the most realistic and eventually the most appropriate means of training, but the need for clinical training and clinical governance must be balanced with patient safety and outcome. Therefore some form of training with the means described above is used prior to and alongside clinical exposure.

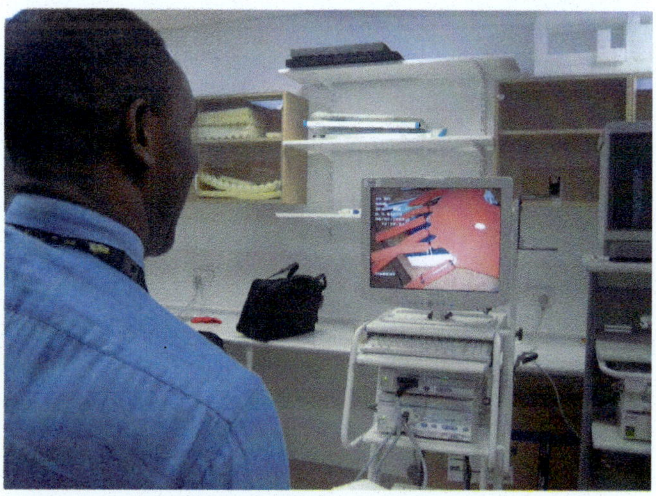

Fig. 1.1 A simple box trainer used in one of our centres

1.3.2 Mentorship

This can be seen as the final training step for those wishing to embark on MIS who have not acquired full training in MIS or in a specific operation. An expert can mentor a senior surgeon in the acquisition of the final stages of the needed skill and experience. This mentorship allows for a good mix of training, governance, and safety.

1.3.3 Continued Development

Even experts in MIS needs to continually develop and modify techniques and skills, keeping abreast of advances in the field. This development is often best done by attending large international conferences that either focus on MIS (e.g., International Pediatric Endosurgery Group/Society of American Gastrointestinal and Endoscopic Surgeons, IPEG/SAGES) or include MIS in their programme (e.g., British Association of Paediatric Surgeons/British Association of Urological Surgeons, BAPS/BAUS), often with manual training running alongside academic sessions.

It is also prudent to audit and frequently evaluate the outcome of MIS cases to document and evaluate outcomes that may need addressing or can help direct continued professional development. Presenting such data at conferences is also a means of peer review and feedback, which can help continued professional development.

1.4 Where Should MIS in Children Take Place?

1.4.1 Centres

If embarking on MIS in children for the first time, it may be advantageous to be in a centre that performs MIS in adults, to allow for some mentorship and sharing of experience between surgeons and teams. It also cuts down on setup cost if most of the basic equipment is already available.

Dedicated paediatric anaesthetists, who appreciate the different physiological changes incurred by MIS, are crucial in allowing safe anaesthesia for children undergoing MIS. The respiratory management is especially challenging in thoracoscopic operations in smaller children; an experienced anaesthetist is needed. There are also different cardiovascular changes introduced by MIS. A dedicated team of personnel (scrub nurses, managers) interested in development and support is needed for the success of MIS in children.

1.4.2 Research

It is also helpful to have some interest in research (both clinical and basic science) and development of MIS in the centre. Although not essential, the research interest will help with the development of personal governance and can assist with input into the general evidence base to inform and consolidate the case for MIS in children.

1.4.3 Operating Rooms

Most regular operating room (ORs) that are functional and suited for paediatric surgery are easily transformed into a suitable MIS suite for most routine procedures by the addition of a simple MIS stack. Commercially available start-up sets are available from well-known companies with a track record in paediatric MIS (e.g., Storz and Olympus). The basic stack and a very simple basic tray (see individual chapters) may suffice for operations like pyloromyotomies, hernia repairs, and appendectomies.

More advanced, integrated MIS suites are being used in centres that perform high volume and/or high-complexity MIS operations. Collaboration between hospitals and endoscopic companies like Storz and Olympus have allowed development of integrated theatre suites (OR1™ and ENDOALPHA™), which combine advanced concepts of MIS to include controls by a common sterile interface for operating lights, insufflators, and electrosurgical equipment; "built-in" LCD monitors; "all-in-one" camera heads; and other advanced solutions for integration. These impressive suites come at a cost but are worth the investment for those doing high volume and dedicated MIS procedures.

1.5 The "How" of Laparoscopic Surgery in Children: Techniques

The bulk of the rest of this book deals with the techniques involved in specific MIS operations in children. Precise, step-by-step approaches to common operations are outlined in individual chapters, but the individual skills required to put these techniques together are not the subject of this atlas, as important and fundamental as they are. Many intuitive steps during open surgery are not as intuitive during laparoscopic surgery; sometimes a complete change in the thought process is required. Others need to be learned in the context of laparoscopy. Familiarity with the equipment unique to MIS is also required; some is addressed in chapters on ergonomics and equipment.

Laparoscopic training sets (both commercial and individualised) are available for practise and to familiarise the surgeon with the techniques such as triangulation, ligation, clipping, and intracorporeal and extracorporeal suturing. These sets are a worthwhile investment for training and building up of necessary skills.

Ergonomics is covered in Chap. 2. Appreciating and mastering some of the ergonomic challenges of laparoscopy is essential to the safe and efficient performance of individual tasks to construct a successful operation.

1.5.1 Port Insertion

Most primary port insertions tend to be at the umbilicus (either infraumbilical or supraumbilical depending on the operation and patient size). Primary ports can be inserted either by open cut-down technique or blind insertion (e.g., Veress needle). Given the small size of many paediatric patients and the inherent safety provided by the cut-down technique, that is the method used and recommended.

A small stab incision is made and tissue planes sequentially dissected onto muscle fascia. This can be stabilised between stay sutures or clips and then incised down to the peritoneum, which is then opened under direct vision. Various port types are available, with various ways to fix them in place after insertion. Some ports have been designed to prevent migration once inserted (e.g., screwing-in shaft, radially expanding sheath, inflatable balloons). A Hasson-type port is commonly used, and allows for fixation and sealing in its mechanism. Specialised ports with a transparent trocar tip that allows insertion of an endoscope to visualise entry during insertion are also available, and may be suitable for the older age group.

Secondary ports also come in various forms. An initial stab incision can be used at the intended site. This incision can be superficial if an introducer is being used with the port. The commonest introducers are pyramidal or conical (sharp or blunt), or ports with a retractable blade. The introduction is always performed under laparoscopic vision, pointing the tip away from viscera and vital structures (point into space). If blunt introducers are being used or instruments are to be inserted directly through the abdominal wall without ports, the stab incision is through all layers with direct laparoscopic vision.

Tips and tricks for port insertion:

- Local anaesthetic can be used to infiltrate the incision site prior to incising.
- Ensure that there is sufficient intra-abdominal pressure to give counter-pressure during insertion. (The pressure can always be temporarily increased if necessary and possible.)
- A direct cut-down of the secondary ports is also possible if desired.

1.5.2 Converting to an Open Approach

During MIS, it sometimes becomes necessary to convert to an open approach. This need is not in itself a complication. Converting to an open approach may indeed avoid complications in situations where it is difficult to carry on using MIS, such as equipment problems, poor visualisation, or bleeding. Many of the techniques used during MIS and open surgery are transferable (e.g., operative principles, tissue handling, and operative steps), whereas others are not (suturing, haemorrhage control, and organ extraction). The surgeon should be familiar with the corresponding techniques and skills for the open operation. Usually this is not difficult, as most surgeons have been exposed to both the open and laparoscopic techniques during training. But as more and more operations are being done laparoscopically, this difficulty may increase. For instance, very few appendectomies are primarily being done by open surgery, so trainees may not routinely gain exposure to these open techniques.

Now that most surgeons in training are being taught MIS procedures without much exposure to the open operation for comparison and reference, the challenge of familiarity with open techniques is not easily addressed. Continued exposure to open operations—even if not the same operation as the one being converted—can keep the repertoire of open techniques in the forefront of the surgeon's mind and psychomotor skills.

References

1. Dingemann J, Ure BM. Systematic review of level 1 evidence for laparoscopic pediatric surgery: do our procedures comply with the requirements of evidence-based medicine? Eur J Pediatr Surg. 2013;23:474–9.
2. Kehlet H. Surgical stress response: does endoscopic surgery confer an advantage? World J Surg. 1999;23:801–7.
3. Cuschieri A. Whither minimal access surgery: tribulations and expectations. Am J Surg. 1995;169:9–19.
4. Petty JK, Bensard DD, Partrick DA, Hendrickson RJ, Albano EA, Karrer FM. Resection of neurogenic tumors in children: is thoracoscopy superior to thoracotomy? J Am Coll Surg. 2006;203:699–703.
5. Laborde F, Folliguet T, Batisse A, Dibie A, da Cruz E, Carbognani D. Video-assisted thoracoscopic surgical interruption: the technique of choice for patent ductus arteriosus. Routine experience in 230 pediatric cases. J Thorac Cardiovasc Surg. 1995;110:1681–4.
6. Lawal TA, Gosemann JH, Kuebler JF, Gluer S, Ure BM. Thoracoscopy versus thoracotomy improves midterm musculoskeletal status and cosmesis in infants and children. Ann Thorac Surg. 2009;87:224–8.
7. Yu L, Shan M, Jiang J, Jing Y, Zang N, Wang T, et al. Combined transcervical and unilateral-thoracoscopic thymectomy for myasthenia gravis: 2 years of follow-up. Surg Laparosc Endosc Percutan Tech. 2008;18:489–92.
8. Friscia ME, Zhu J, Kolff JW, Chen Z, Kaiser LR, Deutschman CS, et al. Cytokine response is lower after lung volume reduction through bilateral thoracoscopy versus sternotomy. Ann Thorac Surg. 2007;83:252–6.
9. Wang L, Qin W, Tian F, Zhang G, Yuan J, Wang H. Cytokine responses following laparoscopic or open pyeloplasty in children. Surg Endosc. 2009;23:544–9.
10. McHoney M, Eaton S, Wade A, Klein NJ, Stefanutti G, Booth C, et al. Inflammatory response in children after laparoscopic vs open Nissen fundoplication: randomized controlled trial. J Pediatr Surg. 2005;40:908–13.
11. Anand KJ, Aynsley-Green A. Measuring the severity of surgical stress in newborn infants. J Pediatr Surg. 1988;23:297–305.
12. Chwals WJ, Letton RW, Jamie A, Charles B. Stratification of injury severity using energy expenditure response in surgical infants. J Pediatr Surg. 1995;30:1161–4.
13. McHoney M, Klein NJ, Eaton S, Pierro A. Decreased monocyte class II MHC expression following major abdominal surgery in children is related to operative stress. Pediatr Surg Int. 2006;22:330–4.
14. Campbell RM, Cuthbertson DP. Effect of environmental temperature on the metabolic response to injury. Q J Exp Physiol Cogn Med Sci. 1967;52:114–29.
15. McHoney M, Mackinlay G, Munro F, Capek A, Aldridge L. Effect of patient weight and anesthetic technique on CO(2) excretion during thoracoscopy in children assessed by end-tidal CO(2). J Laparoendosc Adv Surg Tech A. 2008;18:147–51.
16. Sugi K, Katoh T, Gohra H, Hamano K, Fujimura Y, Esato K. Progressive hyperthermia during thoracoscopic procedures in infants and children. Paediatr Anaesth. 1998;8:211–4.
17. McHoney M, Corizia L, Eaton S, Wade A, Spitz L, Drake D, et al. Laparoscopic surgery in children is associated with an intraoperative hypermetabolic response. Surg Endosc. 2006;20:452–7.
18. Luo K, Li JS, Li LT, Wang KH, Shun JM. Operative stress response and energy metabolism after laparoscopic cholecystectomy compared to open surgery. World J Gastroenterol. 2003;9:847–50.
19. McHoney M, Eaton S, Wade A, Carnielli V, Kiely E, Drake D, et al. Effect of laparoscopy and laparotomy on energy and protein metabolism in children: a randomized controlled trial. J Pediatr. 2010;157:439–44.
20. Mullett CE, Viale JP, Sagnard PE, Miellet CC, Ruynat LG, Counioux HC, et al. Pulmonary CO2 elimination during surgical procedures using intra- or extraperitoneal CO2 insufflation. Anesth Analg. 1993;76:622–6.
21. McHoney M, Corizia L, Eaton S, Kiely EM, Drake DP, Tan HL, et al. Carbon dioxide elimination during laparoscopy in children is age dependent. J Pediatr Surg. 2003;38:105–10.
22. Eaton S, McHoney M, Giacomello L, Pacilli M, Bishay M, De Coppi P, et al. Carbon dioxide absorption and elimination in breath during minimally invasive surgery. J Breath Res. 2009;3:047005.
23. Bishay M, Giacomello L, Retrosi G, Thyoka M, Nah SA, McHoney M, et al. Decreased cerebral oxygen saturation during thoracoscopic repair of congenital diaphragmatic hernia and esophageal atresia in infants. J Pediatr Surg. 2011;46:47–51.
24. McHoney M, Giacomello L, Nah SA, De CP, Kiely EM, Curry JI, et al. Thoracoscopic repair of congenital diaphragmatic hernia: intraoperative ventilation and recurrence. J Pediatr Surg. 2010;45:355–9.
25. Bishay M, Giacomello L, Retrosi G, Thyoka M, Garriboli M, Brierley J, et al. Hypercapnia and acidosis during open and thoracoscopic repair of congenital diaphragmatic hernia and esophageal atresia: results of a pilot randomized controlled trial. Ann Surg. 2013;258:895–900.
26. Sakka SG, Huettemann E, Petrat G, Meier-Hellmann A, Schier F, Reinhart K. Transoesophageal echocardiographic assessment of haemodynamic changes during laparoscopic herniorrhaphy in small children. Br J Anaesth. 2000;84:330–4.
27. Pacilli M, Pierro A, Kingsley C, Curry JI, Herod J, Eaton S. Absorption of carbon dioxide during laparoscopy in children measured using a novel mass spectrometric technique. Br J Anaesth. 2006;97:215–9.
28. Cheng KI, Tang CS, Tsai EM, Wu CH, Lee JN. Correlation of arterial and end-tidal carbon dioxide in spontaneously breathing patients during ambulatory gynecologic laparoscopy. J Formos Med Assoc. 1999;98:814–9.
29. Laffon M, Gouchet A, Sitbon P, Guicheteau V, Biyick E, Duchalais A, et al. Difference between arterial and end-tidal carbon dioxide pressures during laparoscopy in paediatric patients. Can J Anaesth. 1998;45:561–3.
30. Gehring H, Kuhmann K, Klotz KF, Ocklitz E, Roth-Isigkeit A, Sedemund-Adib B, et al. Effects of propofol vs isoflurane on respiratory gas exchange during laparoscopic cholecystectomy. Acta Anaesthesiol Scand. 1998;42:189–94.
31. Giebler RM, Kabatnik M, Stegen BH, Scherer RU, Thomas M, Peters J. Retroperitoneal and intraperitoneal CO2 insufflation have markedly different cardiovascular effects. J Surg Res. 1997;68:153–60.
32. Glascock JM, Winfield HN, Lund GO, Donovan JF, Ping ST, Griffiths DL. Carbon dioxide homeostasis during transperitoneal or extraperitoneal laparoscopic pelvic lymphadenectomy: a real-time intraoperative comparison. J Endourol. 1996;10:319–23.
33. Luiz T, Huber T, Hartung HJ. [Ventilatory changes during laparoscopic cholecystectomy]. Anaesthesist. 1992;41:520–6. German.
34. Haynes SR, Bonner S. Review article: anaesthesia for thoracic surgery in children. Paediatr Anaesth. 2000;10:237–51.
35. Bertozzi M, Melissa B, Magrini E, Bini V, Appignani A. Laparoscopic herniorrhaphy in the pediatric age group: what about the learning curve? J Endourol. 2013;27:840–4.
36. Yoshizawa J, Ashizuka S, Kuwashima N, Kurobe M, Tanaka K, Ohashi S, et al. Laparoscopic percutaneous extraperitoneal closure for inguinal hernia: learning curve for attending surgeons and residents. Pediatr Surg Int. 2013;29:1281–5.
37. Aggarwal R, Grantcharov TP, Eriksen JR, Blirup D, Kristiansen VB, Funch-Jensen P, et al. An evidence-based virtual reality training program for novice laparoscopic surgeons. Ann Surg. 2006;244:310–4.

2 Equipment and Ergonomics

Alex C.H. Lee and Hugh W. Grant

Abstract

Minimal access surgery (MAS) differs from traditional open surgery in that it accesses and visualizes the operative field via small skin incisions. The small access points minimize the morbidity and unsightly scars caused by larger open wounds. The endoscopic visualization can also offer additional views by reaching deeper within the body cavity. The procedure can still be invasive and traumatic, and therefore it is more appropriate to describe the approach as "minimal access" instead of "minimally invasive."

Keywords

Ergonomics • Operative field • Task performance

A.C.H. Lee, MB ChB, PhD, FRCSEd (Paed Surg), FRCS (✉)
H.W. Grant, MB ChB, MD, FRCSEd, FRCS
Department of Paediatric Surgery, Oxford University Hospitals
NHS Foundation Trust, Oxford Children's Hospital, John Radcliffe
Hospital, Oxford, UK

M. McHoney et al. (eds.), *Color Atlas of Pediatric Anatomy, Laparoscopy, and Thoracoscopy*,
DOI 10.1007/978-3-662-53085-6_2

2.1 General Information

Minimal access surgery (MAS) differs from traditional open surgery in that it accesses and visualizes the operative field via small skin incisions. The small access points minimize the morbidity and unsightly scars caused by larger open wounds. The endoscopic visualization can also offer additional views by reaching deeper within the body cavity. The procedure can still be invasive and traumatic, and therefore it is more appropriate to describe the approach as "minimal access" instead of "minimally invasive."

The three interrelated performance-enhancing elements in optimizing operative surgery are technology, ergonomics, and training (Fig. 2.1). The innate abilities of the surgeon and patient factors also influence surgical outcomes. The interaction ergonomics and psychomotor skills are even more evident in pediatric MAS, where there is often a limited operative workspace.

In this chapter we provide an overview of the basic equipment used in pediatric MAS, the ergonomic constraints, and optimization strategies.

Specific requirements in operative procedures and developments in techniques such as single incision laparoscopic surgery (SILS) and robotic surgery are described in other chapters.

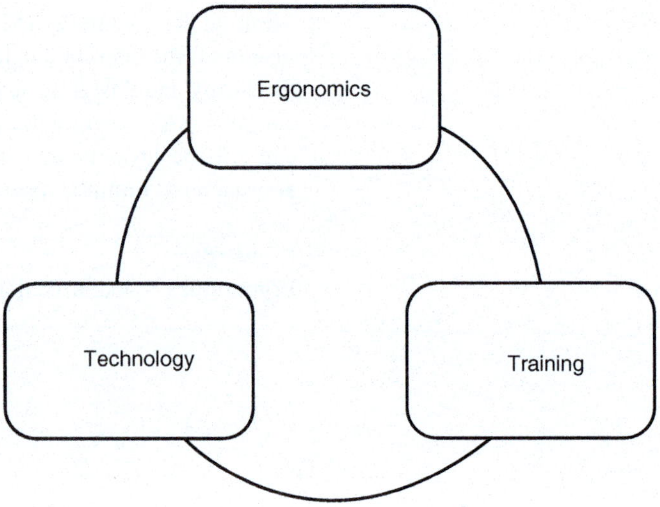

Fig. 2.1 Three interrelated performance enhancing elements

2.2 Working Instruments

The equipment and instruments used in pediatric MAS are mostly the same as those used in adult MAS with some specifically designed for surgery in infants. It is important for the surgeon to understand how the equipment works and to know how to trouble-shoot basic problems.

2.2.1 Creation of the Operative Workspace

An insufflator is used to create, maintain, and control an adequate operative workspace during MAS. The machine regulates and monitors the flow rate, volume, and pressure of CO_2 transported into the body cavity from the CO_2 cylinder. A filter is used to prevent back flow of fluid from the patient. The ambient air within the tube should be purged and filled with CO_2 prior to connecting it to the patient. The desired pressure and flow rate can be set by the user (Fig. 2.2). The insufflations are given in short pulses and not continuously (although set as liters per minute). When the measured pressure is less than the set pressure, another pulse of insufflation is given, and the process is repeated until the set pressure is reached. Most machines are designed for adult use. In a large body cavity, leakage can be easily compensated for by setting the machine at a higher flow rate to maintain pressure. Leakage poses a problem in small infants because each pulse of insufflation (if set at a high rate) could result in a pressure surpassing the set pressure before the negative feedback occurs that stops further insufflation. This can be particularly dangerous when insufflation is used in the neonatal chest. The initial pressure and flow rate settings should be at low levels to start (e.g., a pressure of 6–8 mmHg and a flow rate of 1 L/min).

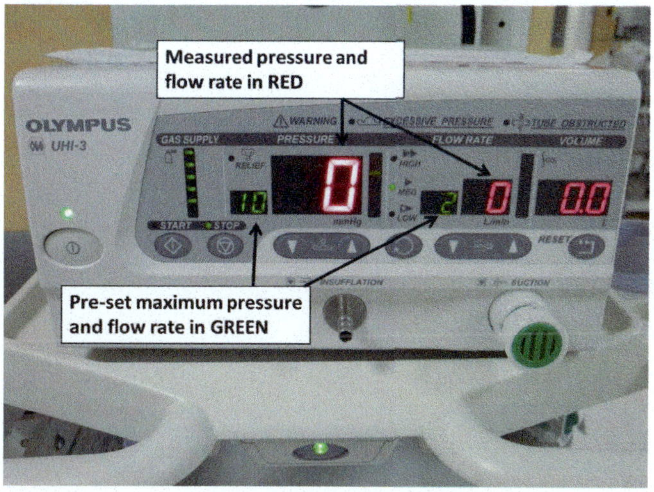

Fig. 2.2 Insufflator displaying preset and measured pressure and flow rate. Some machines only display the preset values temporarily after adjustments are made

2.3 Visualization of the Operative Field

It is crucial that the surgeon understands the imaging chain (Fig. 2.3). Any disruption of this chain results in suboptimal visualization.

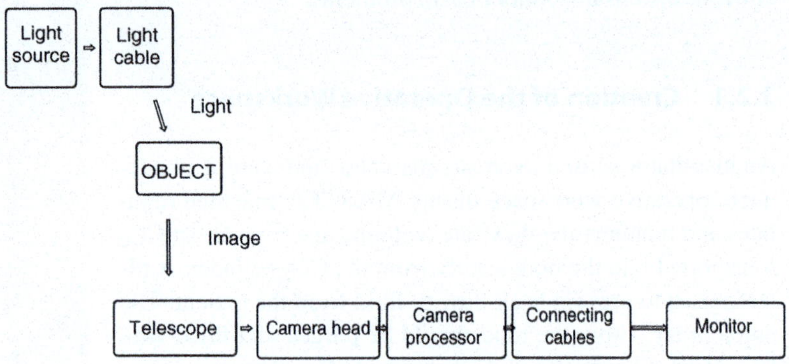

Fig. 2.3 Imaging chain. A clear understanding of this interlinked chain would allow trouble-shooting when the displayed image is absent or poor

2.3.1 Light

Most modern light sources use 300 watt xenon bulbs that emit white light transmitted via a fiberoptic light cable to the light post of the endoscope. Problems arise if there is a size mismatch between the light cable and the endoscope size, loose connections, or plus or minus broken fibers within the light cable (Fig. 2.4). Light is transmitted along the optical fibers within the endoscope, which provides illumination from its tip. These were called "cold" light sources because of the color temperature of 5000–6500 K. They generate a significant amount of heat, which can cause thermal damage to tissue; the temperature at the distal end of the tip of the endoscope can reach up to 95 °C.

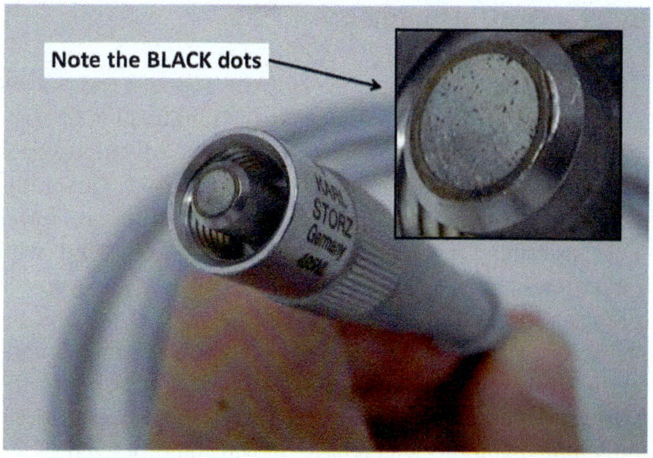

Fig. 2.4 Broken fibers in the light cable are shown as black dots on close inspection

2.3.2 Optical Image

The traditional endoscope, the Hopkins rod lens endoscope, contains optical lenses in the center and illumination fiberoptics in the periphery. The endoscopes come in various sizes and lengths, with a viewing angle of 0–70°. Generally, smaller endoscopes have lower optical resolution, lower light transmission, and greater distortion compared with larger ones (Fig. 2.5a, b). The optical image at the eyepiece of the endoscope is captured by the camera head, which contains the charge coupled device (CCD). It converts the optical information into electrical signals for processing in the camera box. Single chip cameras have been replaced by three-chip cameras, which have one channel for each of the three primary colors. Some cameras are also equipped with a parfocal zoom, which allows enlargement of the image without moving the endoscope. However, zooming results in less resolution, illumination, and perception of depth.

Prior to use, white balancing should be performed by keeping a white object in front of the endoscope and activating the appropriate button on the camera box or camera head. This is used as a white reference to adjust to the three primary colors.

With advances and miniaturization of imaging technology, some manufacturers place the CCD at the tip of the endoscope ("chip-at-tip") rather than at the eyepiece end without the need of Hopkins rod lenses. This construction also allows pivoting at the tip in larger endoscopes.

Whatever type of endoscope is used, the processed electronic signal of the image is then transmitted to the viewing monitor. Previously large cathode-ray tube (CRT) monitors were placed on top of other equipment in the MAS tower. Most hospitals now use flat panel monitors attached to the MAS tower or suspended from the ceiling; they are usually adjustable to allow changes in position. High definition (HD) camera systems and monitor displays are increasingly used in the operating theater. It is important to note that a compatible processor connecting cables and monitor is essential for the superior imaging from three-chip CCD or HD cameras.

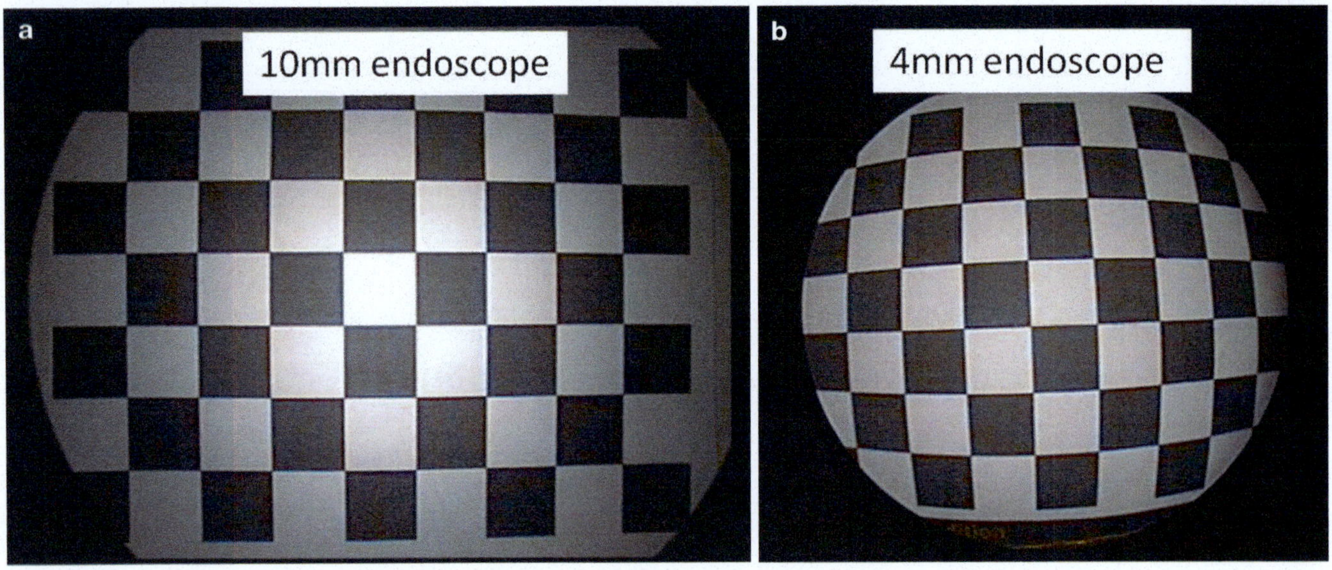

Fig. 2.5 (**a**, **b**) Barrel distortion is less marked using a larger endoscope

2.4 Instruments

Instruments are available in disposable or reusable forms. Disposable instruments are always new, clean, sterile, and work well as manufactured. However, they are expensive. Reusable instruments are generally more economical but have to be cleaned, sterilized, packed, and serviced. It is essential that the cleaning/sterilization department knows the exact requirements of each instrument.

Instrument access into a body cavity in MAS is usually via a port that consists of the cannula and trocar. The cannula is also commonly referred to as the port. There are various types of trocar tips (Fig. 2.6), the commonest being pyramidal, conical (sharp or blunt), or with a retractable blade. Insertion using a pyramidal or blade tip should avoid any twisting action in order to minimize tissue damage. Specialized ports such as those used for bariatric surgery have a bladeless trocar with a transparent trocar tip, allowing for insertion of an endoscope to visualize entry during insertion. Disposable radially expandable sheath ports are popular with some surgeons.

The size of the port depends on the instruments to be used. Most ports have a side stopcock for insufflation and an internal valve to prevent gas leakage when the instrument is removed (Fig. 2.7); some allow instruments of different sizes to be used without the need for adaptors/reducers. The rubber bung at the outer end maintains the gas seal when the instrument is inserted.

The length of the port is important: Long ones are heavy and can limit the surgeon if they are inserted too deeply into the body cavity. On the other hand, short ports increase the risk of dislodging owing to the thin body wall of infants.

There are various ways to fix a port after insertion into the body, and some ports have been designed to prevent slippage, such as the screwing-in shaft, the radially expanding sheath, or the Hasson port (Fig. 2.8). In small children and infants, instruments can be inserted without a port, especially if frequent instrument changes are unlikely, such as in a pyloromyotomy. The incision needs to be small and tight around the instrument to minimize the leakage of gas.

Commonly used working instruments in MAS include graspers/dissectors, scissors, retractors, clippers/staplers, ligature placing devices, suction/irrigation devices, energy supplying devices, and tissue retrieving bags. Some of these instruments are only available in disposable form (e.g., staplers). Several manufacturers have modular design instruments to allow interchangeable handles (e.g., various ratchets) with different tips. The diameter of the shaft is usually 3 mm or 5 mm. Generally, 3-mm instruments are preferable in patients weighing less than 10 kg.

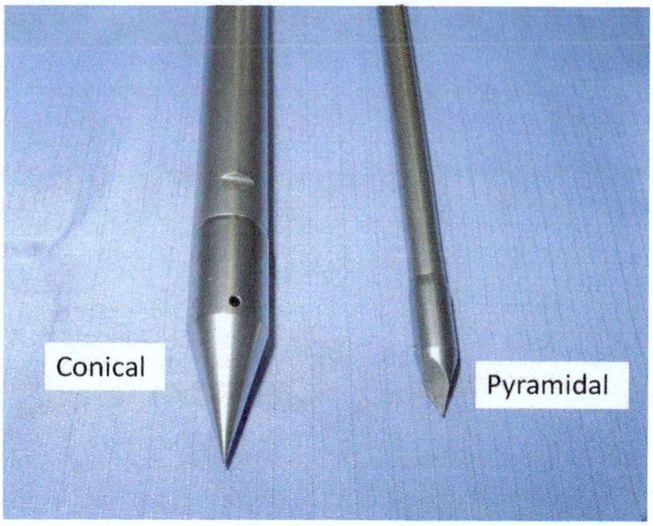

Fig. 2.6 Trocars with conical (10 mm) and pyramidal (5 mm) tips are shown

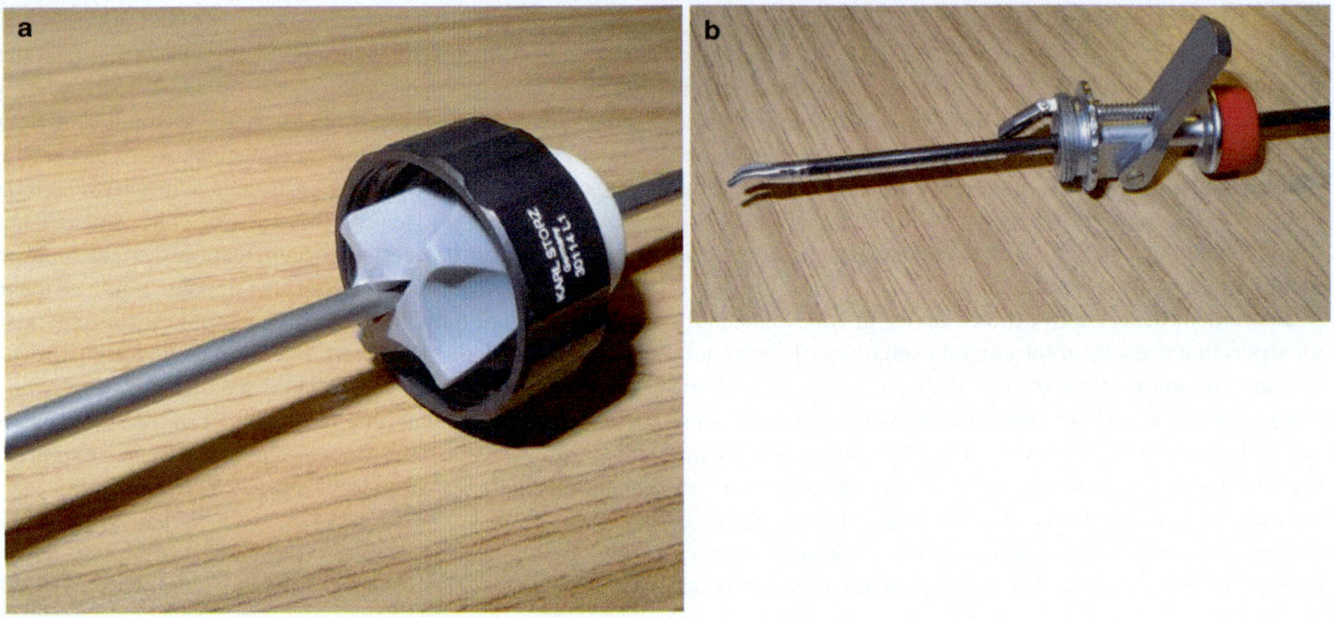

Fig. 2.7 Internal valve mechanisms when an instrument is inserted. (**a**) Silicon valve. (**b**) Metal valve

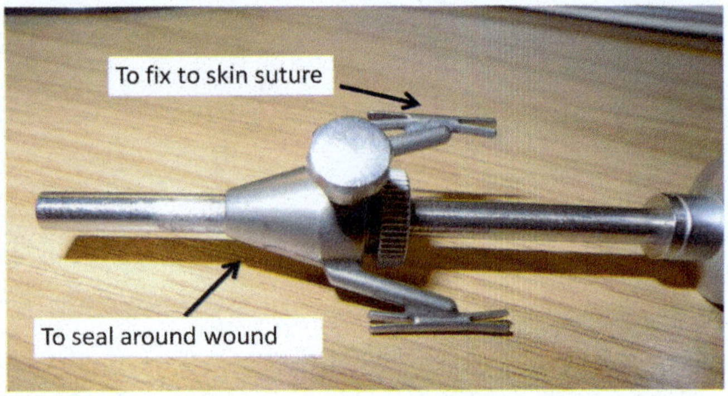

Fig. 2.8 Hassan port system for port fixation. It can be secured at variable internal lengths

2.5 Energy Devices

Electrosurgical devices are used extensively for hemostasis and dissection in MAS. Minor bleeding can obscure the view and reduce light reflection within the operative field. The general principles of monopolar and bipolar diathermy are the same as those for open surgery and are controlled by a foot pedal. Extra care must be taken when using monopolar diathermy to avoid hazards caused by insulation failure, capacitive coupling, and inadvertent direct (coupling) touching of another metal instrument within the operative field. All plastic or all metal port systems can be used but avoid ports that are made from a combination of materials (i.e., hybrid ports). The hook monopolar diathermy instrument is most commonly used for dissection. Bipolar diathermy uses special forceps without the need for use of the patient return plate in monopolar diathermy. In general, bipolar instruments are preferable because the electrical circuit passes between the tips of the instruments, not through the patient's body.

Ultrasonic scalpels (e.g., the harmonic scalpel and Sonosurg [Olympus America, Center Valley, NJ]) convert ultrasonic vibrations into energy for precision cutting and coagulation without the need for an electrical circuit through the patient. Beware of collateral injury caused by thermal spread by the heated instrument tip during or after use. Vessel-sealing technology (e.g., LigaSure [Medtronic, Minneapolis, MN]) uses an optimized combination of pressure and energy to create seals by denaturing the collagen and elastin in vessel walls. It seals vessels up to 7 mm in diameter. When the seal is complete, the computer-controlled feedback ceases the energy. Some hand-held devices also come with a knife mechanism for division of tissue that has just been sealed.

2.6 Other Equipment Considerations in Pediatric MAS

- The Nathanson retractor (Cook Medical; Bloomington, IN, USA) is very useful for liver retraction in upper abdominal surgery in place of a second assistant (e.g., fundoplication). There are other hand-held endoscopic retractors that open up once inserted into the body cavity into snake or fan shapes.
- Pretied surgical loops (e.g., Endoloops [Ethicon, Medline Industries. Mundelein, IL]) are useful for resection procedures such as appendicectomy.
- Specimen retrieval bags of various sizes and designs are available to avoid contamination during organ/tissue removal. These are usually too large for use in infants.
- Knot pushers of various designs are used in extracorporeal knot tying.
- Suction-irrigation devices are useful when there is spillage in the operative field. They can also be used in blunt dissection. They require pressurized fluid for irrigation. A specimen trap can be set up for collection of the suctioned fluid.
- Stapling devices are bulky and suitable for older children only.
- Titanium clips for ligating vessels are available in 5 mm.
- A pyloromyotomy spreader has serrations in the outer and inner surfaces of the instrument tip; the tip comes in straight or hockey-stick configurations, depending on the surgeon's preference.
- Veres needles are used by some surgeons routinely. There is a greater risk of collateral injury in children because of their more elastic abdominal walls. The risk of such injury is avoided by the "open" or "Hassan" entry method. The authors prefer to use the open technique for initial entry in children in all cases.

2.7 Ergonomic Considerations in Pediatric MAS

2.7.1 Definition

The word ergonomics comes from the Greek words *ergos* (work/labor) and *nomos* (natural law). Ergonomics is the scientific study of the interaction between humans and their working environment. It aims to achieve the optimum outcome by fitting the job to the worker and the product to the user (Table 2.1).

There are many mental and physical similarities between surgery and safety-critical industries such as aviation, where optimizing ergonomics plays a crucial role for safe and efficient deployment. The ergonomic challenges in the operating room have become more evident and critical with advances in surgical technology.

2.7.2 Ergonomic Constraints in Minimal Access Surgery

MAS carries a set of mechanical and visual constraints in the execution of surgical tasks that cause degraded task performance and surgeon discomfort.

Table 2.1 Ergonomics can be broadly considered in three domains: physical, cognitive, and organizational

Domain	Considerations	Relevance/Applications
Physical ergonomics	Anatomic, anthropometric, physiologic, and biomechanical characteristics	Work postures Workplace layout Equipment handling Work-related musculoskeletal disorders in surgeon
Cognitive ergonomics	Mental processes: Perception Memory Reasoning Motor response	Decision-making Mental workload Skill performance Human-computer interaction Human reliability Training
Organizational ergonomics	Organizational structures, policies, and processes	Communication Teamwork Team resource management Design of roster and work patterns

2.7.3 Mechanical Constraints

Each degree of freedom (DOF) of movement allows the instrument to move in an independent direction (Fig. 2.9). In MAS, there are four DOFs of movement, namely, across in the X axis, up and down in the Y axis, in and out in the Z axis, and rotational movements. This compares unfavorably to the 36 DOFs of movement in open surgery, where all joints of the upper limbs from shoulder to finger tips provide movement. The movements of MAS instruments are restricted by the entry point in the body wall, and the resulting fulcrum effect leads to paradoxical movements, such as when the surgeon's hand moves to the right, the instrument tip moves to the left.

Direct tactile feedback is lost in MAS, and indirect tactile feedback from the instrument's tip to the handles is reduced. Proprioceptive feedback and the ability to identify the nature of tissue components and planes are therefore diminished.

MAS instruments were adapted from equivalent instruments used in open surgery with modifications to enable them to be inserted into the body cavity. The diameter of the endoscopic instruments is limited by the size of the access port. Long thin instruments have poor mechanical advantage, and the narrow distal tip may cause damage to tissues. Because of the extracorporeal shaft length, the use of a long instrument will result in a large external arc of arm movement by the surgeon when the corresponding intracorporeal shaft length is short.

The locations of the ports are determined by the surgeon according to the available space on the surface of the body. These are known to have crucial effects on task performance. The manipulation angle is defined as the angle between the two instruments (active and assisting) (Fig. 2.10). The elevation angle is the angle of the instrument against the horizontal plane. The best task performance is obtained when the manipulation angle is between 45 and 75°, the ideal being 60°. Wide manipulation angles necessitate wide elevation angles for optimal performance and task efficiency. Generally elevation angles of 45–60° are most suitable.

Most MAS instruments are not well suited for their use; they often require flexion and ulnar deviation at the wrist (which decrease maximum grip force); the handle configuration often requires the surgeon to use opposing thenar and hypothenar muscles for gripping rather than the more powerful deep flexor muscles of the forearm; the muscle contraction force for MAS grasping is three to five times higher than that for open instruments. The inefficient transfer of the mechanical force from the handles to the tip of the instrument causes further discomfort and fatigue. Many laparoscopic instruments are not designed to accommodate the fingers comfortably and can cause neuropraxia.

Surgeons have limited freedom of movement to adjust their body posture and arm positions as a result of the fixed point of insertion through the body wall. In addition, prolonged shoulder abduction is necessary, especially when the table height is not suitably lowered. The presence of the assistant holding the camera in front of the surgeon may further prevent the surgeon as well as the assistant from adopting a comfortable posture. Surgeons often adopt static postures for prolonged periods during MAS; in particular, there is reduced mobility of the head and neck regions and anteroposterior weight shifting. The restricted posture limits the natural posture changes that occur in open surgery. This is worsened when MAS is performed from the side of the patient, requiring rotation of the surgeon along the head-neck spine axis.

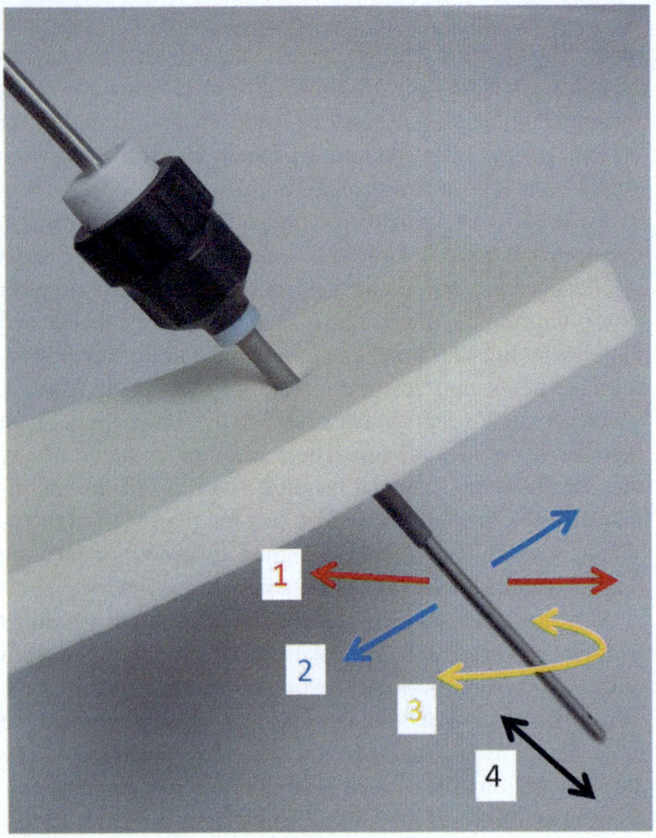

Fig. 2.9 The four degrees of freedom (DOF) of movement along the X axis (*up and down*), Y axis (*left and right*), Z axis (*in and out*) and rotational

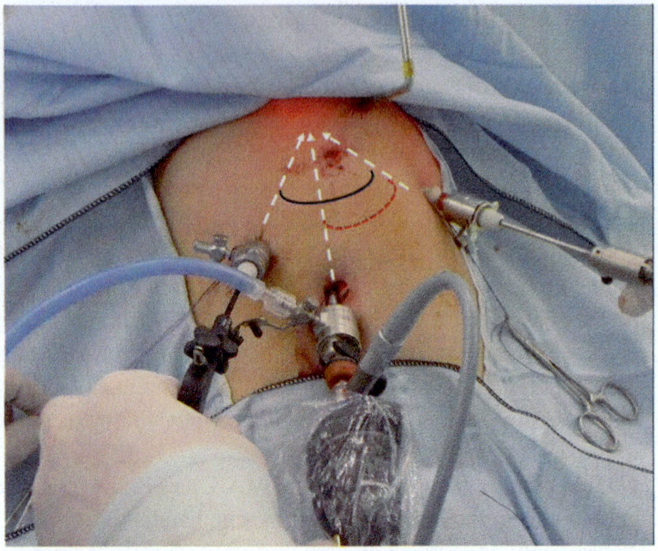

Fig. 2.10 The manipulation angle of the working instruments (*solid line*) ideally should be around 60°. The azimuth angle between the optical and working instrument (*fine dotted line*) ideally should be equal on both sides

2.7.4 Visual Constraints

Instead of having direct vision of the operative field, the use of an image display system between the surgeon and the operative field results in several visual limitations in MAS.

The endoscopic field of vision is determined by the direction of the endoscope; its peripheral field is limited by the optical component of the endoscope. The viewing angle refers to the angle formed by the outer limits of the endoscopic field. The field of view is the area as seen by the eyepiece of the endoscope, the full extent of which may not be completely picked up by the camera head, particularly when zoomed-in. This restricted field of endoscopic vision contributes to incidental tissue injury when instruments move outside the view.

The best detail can be obtained when the distal end of the endoscope is positioned close to the surgical field or "up close." However, as magnification increases, the apparent speed of movement (including tremor) also increases; therefore the operator has to mentally adjust the speed and magnitude of movement accordingly.

In normal binocular vision, the occipital cortex receives two slightly different views from each eye. These two images are processed into a three-dimensional pictorial image in the brain. In standard two-dimensional MAS both retinas receive the same information from the screen, and there is a loss of stereopsis (or binocular disparity). Other physiologic cues from accommodation and vergence are also lost because the image is in a fixed position as opposed to the real objects and background, which are at variable distances. To compensate for the lack of three-dimensional vision, surgeons need to rely on other indirect visual cues to reconstruct a three-dimensional anatomy from the two-dimensional image on the screen. This requires intense perceptual processing, which leads to fatigue. The lack of depth perception accounts for the overshooting/undershooting and imprecise "wobble" by surgeons as they try to determine the location of anatomic structures during MAS. Shadows are important pictorial cues in real life, but these are not available with the current rigid endoscope due to its coaxial arrangement for the optics and illumination. As a result, the target object in view is directly illuminated with its associated cast shadow behind the object. Experimental studies have shown that illumination systems with cast shadows improve task performance.

Studies have demonstrated that a distance of 75–150 mm between the endoscope and target is optimal for task performance. At this distance, instrument clashing is minimized. However, this is not possible in infants, where space is very limited.

Optimum task performance is obtained when the optical axis of the endoscope is perpendicular to the target plane, that is, optical axis to target view angle (OATV) equals 90° (Fig. 2.11). The direction of view of the endoscope (0°, 30°, 45°) has no significant effect on task performance when the optical axis subtends the same angle with the target surface. It is therefore preferable to use an oblique-viewing endoscope or a chip-on-tip endoscope with flexible tip, because a 0° forward-viewing endoscope will only provide a perpendicular view at one axial position. The oblique viewing endoscopes also provide more visual information to allow viewing from different angles by rotating the endoscope and consequently may enhance both the correct interpretation of the anatomy and the execution of advanced laparoscopic maneuvers.

The azimuth angle is the angle between the instrument and the optical axis (*see* Fig. 2.10). Improved task efficiency is achieved with equal azimuth angles on either side of the optical axis. If this is not possible, off axis endoscopic viewing (Fig. 2.12) from the nondominant side gives better performance.

When the optical axis is above the instrumental plane (Fig. 2.13a), the instruments appear to enter the operative field from the sides (real side) (Fig. 2.13b). Conversely, when the optical axis is below the instrumental plane, the instruments appear to enter the operative field from the opposite side (Fig. 2.13c), which is counterintuitive.

When the ports for the endoscope and instruments are aligned in the same direction, the paradoxical movement of the instrument tip is shown as the same direction on the viewing monitor (i.e., first-order paradoxical movement) (Fig. 2.14a). However, if the instruments are aligned against the opposite direction of the camera, manipulation becomes difficult. In this situation the horizontal movements are shown as moving opposite to the tip (second order paradoxical movements), and yet the vertical movements have remained in first order paradox (Fig. 2.14b). Although some individuals can adapt to reverse alignment, this adaptation requires increased mental processing and accelerates mental fatigue.

The best task performance is obtained when the monitor is placed in front of the surgeon just above the patient's body to allow a "gaze-down" view of the operative field. It should ideally be about one meter from the surgeon's eyes. Changing the position of the monitor in line with the surgeon has been shown to improve operating time in laparoscopic appendicectomy in children by over 10%.

2.7.5 Specific Considerations in Minimal Access Surgery in Infants

The operative field is significantly reduced in infants because of the small operative workspace and minimal distention created by the low-pressure pneumoperitoneum. Availability of small instruments may be limited. Very small instruments smaller than 3 mm are fragile and bend easily. Inadvertent injuries may also occur due to their sharper ends. In addition, excessively firm grasping may cause tissue injuries by the working distal end of the instrument on small surface areas.

In video-assisted thoracoscopic surgery, the use of unilateral lung ventilation can be difficult to achieve and poorly tolerated in small children. The lung often needs to be collapsed (by positive pressure) to create an operative space. Movements of the incompletely collapsed lung may obscure the operative field, causing problems in visualization.

In addition to the small internal space, the external space over an infant is also limited. Excessive cluttering of instruments can cause inefficiency and potential dangers, particularly with cables and electrical wires.

The thin body wall in infants provides less grip for anchoring the ports. Since the internal space is limited, the length inserted should be kept to a minimum. This predisposes to dislodgement, and the port should be suitably secured.

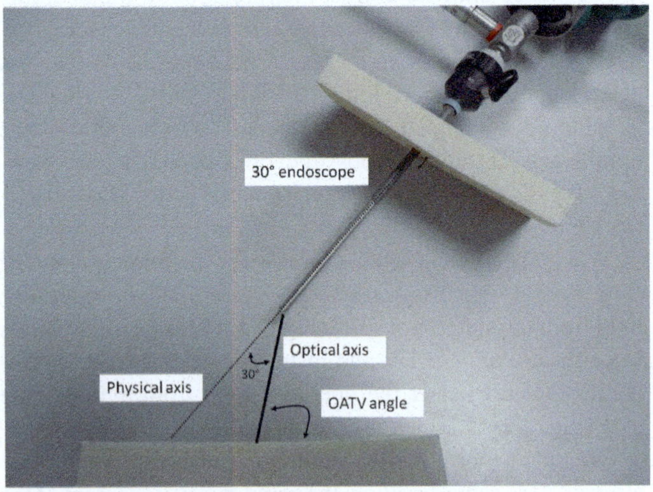

Fig. 2.11 The optical axis to target view angle (OATV) should ideally be 90°

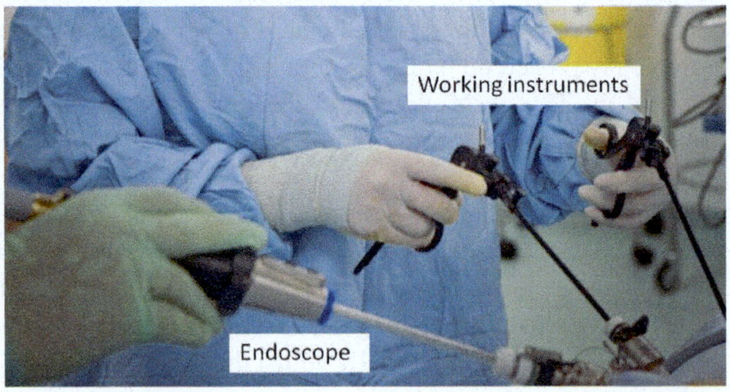

Fig. 2.12 Off axis viewing with both working instruments placed on one side of the endoscope

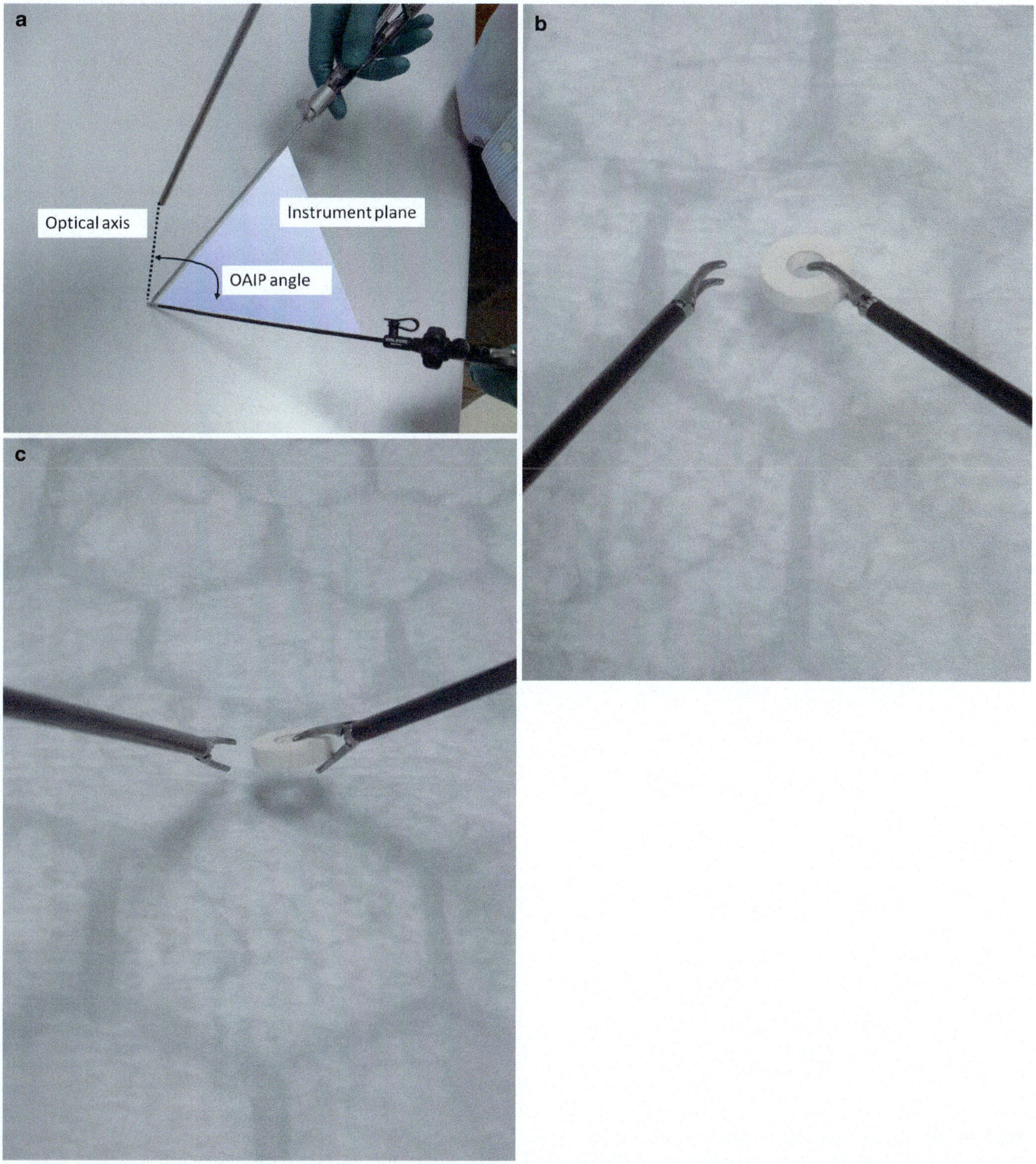

Fig. 2.13 (**a**) The optical axis to instrument plane (OAIP) angle subtends between the optical axis and the instrument plane. This should ideally be positive (**b**)—looking from above. Negative OAIP (**c**) is counterintuitive

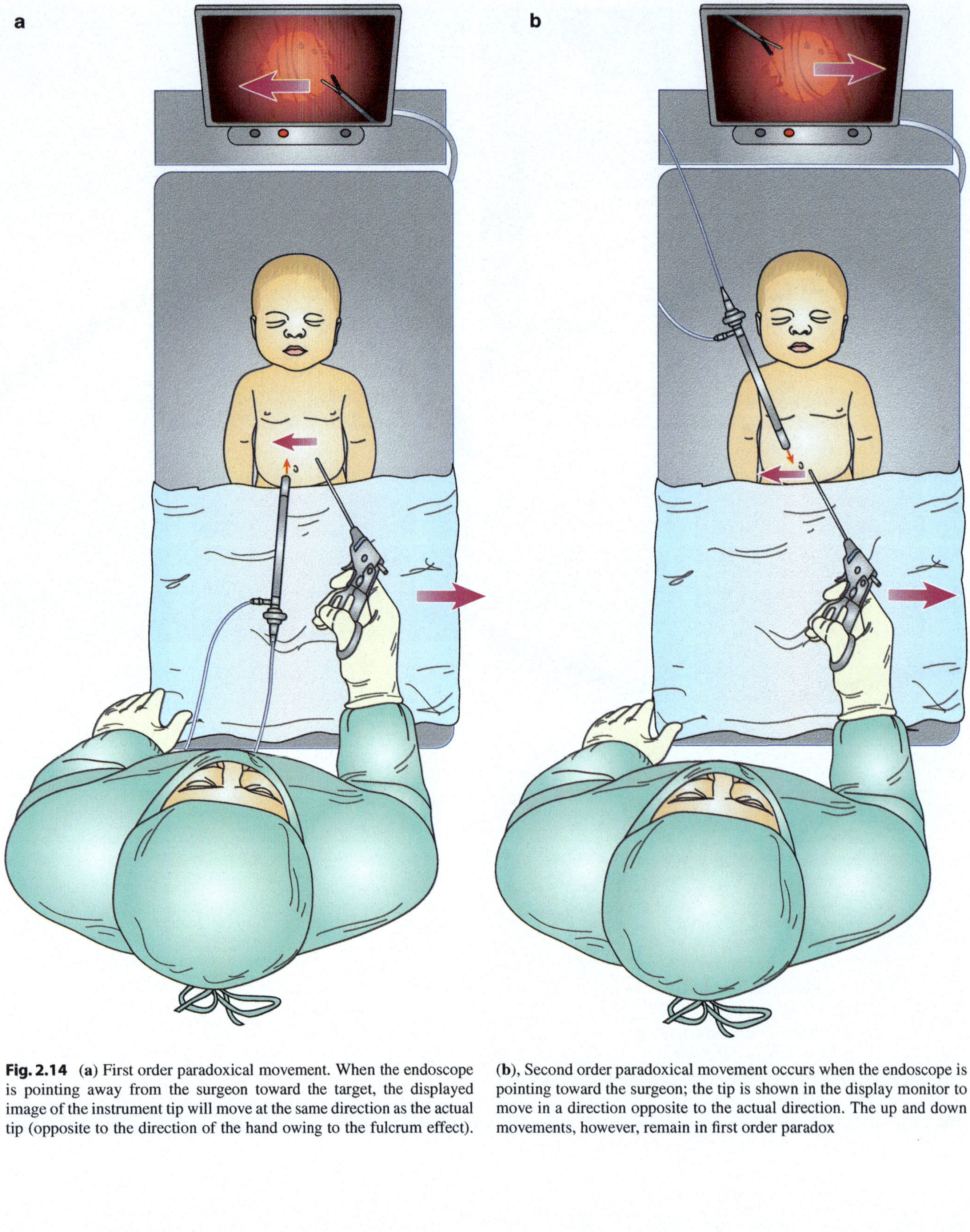

Fig. 2.14 (**a**) First order paradoxical movement. When the endoscope is pointing away from the surgeon toward the target, the displayed image of the instrument tip will move at the same direction as the actual tip (opposite to the direction of the hand owing to the fulcrum effect). (**b**), Second order paradoxical movement occurs when the endoscope is pointing toward the surgeon; the tip is shown in the display monitor to move in a direction opposite to the actual direction. The up and down movements, however, remain in first order paradox

2.8 Strategies to Improve Ergonomics in Pediatric MAS

To avoid second degree paradoxical movements operators should position the crew as close to the SCOPe position as possible (Fig. 2.15) by aligning the Surgeon, Camera, Organ, and Picture in the same direction. For the assistants/camera-holders to help effectively, they also need to be near to the SCOPe position.

It is more comfortable for the camera-holder's arm to be below the operator's arm. Depending on the relative height and size of the surgeon and the assistant, the assistant may need to sit down or the operator to stand on a platform for the best match.

It is useful to position the child at the end of the operating table with the surgeon standing between the legs of an older patient held in stirrups, for example for MAS in the upper abdomen. Beware of the flexed hips and knees, which may clash with the instrument handles.

Place the viewing monitor in front of and below head level as close to the operative field as possible. This aligns the visual and motor axes and puts together the sensory and motor information. This is best achieved by ceiling mounted display monitors or alternatively from moveable displays hinged from a MAS tower.

Table height should be reduced by 20–30 % to accommodate the length of the MAS instrument. This avoids prolonged abduction, which can cause shoulder or neck discomfort or pain.

By considering the position of the main target where most complex maneuvering is required (e.g., suturing), the ports should be placed to achieve an optical axis-to-target view (OATV) angle as close to 90° as possible (*see* Fig. 2.11). Ideally, the optical axis-to-plane (OAIP) angle (Fig. 2.13a) should be 0–15° to look down from above the instruments, as these angles are associated with enhanced task performance. Angled endoscopes are more versatile to achieve these OATV and OAIP angles, in addition to the ability to visualize at an angle from the sides.

Regardless of the optical position, when two working instruments are used, the best task performance can be achieved when these instruments approach the main target position at an angle of 45–75°, preferably with a similar elevation angle.

An intracorporeal to extracorporeal shaft ratio of 1:1 to 2:1 has been shown to improve task performance in experimental studies. The ports may need to be positioned farther away from the target; therefore the OATV angle may also be affected. The use of shorter and smaller instruments, especially in small infants, is beneficial in these respects.

Integrated theaters have provided improved ergonomic solutions to enhance the surgeon-operating room interface.

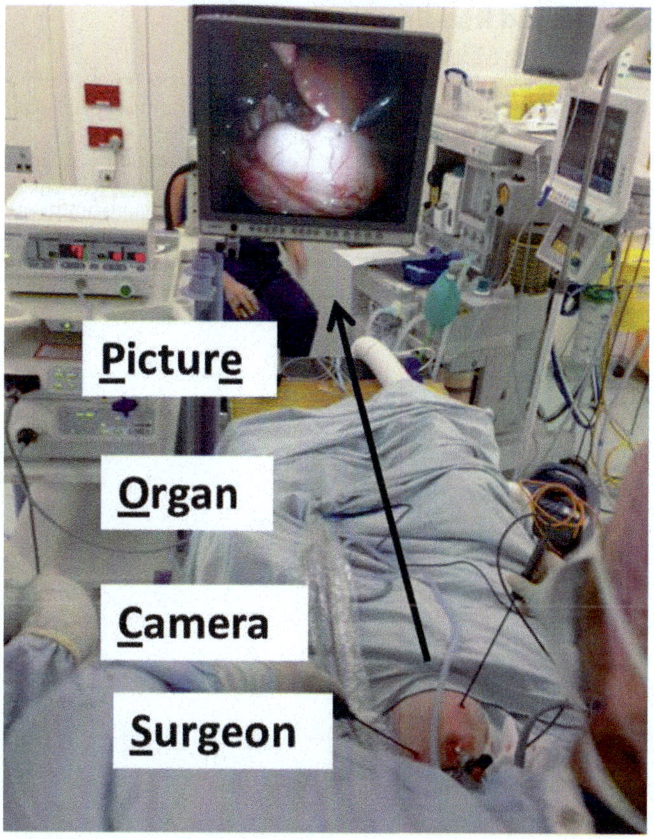

Fig. 2.15 The SCOPe alignment with the Surgeon, Camera, Organ and Picture all in one direction

The adjustable monitor positions allow optimized ergonomic viewing. In addition, the various pieces of equipment can be controlled centrally from within the sterile field. This enables the scrubbed staff to have better control of the procedure. Placing the equipment on a ceiling-mounted pendulum also frees up floor space and minimizes cluttering in the operating theater.

The use of equipment and ergonomics must be customized for each procedure and surgeon. Surgeons also come in various sizes, and therefore user-centered designs should allow the best fit possible for all surgeons, for example in modular handle designs.

Conclusion

It is crucial to understand the ergonomics to optimize task performance. In general one should try to keep the equipment simple and be familiar with how it works.

In the past decade, a lot of effort has been focused on surgical efficiency, task performance, surgical errors, and patient safety events. Further work needs to be done on designing better equipment, introducing safer working practices, and team-working to improve the overall operating room environment and outcomes.

Suggested Reading

Erfanian K, Luks FI, Kurkchubasche AG, Wesselhoeft Jr CW, Tracy Jr TF. In-line image projection accelerates task performance in laparoscopic appendectomy. J Pediatr Surg. 2003;38:1059–62.

Hanna GB, Cuschieri A. Ergonomics of task performance in endoscopic surgery. In: Bax NMA, Georgeson KE, Valla JS, Rothenburg S, Yeung CK, editors. Endoscopic surgery in children. Berlin: Springer; 2008. p. 39–50.

Hanna GB, Shimi SM, Cuschieri A. Task performance in endoscopic surgery is influenced by location of the image display. Ann Surg. 1998;227:481–4.

Lee ACH, Haddad MJ, Hanna GB. Influence of instrument size on endoscopic performance in paediatric intracorporeal knot tying. Surg Endoscopy. 2007;21:2086–90.

Reyes DA, Tang B, Cuschieri A. Minimal access surgery (MAS)-related surgeon morbidity syndromes. Surg Endoscopy. 2006;20:1–13.

Thoracoscopic Debridement of Empyema

3

Fraser D. Munro, Merrill McHoney, and Malcolm Wills

Abstract

The incidence of postpneumonic empyema has been increasing in recent years in most Western countries. The place of surgery in its management is debatable, with many centers advocating initial pleural drainage and instillation of fibrinolytics, reserving surgery for those who fail to respond to these measures. Some, however, suggest that primary surgery leads to a more rapid recovery and a shorter hospital stay. There is little evidence to convincingly show which is the best approach. The aim of surgery is clearance of debris from the pleural space, with reestablishment of a "single" pleural cavity free of loculations and full reexpansion of the lung. Postoperatively, early mobilization and physiotherapy are key. Thoracoscopy has the advantage of excellent visualization of the whole pleural space and minimal postoperative pain, allowing physiotherapy and mobilization to begin almost immediately.

Keywords

Empyema • Thoracoscopy • Pleural infection

3.1 General Information

The incidence of postpneumonic empyema has been increasing in recent years in most Western countries. The place of surgery in its management is debatable, with many centers advocating initial pleural drainage and instillation of fibrinolytics, reserving surgery for those who fail to respond to these measures. Some, however, suggest that primary surgery leads to a more rapid recovery and a shorter hospital stay. There is little evidence to convincingly show which is the best approach. The aim of surgery is clearance of debris from the pleural space, with reestablishment of a "single" pleural cavity free of loculations and full reexpansion of the lung. Postoperatively, early mobilization and physiotherapy are key. Thoracoscopy has the advantage of excellent visualization of the whole pleural space and minimal postoperative pain, allowing physiotherapy and mobilization to begin almost immediately.

3.2 Working Instruments

- 5-mm Camera port
- Two 5-mm accessory ports
- 5-mm 30-degree Telescope
- Two blunt atraumatic graspers with long jaw, e.g., a Johan grasper
- Sucker/irrigator

F.D. Munro • M. McHoney (✉)
Department of Paediatric Surgery, Royal Hospital for Sick Children, Edinburgh, UK

M. Wills
Department of Cardiothoracic Surgery, Royal Infirmary of Edinburgh, UK

M. McHoney et al. (eds.), *Color Atlas of Pediatric Anatomy, Laparoscopy, and Thoracoscopy*,
DOI 10.1007/978-3-662-53085-6_3

3.3 Positioning, Port Siting, and Ergonomic Considerations

The patient is positioned in the lateral position with the affected side uppermost, as for a thoracotomy. Placing a roll under the chest may be helpful in opening the rib spaces a little. One lung ventilation is not normally required, with a low pressure pneumothorax usually giving an adequate view. The patient should be firmly secured to the table so that the lateral roll can be used. It is helpful to have two screens positioned anterior and posterior to the chest as well as access to both sides of the patient, allowing optimum ergonomics while clearing the anterior and posterior parts of the pleural cavity.

If a chest drain is already in place, then the tract can act as the site for placement of the primary port. If not, then this should be placed where the empyema has been shown to be deepest on ultrasound images. Most commonly this is just below the tip of the scapula in the fourth or fifth interspace in mid-axilla. Further port placement will be dictated by the location of the empyema, but for the most common posterobasal empyema, one or two additional ports one or two rib spaces lower and in anterior and posterior axillary lines allow easy access to all areas of the pleural cavity.

3.4 Procedure

Figure 3.1a shows the pleural space in one patient, a loculated pleural space with strands of thin fibrous bands within the pus. On entering the pleural space, infected and loculated exudate and pus are seen (Fig. 3.1b). Initially the scope may be within the fibrinous debris, but by blunt dissection space can be created for insertion of further ports under direct vision.

Further pockets are identified as loculations are broken down (Fig. 3.2). Some bleeding is seen from the pleural surfaces.

Forceps are used to grasp solid purulent material and retrieve it via the ports (Fig. 3.3). Suction is used on liquid portions.

As dissection of the pleural space continues, various cavities containing exudate in various forms (solid, gelatinous, and liquid) can be found in different locations (Fig. 3.4). These are dealt with appropriately.

The forceps can be used to grasp the solid components (peel) in the pleural space, and Swiss rolling allows for removal of large segments at a time (Fig. 3.5).

More of the pleural space is progressively opened up and debrided (Fig. 3.6). Here cavernous spaces are identified and dissected to free the parietal and visceral pleurae from each other.

The peel over the parietal pleura is lifted using forceps (Fig. 3.7). This allows more of the normal lung surface to come into view.

As much of the visceral pleura is debrided as possible to allow visualization of more normal looking lung (Fig. 3.8). At this point it may also be beneficial to request a few forced tidal volumes and see the lung expanding more and more as the peel is removed.

Debridement is performed as extensively as possible from the apex of the lung down to the base (Fig. 3.9). The surgeon should ensure that the base of the lung has been freed from the diaphragm, as shown here. Most of the peel and exudate have been removed to reveal both inflamed surfaces.

Figure 3.10 shows the amount of the solid peel removed from the pleural cavity and surfaces.

At the end of the procedure the lung is inspected and visualized when inflating it with prolonged inspiratory breaths by the anesthetist. A chest drain is left in the pleural space through one of the port sites. If desired, two chest drains may be left: one apical and one basal drain. Chest drains are removed as soon as minimal drainage is obtained (usually less than 50 mL in 24 h).

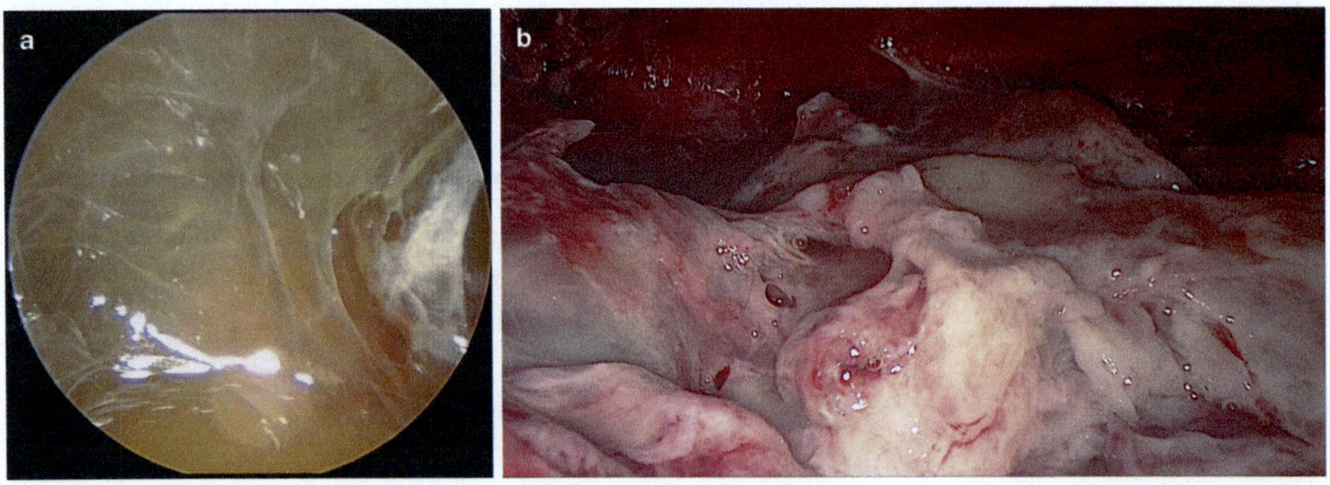

Fig. 3.1 (**a**, **b**) View of empyema

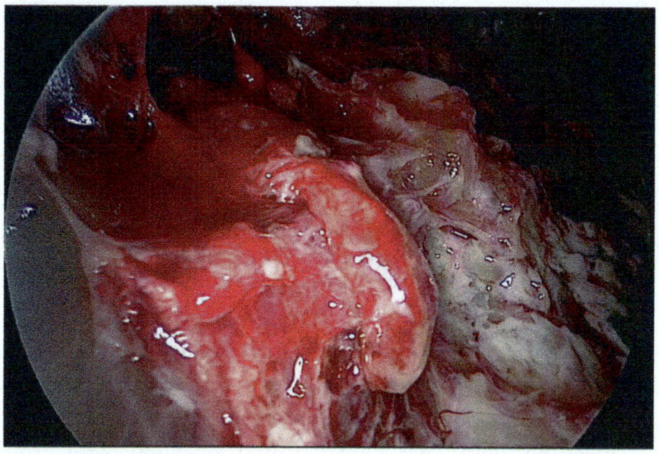

Fig. 3.2 Breaking down of loculations begins

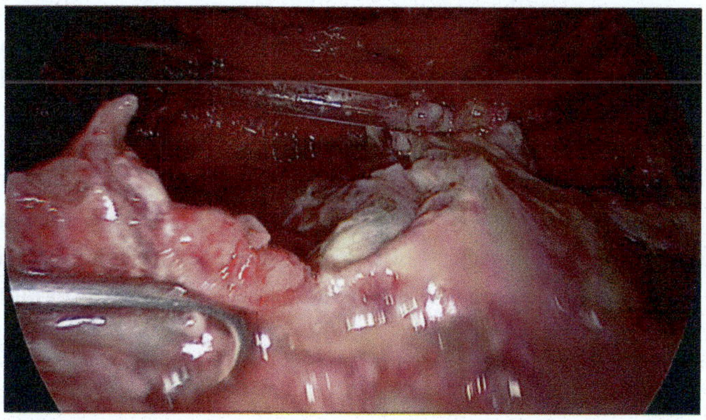

Fig. 3.3 Debridement using forceps and suction

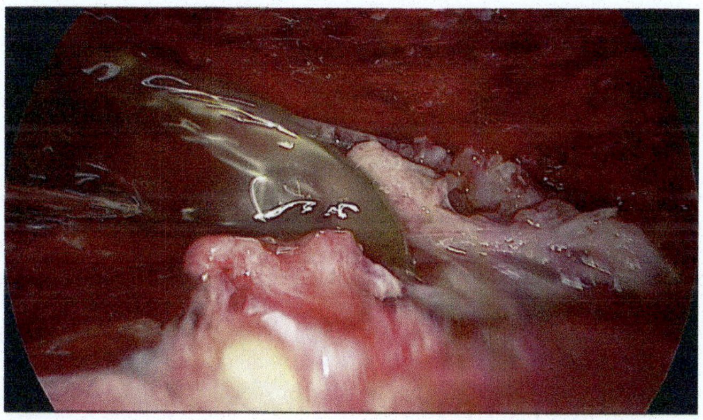

Fig. 3.4 Continued debridement

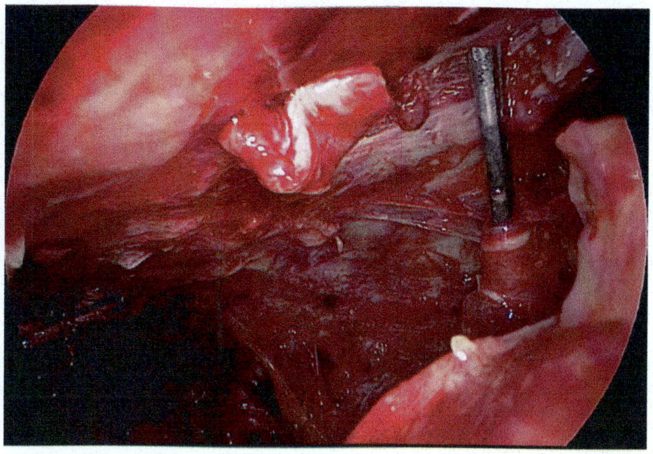

Fig. 3.5 Swiss rolling of the peel

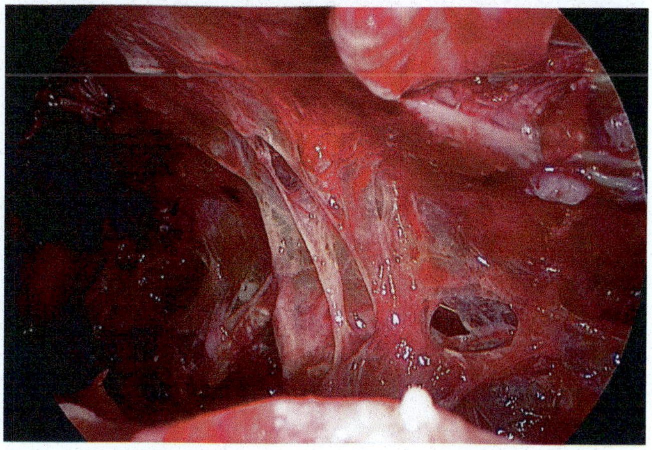

Fig. 3.6 Opening the pleural space more extensively

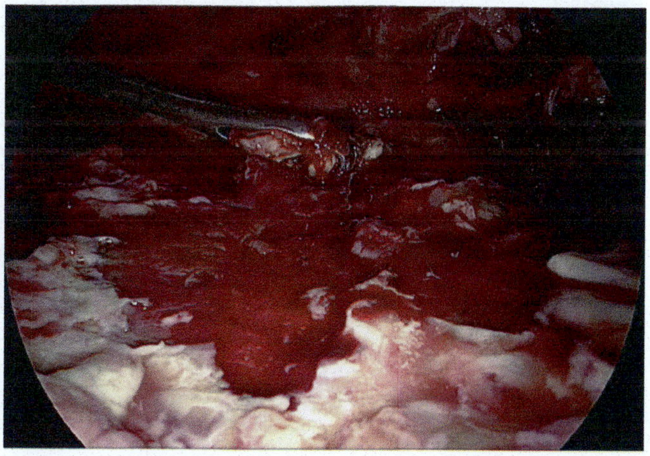

Fig. 3.7 Debridement of the visceral pleura

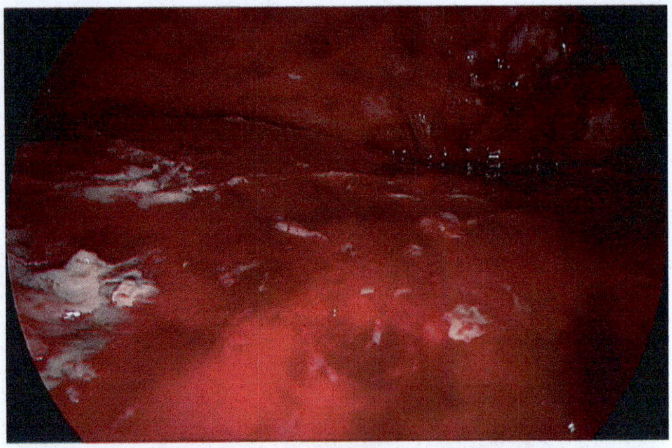

Fig. 3.8 Debrided visceral pleura

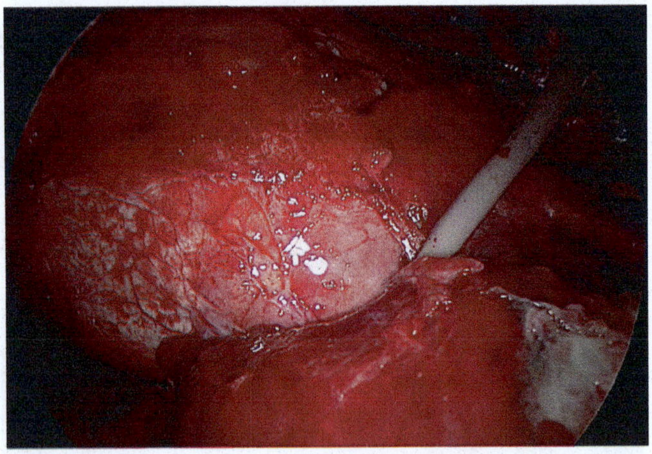

Fig. 3.9 Pleural clearance from apex to base

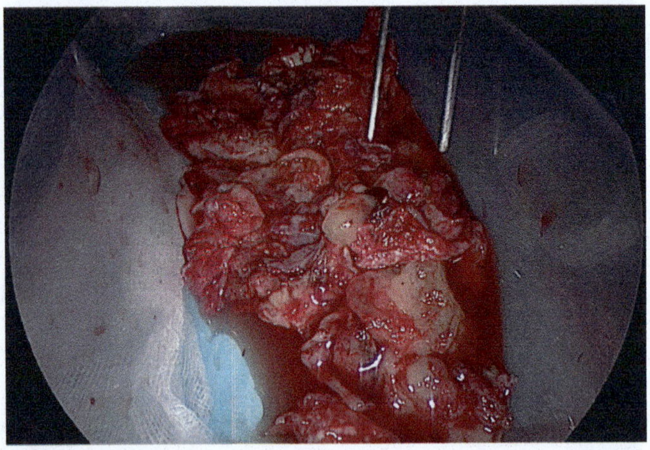

Fig. 3.10 The exudate removed

3.5 Alternatives

In the larger patient, ports may not be necessary, and instruments can be passed via a small stab incision in the intercostal spaces (VATS).

3.5.1 Highlights and Pitfalls

- The valves of ports rapidly fill with fibrin debris as this is removed with graspers. Reusable ports that can be easily dismantled will allow this debris to be cleared during surgery. Long cotton tip applicators may also be useful for cleaning debris and fluid that may obscure vision from inside the ports.
- Bleeding (slight oozing) from the pleural spaces is common and will normally settle, but it can interfere with and obscure vision. Changes in positioning (and the position of the camera if necessary) can assist in preventing blood from dripping onto the camera continuously.
- In general, adequate clearance of the pleural cavity should be achieved in 60–90 min. Prolonged efforts to remove all peel usually result in increasing bleeding and are to be avoided. It is important to keep off the mediastinal pleura to avoid damage to the phrenic nerve and other important structures.
- Postoperative radiographic appearances can be alarming in the first few days, often showing apparent reopacification, but as long as there is improvement in chest wall movement, air entry, and fever, these can be ignored.

Suggested Reading

Avansino JR, Goldman B, Sawin RS, Flum DR. Primary operative versus nonoperative therapy for pediatric empyema: a meta-analysis. Paediatrics. 2004;115:1652–9.

Balfour-Lynn IM, Abrahamson E, Cohen G, Hartley J, King S, Parikh D, et al. BTS guidelines for the management of pleural infection in children. Thorax. 2005;60:1–21.

Sonnappa S, Cohen G, Owens CM, van Doorn C, Cairns J, Staojevic S, et al. Comparison of urokinase and video-assisted thoracoscopic surgery for treatment of childhood empyema. Am J Respir Crit Care Med. 2006;174:221–7.

Spontaneous Pneumothorax

4

Merrill McHoney and Michael Singh

Abstract

True spontaneous pneumothorax results almost exclusively from apical subpleural blebs (or less often, from bullae) in tall, thin adolescent males, sometimes with a smoking history. Secondary pneumothorax may occur in patients with underlying lung disease (e.g., cystic fibrosis, Marfan's disease, asthma). Surgery is indicated in those who do not respond to conservative measures and in those with recurrence. The minimally invasive approach to spontaneous pneumothorax allows possible surgical cure while minimising the invasiveness, postoperative pain, and chest wall deformity from open surgery. It is relatively simple and may be performed with low-cost instrumentation.

Keywords

Spontaneous pneumothorax • Apical bullectomy • Pleurectomy

4.1 General Information

True spontaneous pneumothorax results almost exclusively from apical subpleural blebs (or less often, from bullae) in tall, thin adolescent males, sometimes with a smoking history. Secondary pneumothorax may occur in patients with underlying lung disease (e.g., cystic fibrosis, Marfan's disease, asthma). Surgery is indicated in those who do not respond to conservative measures and in those with recurrence. The minimally invasive approach to spontaneous pneumothorax allows possible surgical cure while minimising the invasiveness, postoperative pain, and chest wall deformity from open surgery. It is relatively simple and may be performed with low-cost instrumentation.

Anatomically, blebs are found most commonly on the apical segment of the upper lobe but also may be located elsewhere. The second most frequent location is the superior segment of the lower lobe.

M. McHoney (✉)
Department of Paediatric Surgery, Royal Hospital for Sick Children, Edinburgh, UK

M. Singh
Department of Paediatric Surgery, Birmingham Children's Hospital, Birmingham, UK

M. McHoney et al. (eds.), *Color Atlas of Pediatric Anatomy, Laparoscopy, and Thoracoscopy*,
DOI 10.1007/978-3-662-53085-6_4

4.2 Working Instruments

- 5-mm port
- Two 3- or 5-mm accessory ports
- 30° telescope
- Two Endoloops (Ethicon; Somerville, NJ, USA)
- Hook diathermy
- Scissors
- Maryland forceps
- (Other instruments described in “Alternatives”)

4.3 Relevant Anatomy

Blebs causing spontaneous pneumothorax almost exclusively involve the apex of the upper lobe (Fig. 4.1). The next commonest site is the superior segment of the lower lobe (Fig. 4.2). Other relevant anatomy is shown in Figs. 4.3 and 4.4.

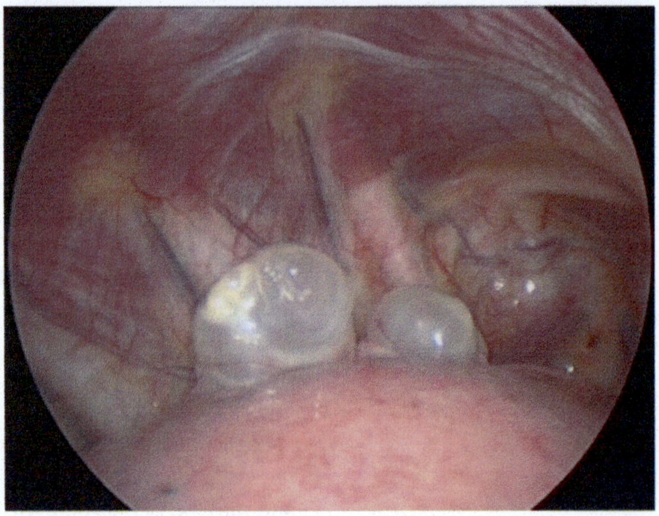

Fig. 4.1 Shown is the apex of the left lung in a patient with an apical bleb

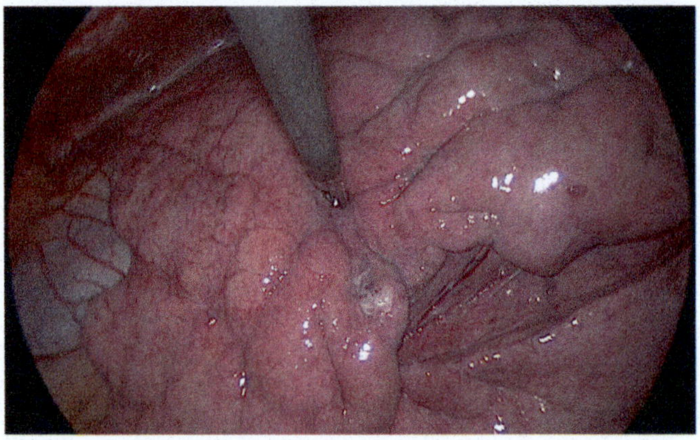

Fig. 4.2 A ruptured bleb is seen in the lingular segment of the left lung in a patient with recurrent pneumothorax

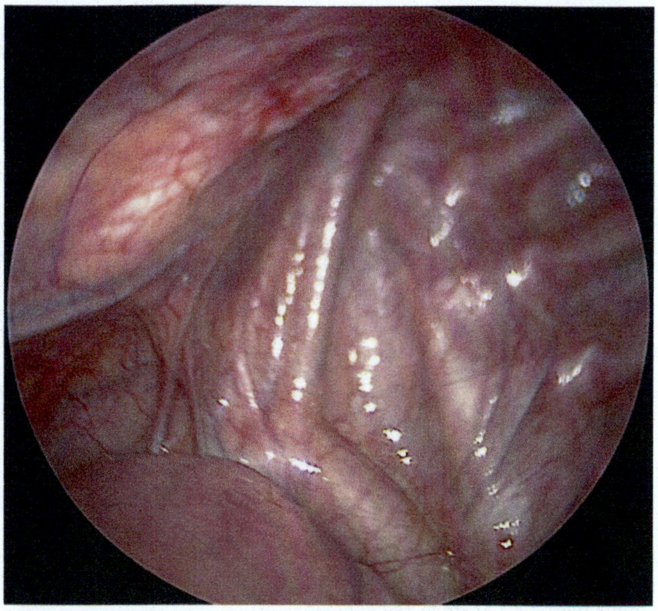

Fig. 4.3 Anatomy of the left mediastinum at the apex of the lung. The aortic arch is seen here giving rise to the left subclavian artery, which subsequently gives rise to the internal mammary (thoracic) artery with its accompanying vein. All are covered by parietal pleura

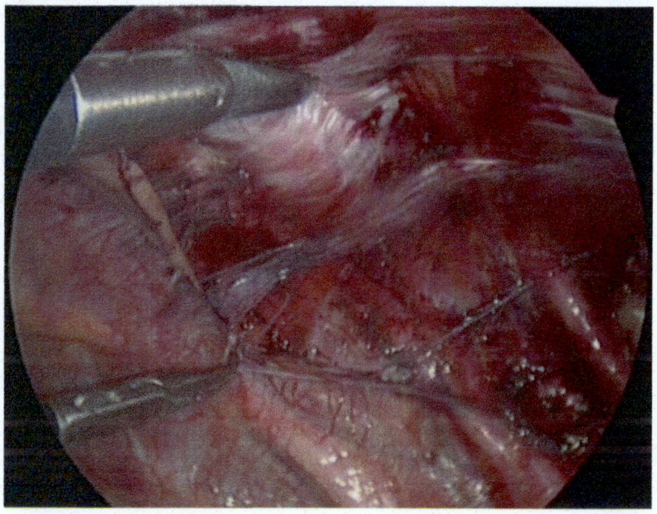

Fig. 4.4 Anatomy of the stripped right chest wall. The structures deep to the parietal pleura can be seen in this figure. The intercostal bundles (vein, artery, and nerve) are seen more clearly now beneath each rib. Shiny fibres of the innermost intercostal muscles are also seen overlying the inner intercostals

4.4 Positioning, Port Siting, and Ergonomic Considerations

The patient is placed in the lateral position (affected side up). Single-lung ventilation, if available, may assist with access by allowing desufflation of the affected side. A 5-mm port is used as the primary port and may be placed in the fourth to sixth intercostal space in the midaxillary line. Subsequent working ports are placed in the axillary folds, or the most ergonomically as determined after primary port insertion.

4.5 Procedure

The surgical steps are shown in the following figures; (Figs 4.5, 4.6, 4.7, 4.8, 4.9, 4.10, 4.11, 4.12, 4.13 and 4.14).

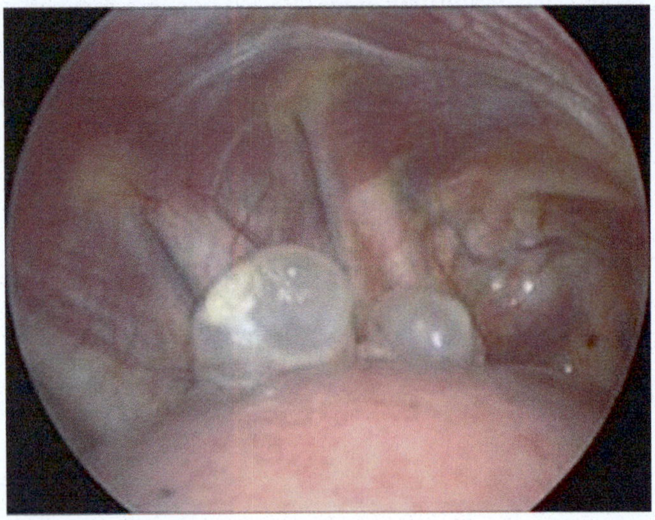

Fig. 4.5 Visualisation of the apical blebs. On entering the pleural cavity, the anatomy and position of the bleb are identified. In this illustration, the apical left lobe is involved

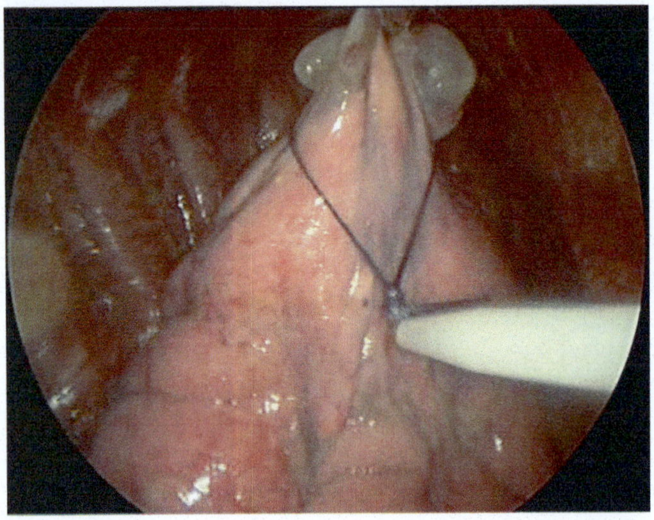

Fig. 4.6 Endoloop ligation of apical blebs. The blebs have been grasped by forceps in the left hand, and a Vicryl Endoloop is being used to encircle the bullae. The Endoloop includes a small portion of normal-looking lung at the base of the blebs. The Endoloop is secured tightly to ligate the involved bleb and adjacent lung

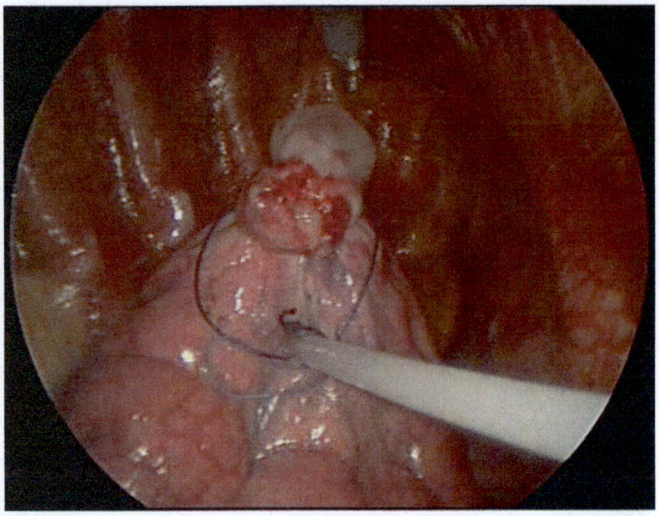

Fig. 4.7 Placing the second Endoloop on the bleb. The first Endoloop has been tied and cut. The second Endoloop then is placed at the base of the bleb. This again is tied and cut

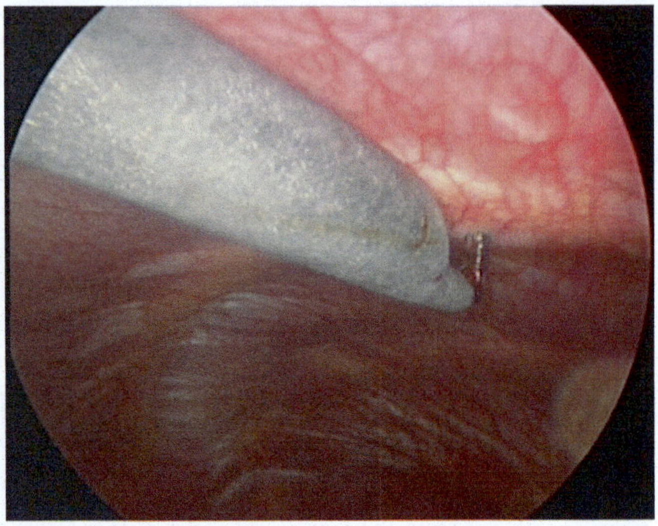

Fig. 4.8 Incision onto the rib bed. An incision (1–2 cm long) is made over the bed of an upper rib using the hook diathermy, remembering to avoid the intercostal bundle, by carefully placing the incision on the rib itself

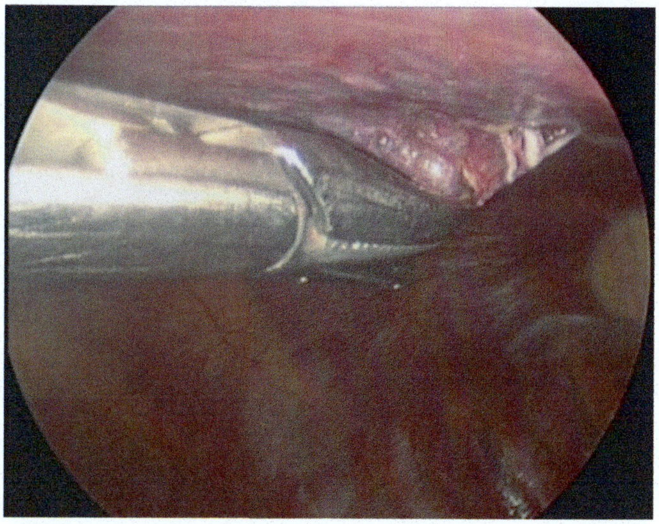

Fig. 4.9 Lifting the pleura off the rib. The incision in the pleura then is lifted off the rib, and the pleural dissection is begun by bluntly inserting graspers (or a pair of sciscors) beneath the pleura. The plane then is developed up to the apex of the hemithorax

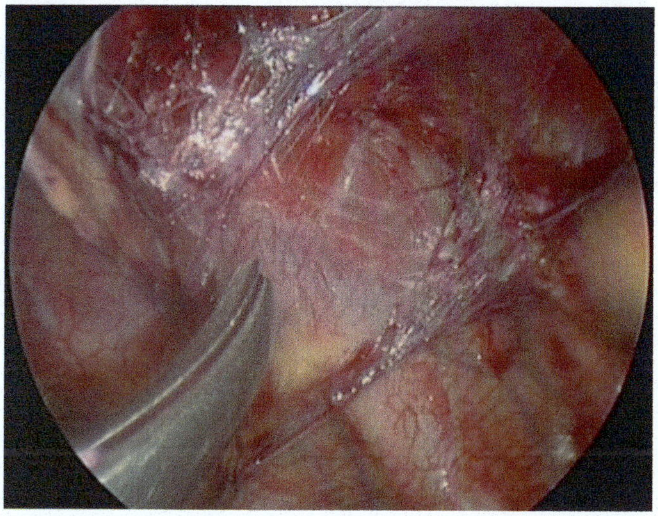

Fig. 4.10 Dissection of the pleura off the chest wall. The pleura is dissected free from the apical chest wall by gentle lifting and by blunt and sharp dissection where needed. As shown here, a definite plane sometimes may be found, which fascilitates dissection

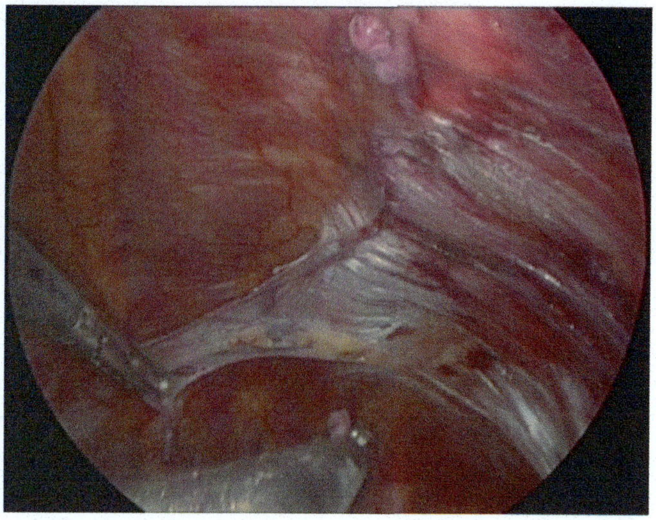

Fig. 4.11 Stripping off the pleura. In areas, the pleura may be dissected off by pulling and stripping it off the chest wall. This sometimes can be done in a Swiss roll manner to remove an appreciable segment of pleura

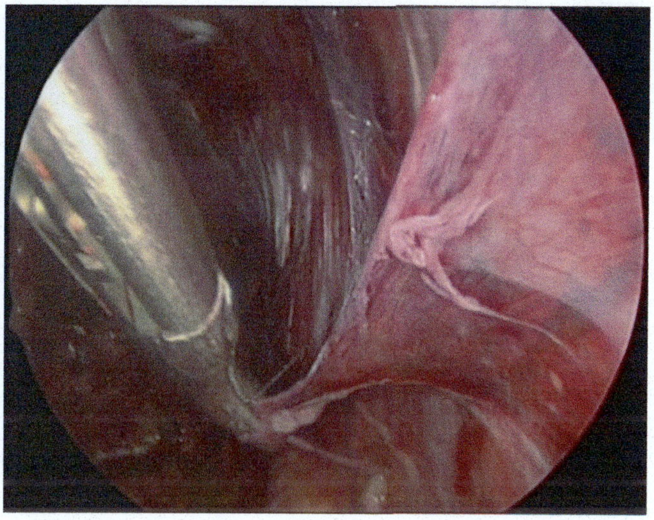

Fig. 4.12 Extending the dissection to the medial chest wall. The medial aspect of the superior chest wall also is stripped of the parietal pleura

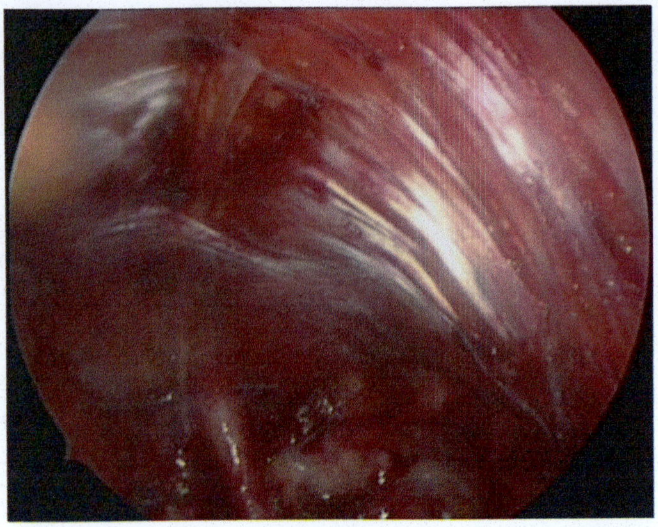

Fig. 4.13 The entire apical area of the hemithorax is stripped of pleura. After the entire apical area has been stripped of pleura, the intercostal muscles, ribs, and intercostal bundles can be seen at the end of the pleurectomy

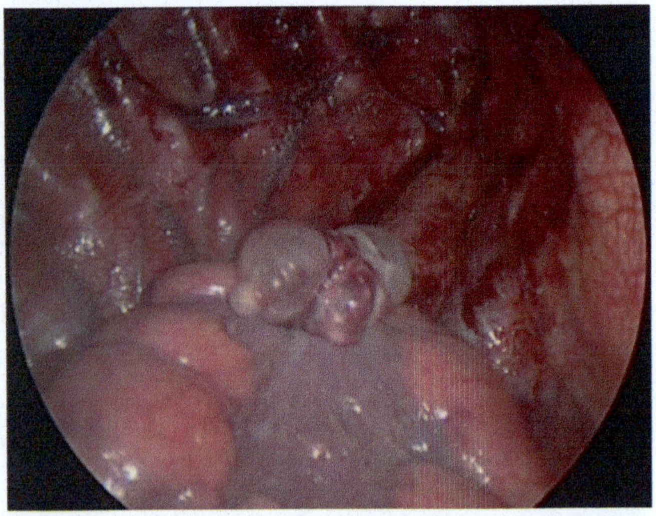

Fig. 4.14 Reinflation of the lung. The lung is then reinflated and seen to expand under vision before removing the scope. A chest drain connected to an underwater drain is inserted in one of the port sites and directed in an apical direction. The wounds are closed, the chest drain is connected to an underwater seal, and chest radiography is done postoperatively. Postoperative chest drain suction is not routine. The chest drain may be removed within 48 h if the lung remains insufflated

4.6 Alternatives (Figs. 4.15 and 4.16)

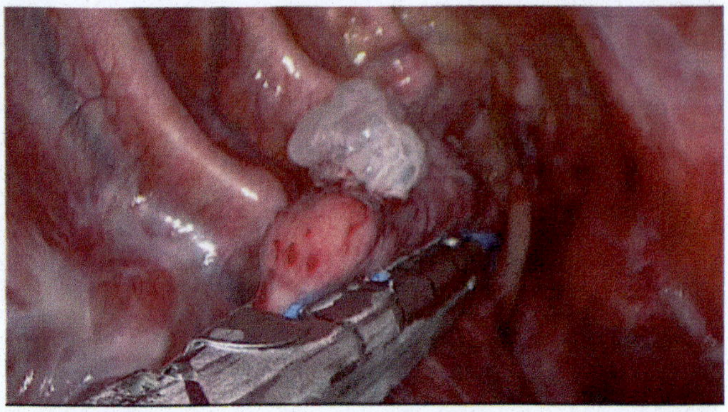

Fig. 4.15 Endoscopic bullectomy. An apical bullectomy may be performed using a blue cartridge endoscopic stapler. The stapler is inserted through a 12-mm port inserted posteriorly. The resection margin must extend into normal lung parenchyma

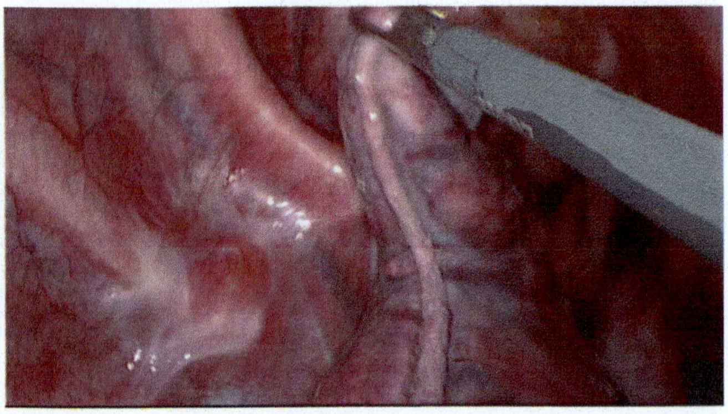

Fig. 4.16 Lung surface after endoscopic stapling. View of the lung surface before placement of a second bite of staples to complete the bullectomy shows the sealed surface at the staple line and an absence of any air leak. After completion of the bullectomy, a pleurectomy is done as described in Figs. 4.8, 4.9, 4.10, 4.11, 4.12 and 4.13. An alternative to pleurectomy is abrasion on the apical pleura using, for example, an abrasive scratch pad to stimulate pleurodesis. This technique usually is somewhat more bloody than the surgical pleurectomy described. Chemical pleurodesis is an alternative to physical abrasion

4.7 Highlights and Pitfalls

During the pleurectomy, once the plane is developed between the pleura on the rib bed, taking care to remain in this plane usually allows almost bloodless dissection.

Suggested Reading

Chen F, Yamada T, Aoyama A, Isowa N, Chihara K. Position of a chest tube at video-assisted thoracoscopic surgery for spontaneous pneumothorax. Respiration. 2006;73:329–33.

Schipper PH, Meyers BF. Surgery for bullous disease. In: Patterson GA, Cooper JD, Deslauriers J, Lerut AEM, Luketich JD, Rice TW, editors. Pearson's thoracic and esophageal surgery. 3rd ed. Philadelphia: Churchill Livingstone; 2008. p. 631–52.

Sepehripour AH, Nasir A, Shah R. Does mechanical pleurodesis result in better outcomes than chemical pleurodesis for recurrent primary spontaneous pneumothorax? R Interact Cardiovasc Thorac Surg. 2012;14(3):307–11.

Thoracoscopic Lung Biopsy and Segmentectomy

5

Kokila Lakhoo

Abstract

The potential advantages of the thoracoscopic approach include avoiding the morbidity associated with a thoracotomy and better visualisation. Planning and final treatment of the condition may be expedited. For any thoracoscopic procedure, full preparation must be made for thoracotomy in case conversion is required urgently.

Keywords

Thoracoscopy • Biopsy • Lung • Segmentectomy

5.1 General Information

The potential advantages of the thoracoscopic approach include avoiding the morbidity associated with a thoracotomy and better visualisation. Planning and final treatment of the condition may be expedited. For any thoracoscopic procedure, full preparation must be made for thoracotomy in case conversion is required urgently.

The preoperative workup includes careful documentation of the number of lesions and their location using specialised imaging, such as high-resolution CT scanning. Thoracoscopy is most useful for peripheral lesions. CT-guided localisation of deep-seated lesions using a needle technique or wire may be used with radiologic support. However, in some deep-seated lung lesions, thoracotomy may not be avoided.

5.2 Working Instruments

- 5-mm ports
- Either a 30° or 0° telescope
- Two 5-mm graspers
- 5-mm LigaSure (infants; Covidien; Boulder, CO, USA)
- 10-mm endovascular reticulated staplers (children)

5.3 Positioning, Port Siting, and Ergonomic Considerations

Single-lung ventilation is ideal; alternatively, insufflation with CO_2 using low flow (1 L/min) and low pressures (4–6 mm Hg in infants and 8–10 mm Hg in older children) may be used. The patient is placed in the lateral decubitus position. For anterior mediastinal lesions, a slight posterior tilt is useful, whereas for a posterior mediastinal lesion, a slight anterior tilt should be considered. The optical port usually is placed in the midaxillary line in the fourth to fifth intercostal space and two other ports are placed one to two spaces above in the posterior (or below the scapular tip) and anterior axillary line for upper-lobe lesions. Working ports for lesions in the lower lobe are placed one to two spaces below in the anterior and posterior axillary line. Port positions are variable and may be best decided after the lesion is identified telescopically. Some surgeons may prefer stab incisions for working instruments (Figs. 5.1, 5.2 and 5.3).

K. Lakhoo, PhD, FRCS, FCS, MRCPCH, MBCHB
Department of Paediatric Surgery, Oxford Children's Hospital, Oxford University Hospitals & University of Oxford, Oxford, UK

M. McHoney et al. (eds.), *Color Atlas of Pediatric Anatomy, Laparoscopy, and Thoracoscopy*,
DOI 10.1007/978-3-662-53085-6_5

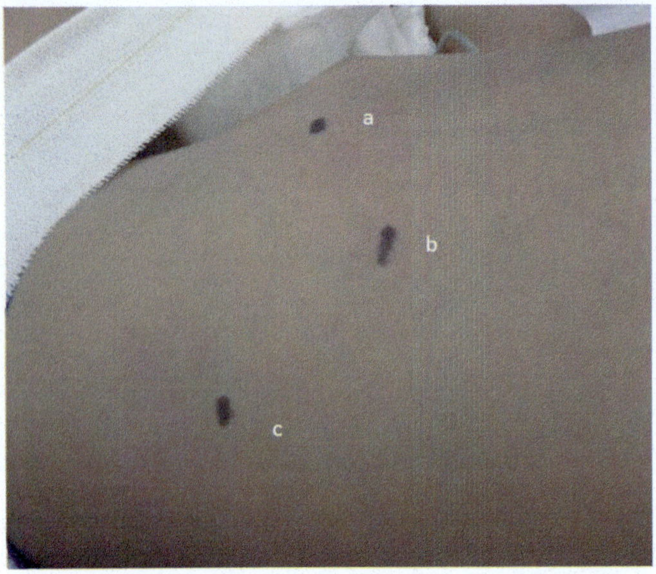

Fig. 5.1 Port position for a right upper-lobe lesion (infant). Working ports (**a** and **c**); optics (**b**)

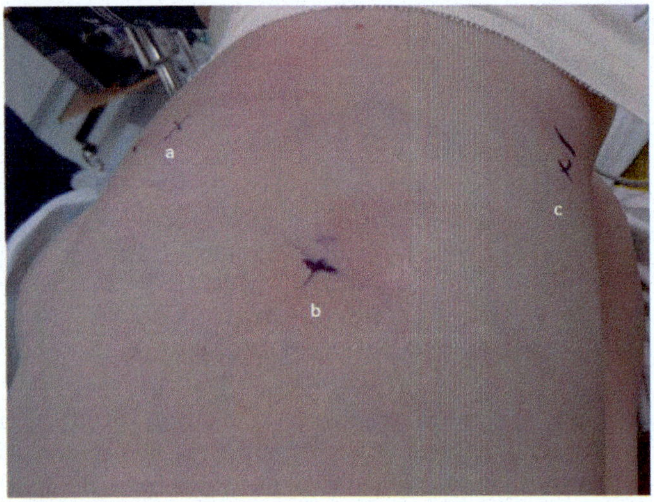

Fig. 5.2 Port position for a right upper-lobe lesion (child). Working ports (**a** and **c**); optics (**b**)

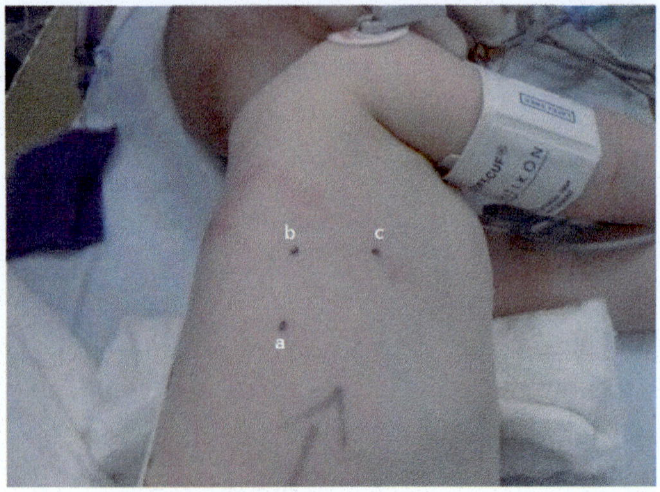

Fig. 5.3 Port position for a right lower-lobe lesion (infant under 6 months). Working ports (**a** and **c**); optics (**b**)

5.4 Relevant Imaging (Figs. 5.4, 5.5 and 5.6)

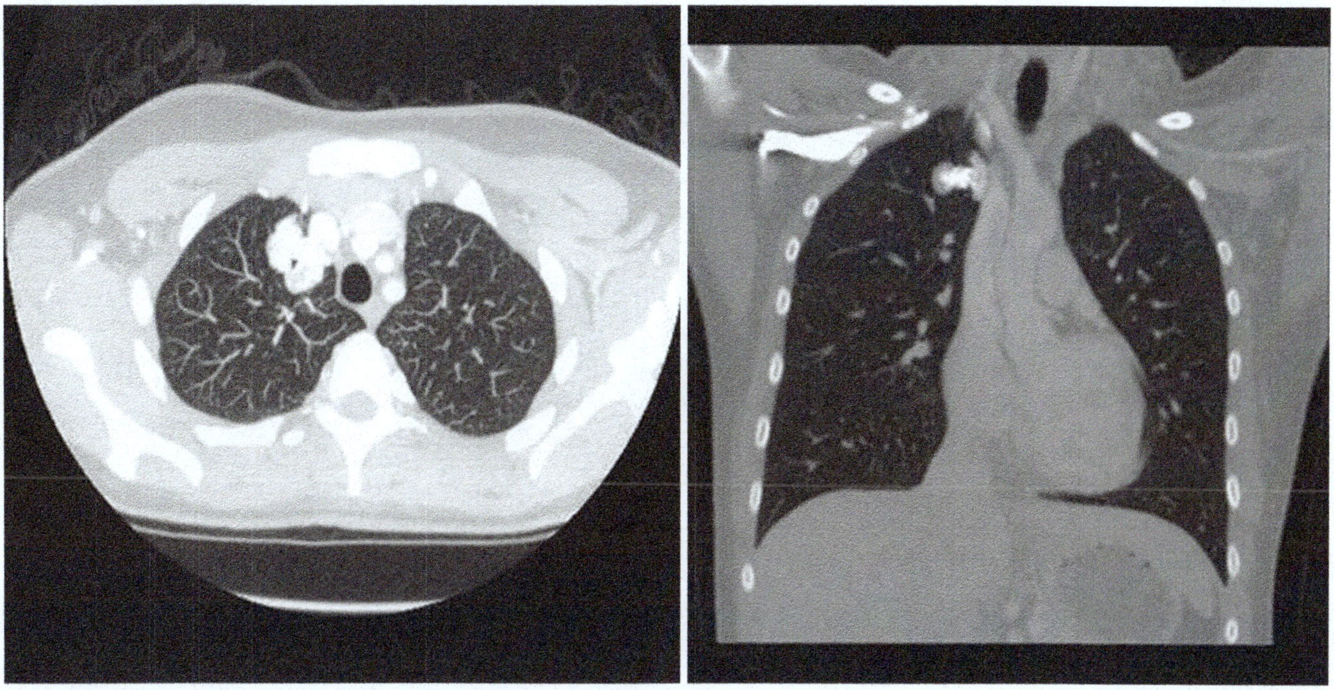

Fig. 5.4 CT scan of a secondary lung tumour. The surface lesion is accessible thoracoscopically

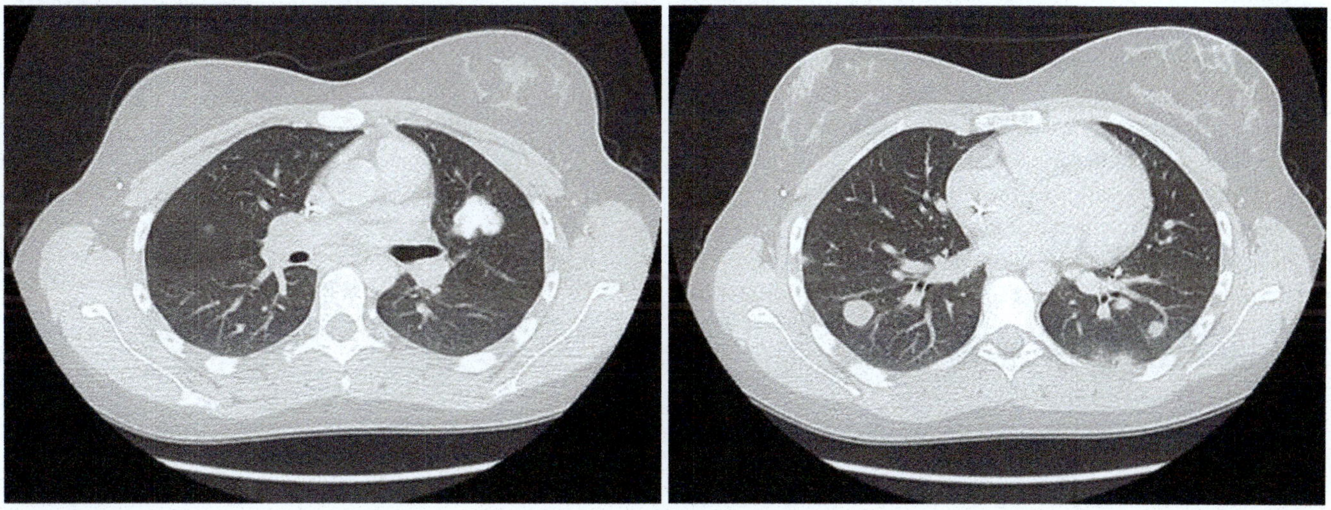

Fig. 5.5 CT scan of a secondary lung tumour. The image shows an intralobar lesion needing open thoracotomy

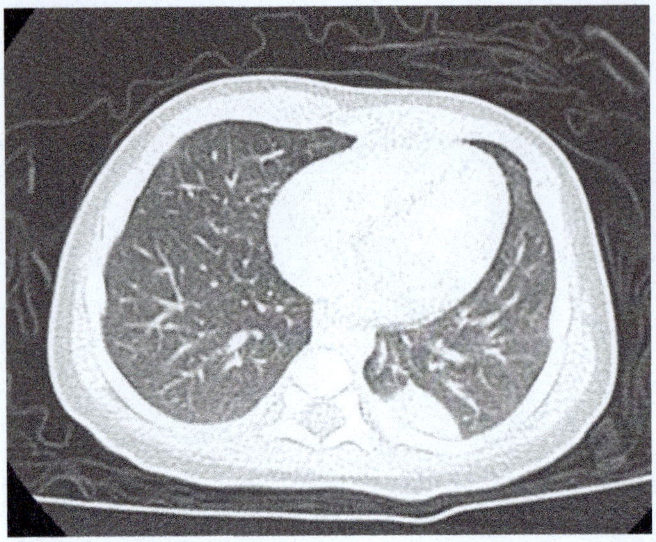

Fig. 5.6 CT scan of a segmental lesion in the left lower lobe

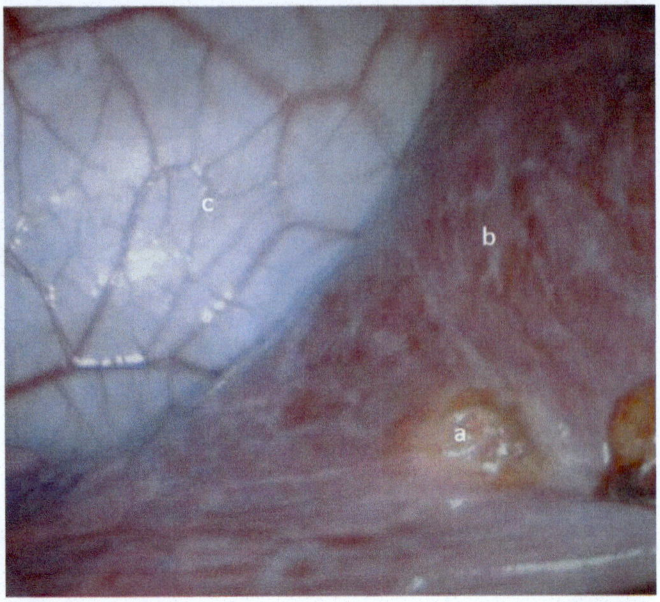

Fig. 5.7 Lung secondary from primary sarcoma. Lung secondary (**a**); lung (**b**); heart (**c**)

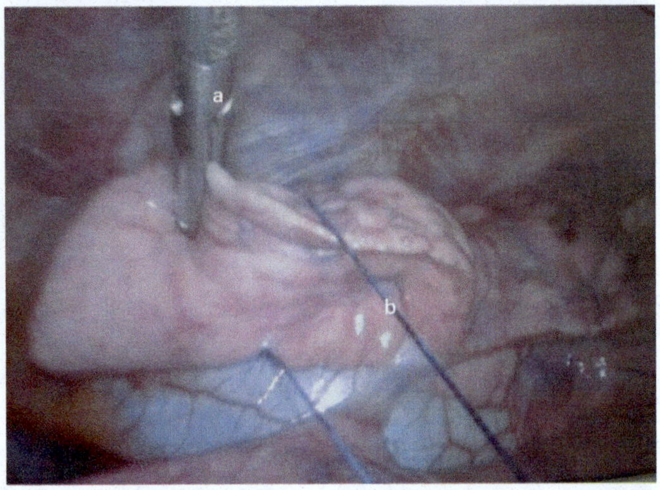

Fig. 5.8 Lung biopsy using an Endoloop. Grasper (**a**); endoloop (**b**)

5.5 Surgical Technique (Figs. 5.7, 5.8, 5.9 and 5.10)

- Identify the lesion.
- Place the desired tissue on stretch and take the lesion with a stapler (child) or LigaSure (infant).
- Sometimes two applications with the stapler or LigaSure are necessary to complete the excision.
- In infants, Endoloops (Ethicon, Somerville, NJ) may be used in place of LigaSure. In this situation two Endoloops are used over the lesion and the specimen is resected between the loops using scissors.
- Take the lesion with the rim of normal lung.
- Remove the specimen using a retrieval bag.
- Reexpand the lung and exclude air leaks.
- Introduce a chest drain using one of the port sites and close the remaining ones.
- A chest drain is not mandatory if the lung is fully expanded, with no air leaks.

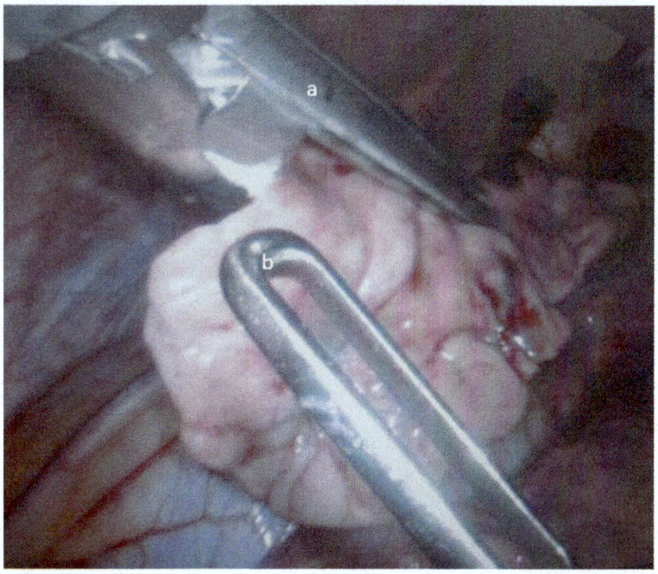

Fig. 5.9 Lung biopsy or segmentectomy using a stapling device. Stapling device (**a**); grasper (**b**)

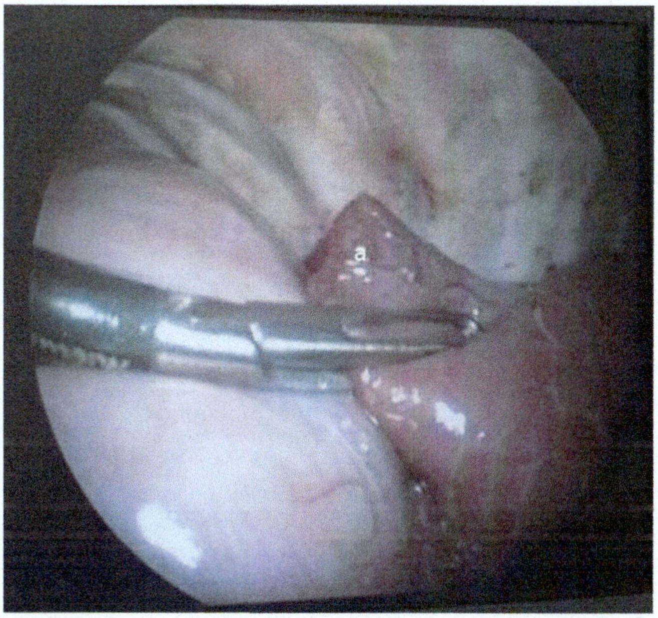

Fig. 5.10 Segmental lesion. Lesion (**a**)

5.6 Highlights and Pitfalls

- Plan the procedure using imaging.
- Discuss the procedure with your anaesthetist.
- Initial collapse of the lung will cause temporary oxygen desaturation.
- Port positioning is vital, thus it may change after insertion of the optics.
- The surgeon and assistant are on the same side of the patient to avoid working backwards.
- Lung lesions within parenchyma may require open thoracotomy.
- Be prepared to convert if the lesion is not visible or the procedure is suboptimal.

Suggested Reading

Holcomb 3rd GW, Georgeson KE, Rothenberg SS. Atlas of paediatric laparoscopy and thoracoscopy. Philadelphia: Elsevier; 2008.
Saxena AK, Hollwarth ME. Essentials of pediatric endoscopic surgery. Berlin: Springer; 2009.

Thoracic Sympathectomy

6

Giampiero Soccorso and Michael Singh

Abstract

The main indication for the thoracoscopic sympathectomy is hyperhidrosis: excessive sweating in amounts greater than physiologically needed for thermoregulation. Adolescents usually complain of palmar hyperhidrosis, but axillary, plantar, and craniofacial sites can also be involved.

Keywords

Hyperhidrosis • Sympathetic chain • Thorascopic

6.1 General Information

The main indication for the thoracoscopic sympathectomy is hyperhidrosis: excessive sweating in amounts greater than physiologically needed for thermoregulation. Adolescents usually complain of palmar hyperhidrosis, but axillary, plantar, and craniofacial sites can also be involved.

For palmar hyperhidrosis, the sympathetic chain over ribs two to four is resected.

6.2 Working Instruments

- 0 degree 5-mm or 30 degree 5-mm telescope
- 3-, 5-mm ports
- 5-mm hook cautery
- 5-mm scissors, Johans (straight) or Kelly (curved) grasping forceps
- Suction

6.3 Positioning, Port Siting, and Ergonomic Considerations

The patient is placed in the lateral position on top of an axillary roll. The monitor is positioned above the patient's head. The first port is placed anterior to the inferior angle of the scapular. A pneumothorax of 6 mmHg is established with flows of 1.5 L/min. The second and third ports are placed in the fourth or fifth intercostal space in the mid-axillary line and posteriorly.

G. Soccorso
Birmingham Children's Hospital NHS Foundation Trust, Birmingham, UK

M. Singh (✉)
Department of Paediatric Surgery, Birmingham Children's Hospital, Birmingham, UK

M. McHoney et al. (eds.), *Color Atlas of Pediatric Anatomy, Laparoscopy, and Thoracoscopy*,
DOI 10.1007/978-3-662-53085-6_6

6.4 Relevant Anatomy

The sympathetic chain runs vertically over the necks of the ribs in the upper costovertebral region under the parietal pleura. It lies just lateral to the superior intercostal vein. The stellate ganglion lies above and below the neck of rib one, deep to a yellow fat pad just inferior to the subclavian artery (Fig. 6.1).

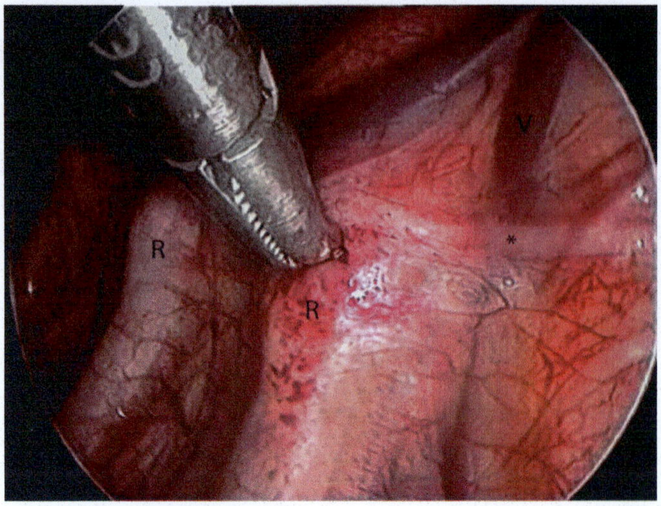

Fig. 6.1 The sympathetic chain (*asterisk*) runs along the neck of the ribs (*R*) and anterior to the intercostal veins (*V*)

6.5 Surgical Technique

We perform bilateral synchronous sympathectomy starting on the right side. Hook diathermy dissection of parietal pleura commences at rib four to expose the sympathetic chain. Once the pleura has been opened, dissection of the sympathetic trunk begins at the level of the neck of rib four (T4) and continues cephalad. The chain is dissected to the level of rib two (T2) (Fig. 6.2). Dissection should be carried out with scissors and no diathermy in order to avoid conduction and thermal injury to the stellate ganglion which can result in Horner syndrome (Fig. 6.3).

Applying traction to the transacted end, the chain is freed posteriorly and on each side by a combination of sharp and blunt dissection. Once the chain has been freed up to rib two, the final stage of resection is carried out with the scissors to avoid damage to the stellate ganglion.

After removal of the resected sympathetic trunk, diathermy cautery for 5 cm on the neck of ribs two to four is carried out to divide the rami communicantes.

A number 16 Fr chest tube is inserted via a port. After expanding the patient's lungs with positive pressure ventilation, the tube is removed from the chest at positive pressure to avoid a residual pneumothorax. Any residual CO_2 will be absorbed over a few hours. No thoracic drain is needed postoperatively. The procedure is then repeated on the left side. An overnight hospital stay may be required for pain control. A chest radiograph is obtained after the following day to ensure complete lung expansion.

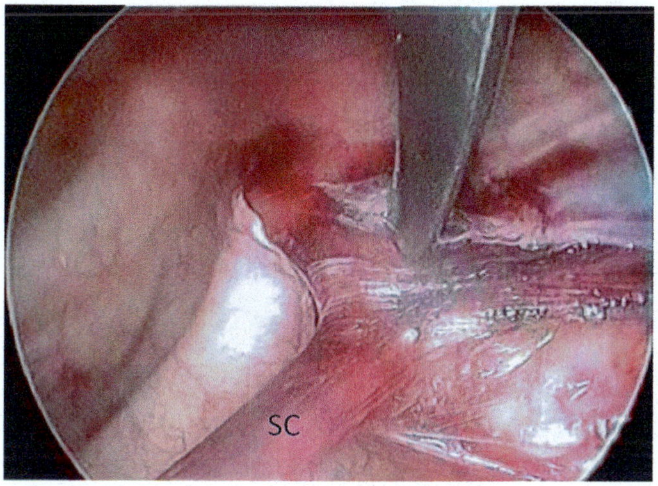

Fig. 6.2 The sympathetic chain (*SC*) is resected from ribs four to two

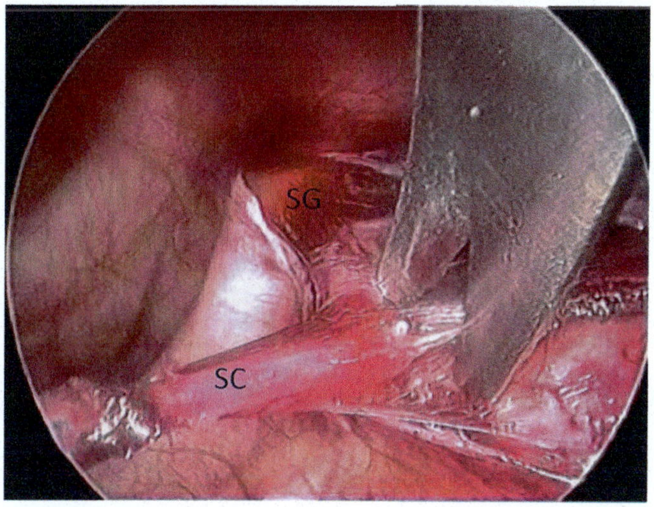

Fig. 6.3 The sympathetic chain (*SC*) is cut without diathermy at the level of the second rib (T2) to avoid injury to the stellate ganglion (*SG*)

6.6 Alternatives

A spatulated diathermy blade instead of a hook has been advocated by some authors to avoid slipping and inadvertent injuries to surrounding vessels. The procedure can be done in the prone position as well.

6.7 Highlights and Pitfalls

- Diathermy use should be kept to a minimum to avoid conduction or thermal injury to the stellate ganglion.
- An unusual double sympathetic trunk, multiple branches of the trunk, or an aberrant nerve bundle of Kuntz should be recognized and divided.
- The sympathetic trunk usually crosses the intercostal veins anteriorly, but one or more of these veins may run anterior to the trunk. In this case dissection (inferior and superior) of the trunk from the posterior aspect of the vein is carried out.
- Avoid blind diathermy of hemorrhage from intercostal vessels to prevent thermal injury to stellate ganglion.
- Injury to the thoracic duct can cause a chylothorax.

Suggested Reading

Cameron AE, Connery CP, De Campos JR, Hashmonai M, Licht PB, Schick CH, Bischof G. Endoscopic thoracic sympathectomy for primary hyperhidrosis: a 16-year follow up in a single UK centre. Int Soc Sympath Surg. 2014;12:59.

Cerfolio RJ, De Campos JR, Bryant AS, Connery CP, Miller DL, DeCamp MM, et al. The Society of Thoracic Surgeons expert consensus for the surgical treatment of hyperhidrosis. Ann Thorac Surg. 2011;91:1642–8.

Thymectomy

7

Melissa Short and Dakshesh H. Parikh

Abstract

Thymectomy in children is performed most commonly for autoimmune myasthenia gravis. Although there are no randomized controlled trials, there is evidence to show that selected patients undergoing thymectomy are more likely to improve, become asymptomatic, or achieve medicine-free remission compared with those who do not [1]. Advances in thoracoscopy have allowed this procedure to be carried out with reduced morbidity compared with open thymectomy.

Keywords

Thoracoscopy • Thymus • Myasthenia gravis

7.1 General Information

Thymectomy in children is performed most commonly for autoimmune myasthenia gravis. Although there are no randomized controlled trials, there is evidence to show that selected patients undergoing thymectomy are more likely to improve, become asymptomatic, or achieve medicine-free remission compared with those who do not [1]. Advances in thoracoscopy have allowed this procedure to be carried out with reduced morbidity compared with open thymectomy.

M. Short
Department of Paediatric Surgery, Birmingham Children's Hospital, Birmingham, UK

D.H. Parikh (✉)
Department of Paediatric Surgery, Birmingham Children's Hospital NHS Foundation Trust, Birmingham, UK

M. McHoney et al. (eds.), *Color Atlas of Pediatric Anatomy, Laparoscopy, and Thoracoscopy*,
DOI 10.1007/978-3-662-53085-6_7

7.2 Relevant Anatomy

The thymus sits in the anterior mediastinum. It has two upper poles that extend into the neck and lower poles that drape over the pericardium of the heart, connected by a small isthmus in the middle. The thymus receives its arterial supply from the internal mammary artery and/or inferior thyroidal artery. A significant variation in size, shape, and extent exists, and the knowledge of the thymus's anatomic variations is vitally important for surgery [2]. We therefore recommend a preoperative CT scan with intravenous (IV) contrast before thymectomy (Fig. 7.1).

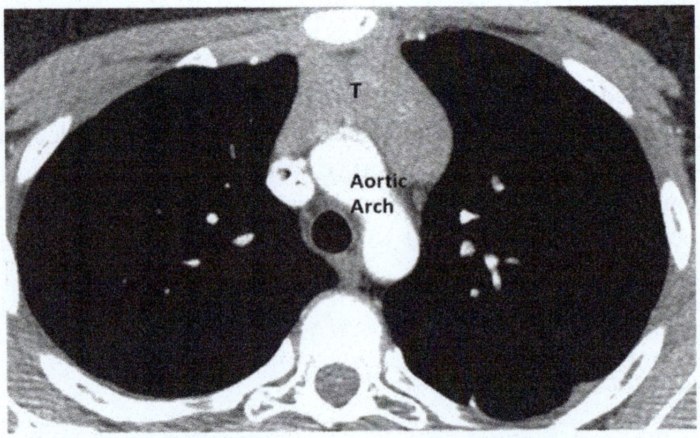

Fig. 7.1 Preoperative CT scan with IV contrast. The scan is obtained to delineate the thymus and to identify any anomalous blood supply or, rarely, associated thymic pathology. There is no ectopic tissue in this scan

7.3 Working Instruments

- 30° scope
- Three ports (5 mm): one camera port and two working ports
- The ultrasonic surgical system is preferred over bipolar electrocoagulation sealing devices
- Monopolar diathermy

7.4 Surgical Technique

7.4.1 Port Positioning with Ergonomic Considerations

Single-lung anaesthesia helps but is not absolutely essential. For right thoracoscopy:

- The patient is supine with a sandbag under the right hemithorax (Fig. 7.2).
- The right arm is abducted to 90°
- The camera port is placed in the mid- or posterior axillary line at approximately the fourth or fifth intercostal space.
- The two working ports are placed at the anterior axillary line, one at approximately the third intercostal space and the other at approximately the fifth or sixth intercostal space (Fig. 7.3).

Pressures should be 4–6 mmHg and the flow 2 L/min. (If single-lung anaesthesia is used, CO_2 insufflation may be stopped once the lung has collapsed on the right side.)

Antibiotics should be given prophylactically and in three doses postoperatively; however, antibiotics thought to exacerbate myasthenia gravis, such as aminoglycosides, quinolones, telithromycin, azithromycin, erythromycin, clindamycin, ampicillin, imipenem, vancomycin, and metronidazole, should be avoided [3].

Dissection is commenced by lifting the inferior pole overlying the pericardium using either the monopolar diathermy hook or scissors to divide the visceral pleura and loose areolar tissue around the thymic capsule (Fig. 7.4).

The posterior surface is dissected off the pericardium gradually with cephalad progression (Fig. 7.5).

The short thymic vein or veins draining into the bracheocephalic vein should be identified next and isolated (Fig. 7.6).

We divide this vein using an ultrasonic device, but if the vein is large, it may require division between the surgical clips. Dissection then should proceed towards the right superior pole. The deep cervical fascia adheres to the superior pole as it passes into the neck; it requires careful dissection with identification and division of the arterial supply with an ultrasonic device.

Once the thymus is dissected completely on both sides, it is removed by extending the anterior port incision (Fig. 7.7).

Port site closure is in layers involving the muscle, subcutaneous and skin with absorbable sutures. The right pleural cavity in intubated to evacuate CO_2 with positive pressure ventilation to reexpand the right lung before closure.

Ectopic thymic tissue beyond the main thymic gland capsule may account for some recurrences after an initial improvement post thymectomy. Fifteen percent of patients with myasthenia gravis will develop a thymoma, which accounts for 1 % of anterior mediastinal childhood tumours. As these tumours may be malignant, they require radical thymectomy [4].

Structures to identify and/or be aware of include (Fig. 7.8):

- Right innominate artery
- Internal mammary artery and veins
- Phrenic nerves
- Pericardium

7.4.2 Alternatives

- Open approach via a median sternotomy
- Robotic surgery
- Transcervical approach

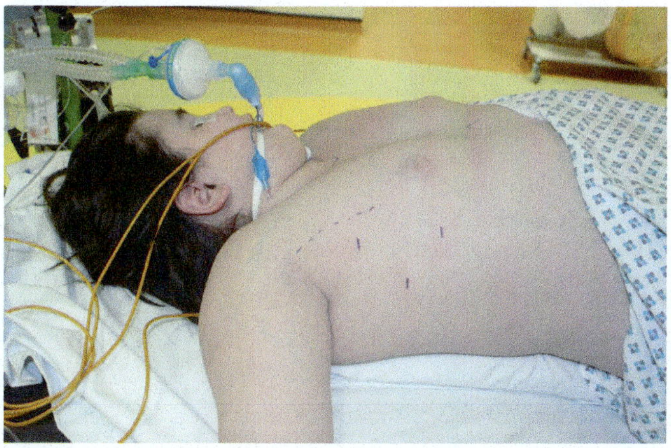

Fig. 7.2 Exposure is from the left nipple to the right posterior axillary line with a sandbag underneath the right hemithorax. The *dotted line* represents the lateral border of the pectoralis major muscle. Port positions are demonstrated by the three *horizontal lines*

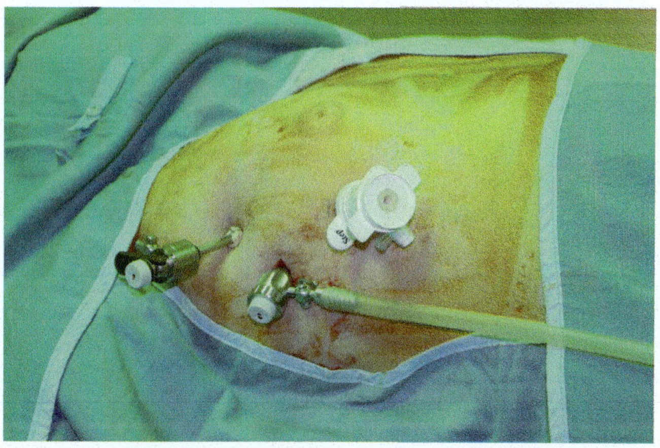

Fig. 7.3 Posterior visual port in the mid-/posterior axillary line. Two anterior working ports are seen in the anterior axillary line

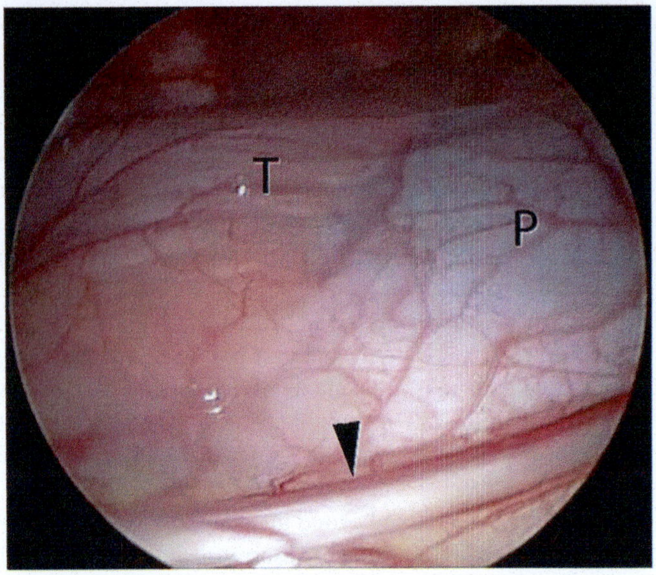

Fig. 7.4 Right lower pole of the thymus (*T*) draped over the pericardium (*P*) with the phrenic nerve (*arrowhead*)

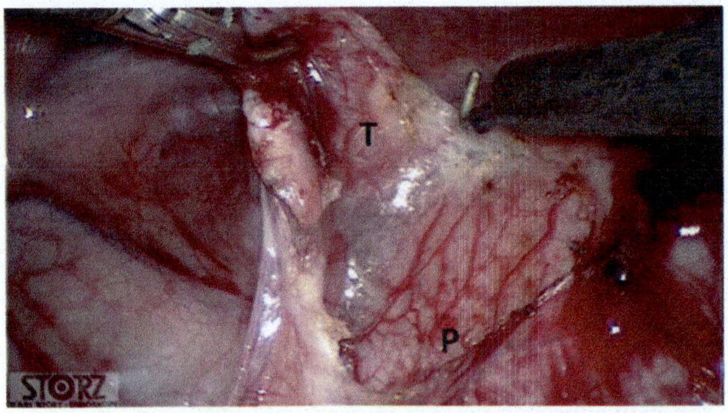

Fig. 7.5 The initial dissection takes place at the inferior border of the thymus over the pericardium. Here, we see loose areolar tissue between the thymus and pericardium being diathermied. The superior vena cava (SVC) also may be seen at the lateral border of the thymus

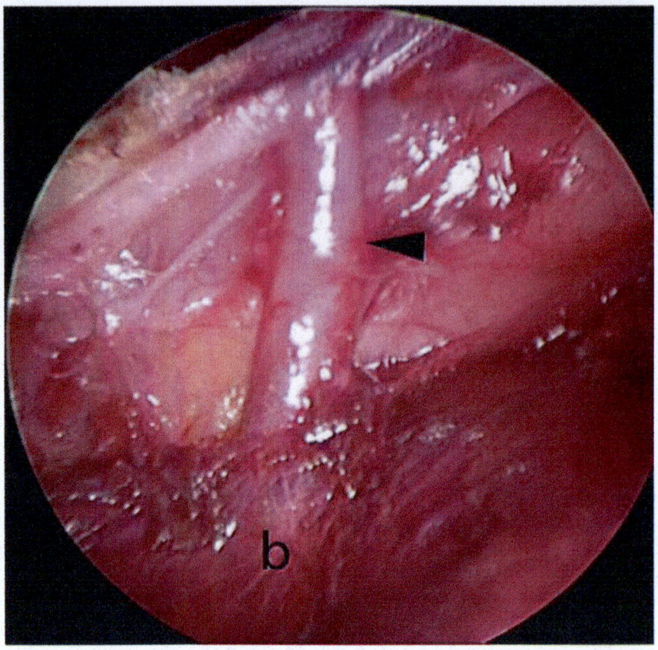

Fig. 7.6 Short thymic veins draining into a generally large brachiocephalic vein (*b*)

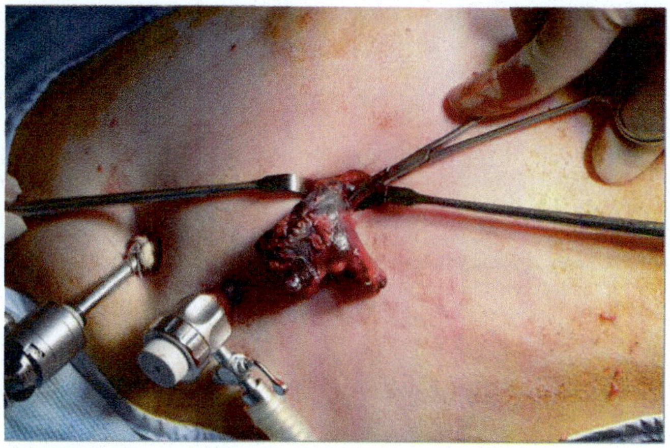

Fig. 7.7 Removal is via the anterior port

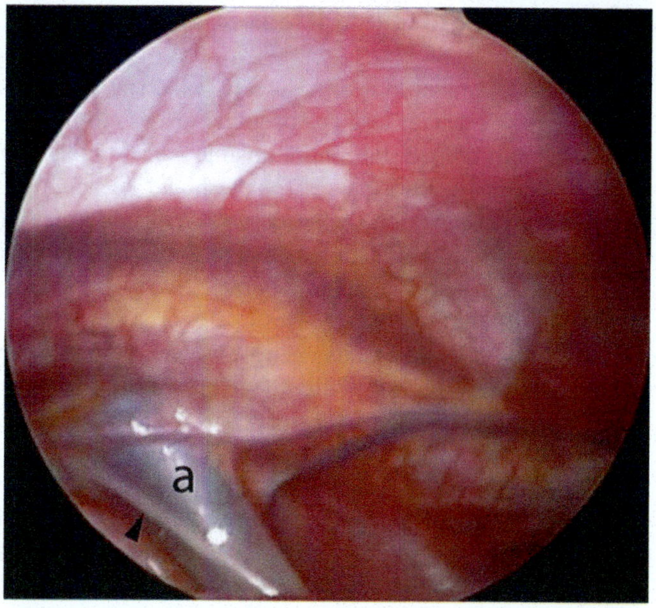

Fig. 7.8 Anterior chest wall with the SVC (*a*) and phrenic nerve (*arrowhead*) entering through the thoracic inlet. This forms the right lateral border of the thymus. The right internal mammary artery (which occasionally gives rise to thymic branches) and veins are visible

7.5 Highlights and Pitfalls

This procedure is technically safe in experienced hands and is associated with less postoperative pain and morbidity. Total excision of the entire thymic gland, including neck extensions of the superior poles should be achieved. Thoracoscopic technique allows good visualisation of important structures, as detailed earlier, especially the short thymic veins and arterial supply. Phrenic nerves are visualised easily; therefore, they are less likely to be injured. No chest drain is needed. Patients generally are discharged the following day. Follow-up is with the referring physician.

Pitfalls include incomplete resection of the thymus. There are descriptions in the adult literature of recurrence due to aberrant thymic tissue, requiring further surgery, with the most likely site being pleural [5]. There is always a risk with thoracosopic surgery that conversion may occur, and this should be achieved preferably through a midline sternotomy.

References

1. Gronseth GS, Barohn RJ. Practice parameter: thymectomy for autoimmune myasthenia gravis (an evidence-based review): report of the quality standards subcommittee of the American Academy of Neurology. Neurology. 2000;55:7–15.
2. Parikh DH, Crabbe DCG. The thymus and myasthenia gravis. In: Crabbe DCG, Parikh DH, Auldist AW, Rothenberg SS, editors. Paediatric thoracic surgery. London: Springer; 2009. p. 579–88.
3. Karcic AA. Drugs that can worsen myasthenia gravis. Postgrad Med. 2000;108(2):25.
4. Romi F. Thymoma in myasthenia gravis: from diagnosis to treatment. Autoimmune Dis. 2011;2011:474512. doi:10.4061/2011/474512.
5. Haniuda M, Kondo R, Numanami H, Makiuchi A, Machida E, Amano J. Recurrence of thymomas: clinicopathological features, re-operation, and outcome. J Surg Oncol. 2001;78:183–8.

Suggested Reading

Pavia R, Mondello B, Monaco F, Pavone A, Micali V, Barresi P, et al. Role of thymectomy in the treatment of myasthenia gravis: considerations and personal cases [in Italian]. G Chir. 2003;24:255–8.

Safieddine N, Keshavjee S. Anatomy of the thymus gland. Thorac Surg Clin. 2011;21:191–5.

Spillane J, Higham E, Kullmann DM. Myasthenia gravis. BMJ. 2012;345:e8497.

8 Mediastinal Cysts

Michael Singh

Abstract

Most mediastinal cysts are diagnosed at antenatal ultrasound scanning; postnatally most patients are asymptomatic. The differential diagnosis includes: bronchogenic cysts, oesophageal duplications, neurenteric cysts, thymic cysts, cystic hygromas, and teratomas. Thoracoscopic excision is the preferred approach for antenatally diagnosed asymptomatic lesions. Infected cysts are more difficult to excise thoracoscopically. This chapter outlines the operative procedure for thoracoscopic excision of a mediastinal cyst.

Keywords

Mediastinal cysts • Oesophageal duplication • Bronchogenic cyst

Most mediastinal cysts are diagnosed at antenatal ultrasound scanning; postnatally most patients are asymptomatic. The differential diagnosis includes: bronchogenic cysts, oesophageal duplications, neurenteric cysts, thymic cysts, cystic hygromas, and teratomas. Thoracoscopic excision is the preferred approach for antenatally diagnosed asymptomatic lesions. Infected cysts are more difficult to excise thoracoscopically. This chapter outlines the operative procedure for thoracoscopic excision of a mediastinal cyst.

8.1 General Information

Most mediastinal cysts are diagnosed at antenatal ultrasound scanning, postnatally most patients are asymptomatic. The differential diagnosis includes: bronchogenic cysts, oesophageal duplications, neuroenteric cysts, thymic cysts, cystic hygromas, and teratomas. Thoracoscopic excision is the preferred approach for antenatally diagnosed asymptomatic lesions. Infected cysts are more difficult to excise thoracoscopically.

M. Singh
Department of Paediatric Surgery, Birmingham Children's Hospital, Birmingham, UK

8.2 Working Instruments

- 3- or 5-mm ports and instruments: Maryland, Kelly, and Johan graspers; Mixter dissector; scissors; needle holder
- Monopolar hook diathermy, ultrasonic scalpel, LigaSure (Valley Lab; Boulder, CO, USA)
- 5-mm 0° telescope
- Suction irrigation

8.3 Positioning, Port Siting, and Ergonomic Considerations

The patient is positioned laterally with the affected side up and an axillary roll underneath, as for a thoracotomy. For superior mediastinial cysts, the monitor is positioned over the patient's head and the surgeon stands at the foot of the table. The converse arrangement is used for cysts in the lower half of the chest. The first port (5-mm, optical) is inserted anterior to the inferior angle of the scapula in the adjoining intercostal space. A pneumothorax of 5–6 mm Hg with flows of 1.5–2 L/min is maintained. After lung collapse, two or three working ports are inserted under direct vision between the anterior and posterior axillary lines to achieve effective triangulation and ergonomic working. Occasionally, a port may have to be inserted more anteriorly or posteriorly to achieve these goals.

M. McHoney et al. (eds.), *Color Atlas of Pediatric Anatomy, Laparoscopy, and Thoracoscopy*,
DOI 10.1007/978-3-662-53085-6_8

8.4 Relevant Anatomy

8.4.1 Right-Sided Lesions

The relevant anatomic structures, from superior to inferior, are the superior vena cava, phrenic nerve, thymus, azygos vein, pericardium, trachea, oesophagus, vagus nerve, hilum of lung, inferior pulmonary ligament and vein, thoracic duct, inferior vena cava, and diaphragm (Fig. 8.1).

8.4.2 Left-Sided Lesions

The relevant anatomic structures, from superior to inferior, are the brachiocephalic vein, phrenic nerve, thymus, vagus nerve, aorta, pulmonary artery, aortopulmonary window, thoracic duct, pericardium, inferior pulmonary ligament and vein, oesophagus, and diaphragm (Fig. 8.2).

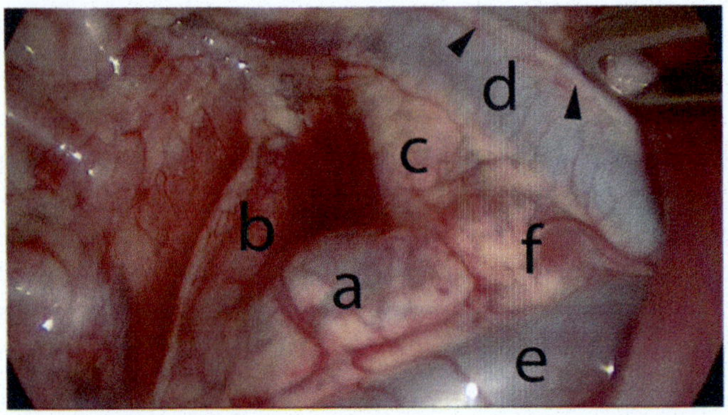

Fig. 8.1 Anatomy encountered during excision of a right-sided bronchogenic cyst. *a* – cyst; *b* – oesophagus; *c* – trachea; *d* – superior vena cava; *e* – azygos vein; *f* – lymph nodes; *arrowhead* indicates phrenic nerve

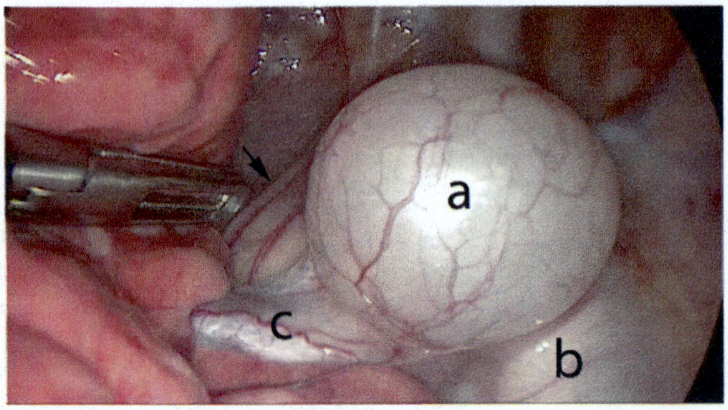

Fig. 8.2 Anatomy encountered during excision of a left-sided bronchogenic cyst. *a* – cyst; *b* – aorta; *c* – pulmonary artery; *arrowhead* indicates vagus nerve

8.5 Surgical Technique

The bronchogenic cyst is grasped and lifted to expose its base. The mediastinial pleura at its base is incised with a monopolar hook (Fig. 8.3). The pleura is divided progressively, and the cyst is mobilised using the monopolar hook to lift the layers away from important structures. Its blood supply is coagulated with monopolar diathermy (Figs. 8.3 and 8.4). Haemostasis may be achieved with judicious use of the LigaSure or ultrasonic scalpel.

Aspirating large cysts may help dissection as it will increase the available working space. The cyst is aspirated and removed through one of the port sites. The port site may be enlarged to aid removal (Fig. 8.5).

Oesophageal duplication cysts may have a common wall with the oesophagus, making it more challenging. A large nasogastric tube or intraoperative oesophagoscopy may help identify the oesophagus. The cyst should be excised, leaving the oesophageal mucosa intact. The muscular defect should then be sutured.

After simple, uncomplicated cyst excisions, the pneumothorax may be evacuated by inserting a 16 F nasogastric tube via a 5-mm port with the opposite end submerged under saline in a dish. The anaesthetist may ventilate the lung manually to help it expand and to evacuate the pneumothorax. Then, the nasogastric tube is removed and the skin is closed.

If the cyst is infected or there is lung, tracheal, or oesophageal injury or repair, it is advisable to leave a chest drain in situ.

A chest X-ray should be done the following day. A small, asymptomatic, residual, pneumothorax does not require treatment. Some patients may be discharged on postoperative day 1.

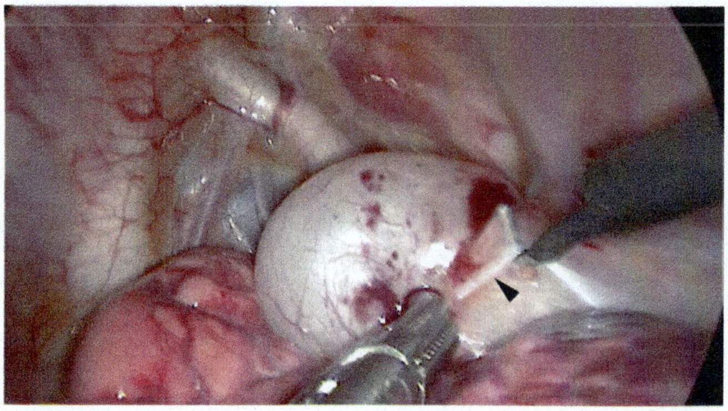

Fig. 8.3 The mediastinal pleura at the base of the cyst is incised with monopolar hook diathermy (*arrow*)

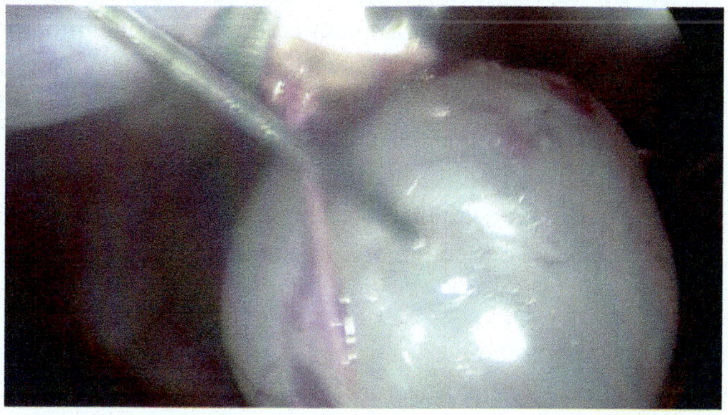

Fig. 8.5 Aspiration of the cyst aids its removal

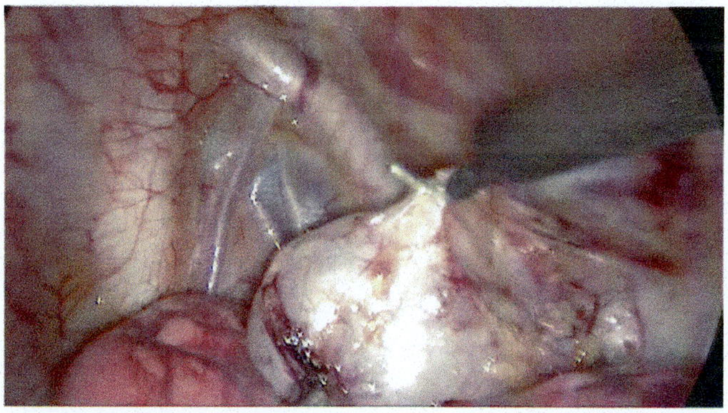

Fig. 8.4 Dissection is aided by lifting the layers with the monopolar hook and keeping close to the cyst surface

8.6 Alternatives: Patient Positioning

An alternative for posterior mediastinal cysts is to have the patient positioned laterally and prone, which allows the lung and mediastinal structures to fall away from the cyst. The monitor is placed towards the patient's back, and the surgeon and assistant stand towards the front. The ports are inserted and triangulated along the posterior or midaxillary line to achieve an ergonomic working environment.

8.7 Highlights and Pitfalls

- Thoracoscopic excisions of mediastinal cysts require an experienced team. The anaesthetist should be competent in using single-lung ventilation (bronchial blockers, double-lumen endotracheal tubes), paravertebral blocks, and thoracic epidural catheters. He or she also should be able to maintain adequate ventilation with a pneumothorax and collapsed lung. The surgeon must have advanced skills in minimally invasive surgery and an intimate knowledge of thoracic pathology and anatomy.
- Careful dissection is important to avoid injury to the thoracic structures, as the cysts may be closely related to them. During dissection, it is advisable to stay close to the cyst's surface.
- The monopolar hook diathermy is an excellent instrument for dissection. However, caution must be taken to avoid electrical or heat conduction injuries to the nerves, oesophagus, and trachea.
- Haemostasis is important for maintaining clear vision during dissection and may be achieved by the precise use of monopolar diathermy. Alternatively, haemostasis may be achieved with the use of an ultrasound dissector or LigaSure. It is important to note that high temperatures are generated with these instruments, and heat injures are possible; hence, their use should be precise and in short bursts. Sufficient time should be given to allow the instrument jaws to cool before holding tissue with it.
- The cyst must be excised completely; otherwise, recurrence is inevitable.

Suggested Reading

Bax KMA, Georgeson KE, Rothenberg S, Valla J-S, Yeung CK, editors. Endoscopic surgery in infants and children. Berlin: Springer; 2008.

Parikh DH, Crabbe D, Auldist A, Rothenberg S, editors. Pediatric thoracic surgery. London: Springer; 2009.

Puri P. Newborn surgery. 3rd ed. London: Hodder Arnold; 2011.

Thoracoscopic Nuss Procedure

9

Joanna Stanwell and Robert Wheeler

Abstract

Minimally invasive pectus excavatum repair involves placement of a retrosternal steel bar through bilateral thoracic incisions, with the attendant risks of pericardial/cardiac puncture, haemothorax, and pneumothorax. These risks are minimised by performing the procedure under thoracoscopic guidance.

Keywords

Minimally invasive pectus excavatum repair • Wire tethering sutures • Bilateral thoracoscopy

9.1 General Information

Minimally invasive pectus excavatum repair involves placement of a retrosternal steel bar through bilateral thoracic incisions, with the attendant risks of pericardial/cardiac puncture, haemothorax, and pneumothorax. These risks are minimised by performing the procedure under thoracoscopic guidance.

9.2 Working Instruments

- Introducer, Nuss bar, and stabilisers
- 5-mm camera port and 30° thoracoscope
- Water seal system to evacuate pneumothorax

J. Stanwell
Department of Paediatric Surgery,
Consultant Paediatric & Neonatal Surgeon, University Hospital
Southampton NHS Foundation Trust, Southampton, UK

R. Wheeler (✉)
Consultant Paediatric & Neonatal Surgeon, University Hospital
Southampton NHS Foundation Trust, Southampton, UK

M. McHoney et al. (eds.), *Color Atlas of Pediatric Anatomy, Laparoscopy, and Thoracoscopy*,
DOI 10.1007/978-3-662-53085-6_9

9.3 Positioning, Port Siting and Ergonomic Considerations (Figs. 9.1 and 9.2)

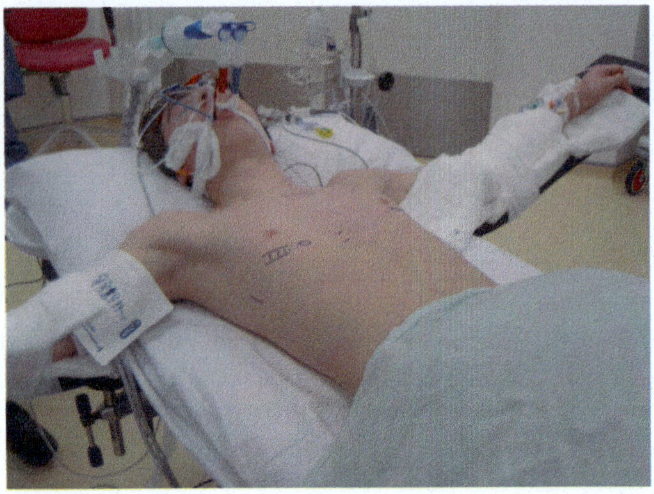

Fig. 9.1 The patient is positioned supine, with the arms outstretched and secured on armboards. Under general anaesthesia, a thoracic epidural catheter is inserted and the right main bronchus is blocked. The surgeon and thoracoscopist stand to the right, with the assistant and scrub nurse to the left of the patient. A Steri-Drape (3 M Healthcare, St. Paul, MN, USA) is used to minimise the risk of implant contamination. Thoracic insufflation using a pressure of 8 mm Hg and low flow of 1 L/min provides a good view of the right thoracic cavity

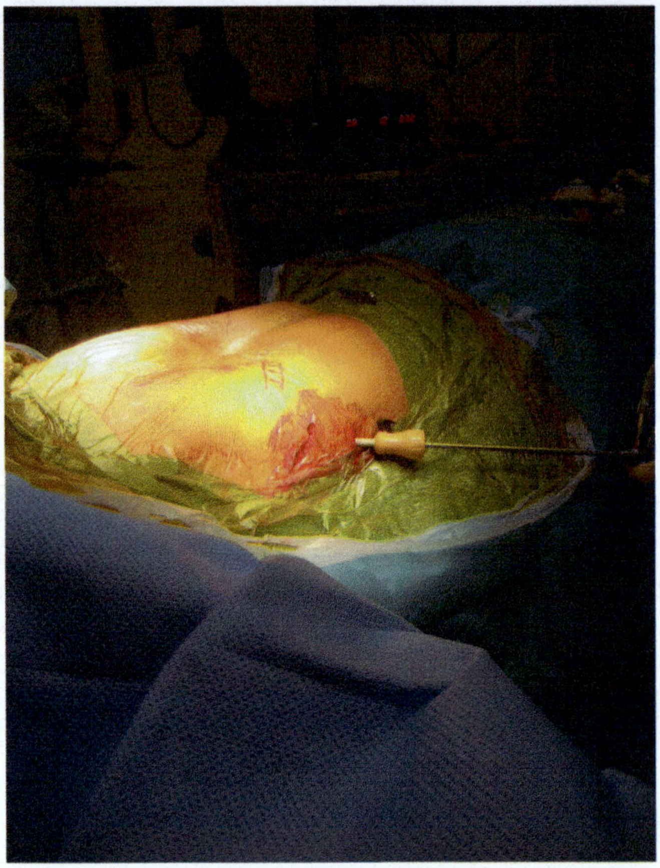

Fig. 9.2 A 5-mm thoracoscopic port is positioned on the *right side*, one or two intercostal spaces below the planned right thoracic incision

9.4 Relevant Anatomy (Figs. 9.3, 9.4 and 9.5)

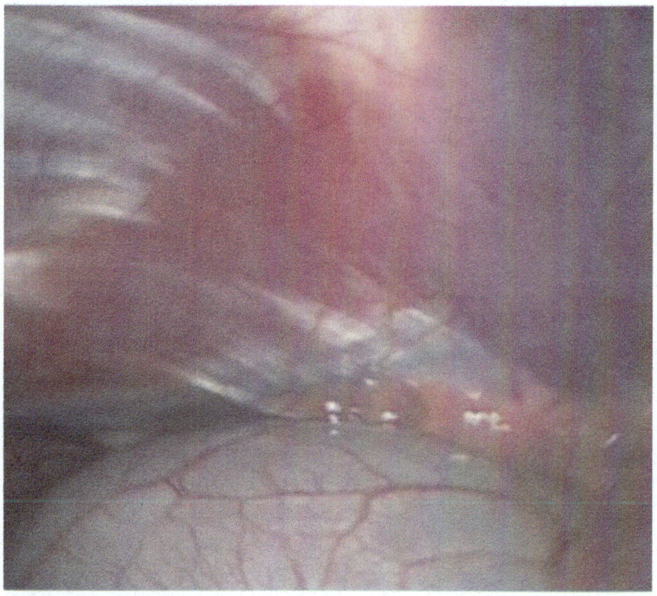

Fig. 9.3 Pectus deformity observed via thoracoscope. The right internal thoracic (mammary) artery and vein are seen

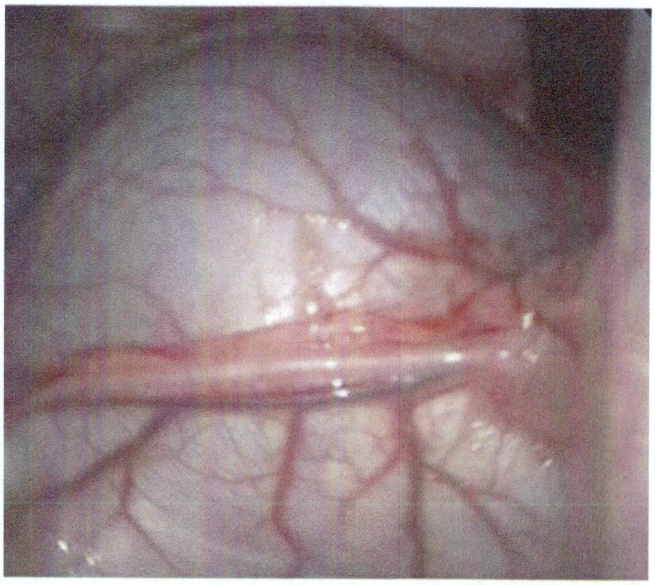

Fig. 9.5 Right phrenic nerve and right pericardiophrenic artery and vein, overlying fibrous pericardium

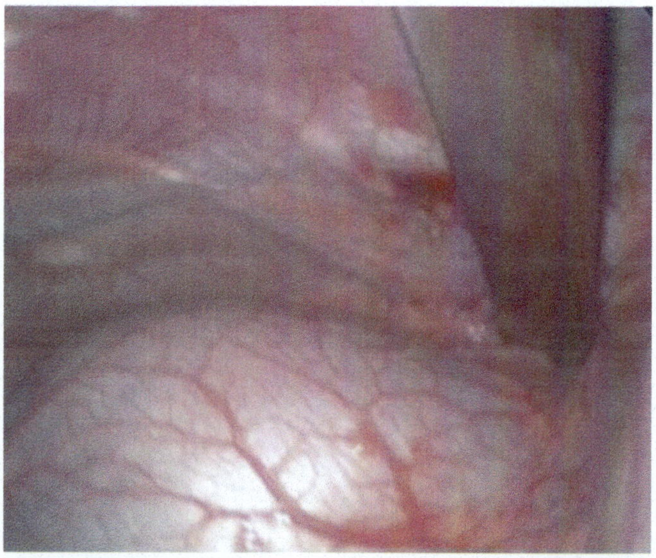

Fig. 9.4 Right hemidiaphragm. Shown are the fibrous pericardial sac, epicardial fat pad, right intercostal branches of the right internal thoracic (mammary) artery, and introducer in situ, elevating the sternum

9.5 Surgical Technique (Figs. 9.6, 9.7, 9.8, 9.9, 9.10, 9.11, 9.12 and 9.13)

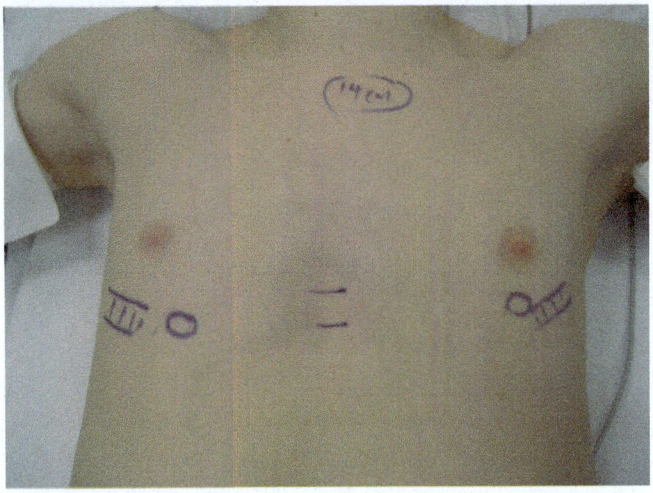

Fig. 9.6 Calculate the required Nuss bar length by measuring the distance from the right to left midaxillary line and subtracting 1–2 cm (1 inch). Mark the deepest point of the pectus excavatum, using a dependent collection of surgical preparation fluid as a guide. If this point is inferior to the sternum, then aim superiorly for the lower end of the sternum just above the xiphoid; this sets the horizontal plane for insertion. Premark the entry point from the subcutaneous tunnel into the thorax by selecting a site directly medial to the most medial rib on which the bar is destined to rest. Make lateral thoracic skin incisions, and create a subcutaneous tunnel anteromedially toward the lateral margin of the pectus deformity

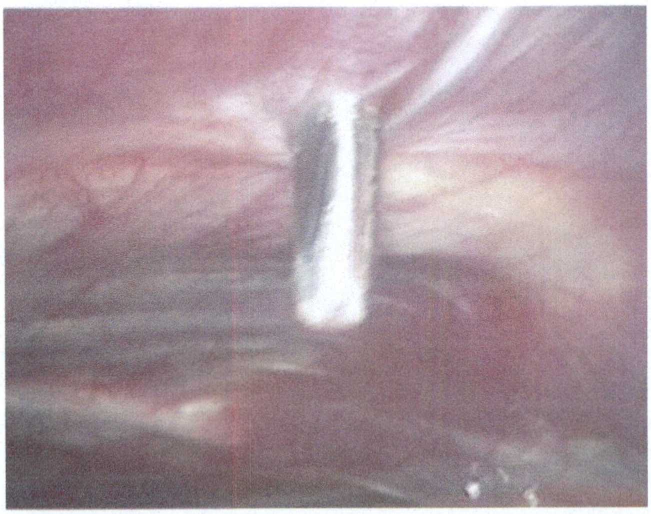

Fig. 9.7 Under thoracoscopic control, make the initial transpleural puncture with scissors and immediately replace with the introducer

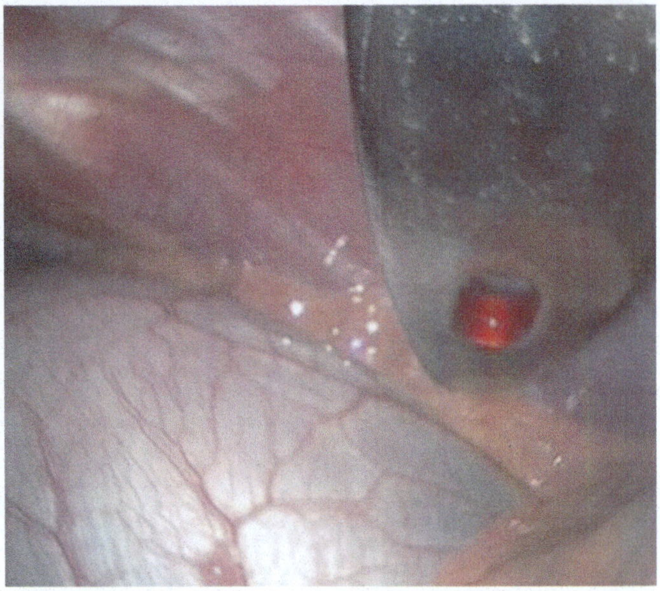

Fig. 9.8 While tunnelling retrosternally through the epicardial fat pad, ensure that the introducer tip is being firmly directed anteriorly against the sternum

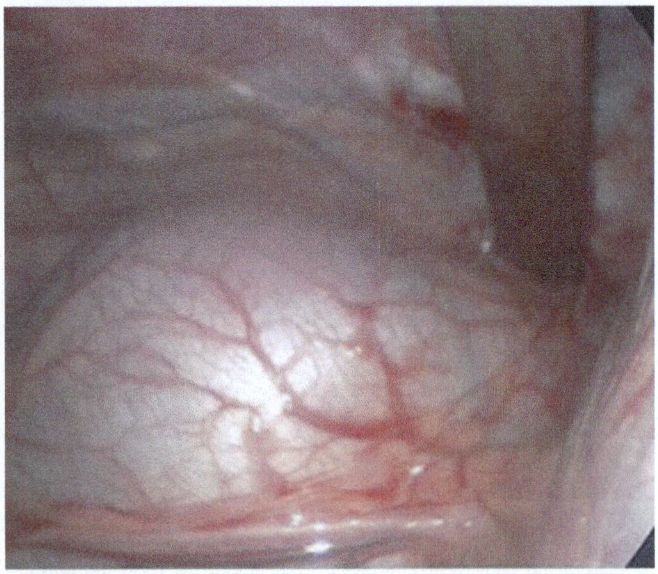

Fig. 9.9 Once through to the *left side*, gently push the tip of the introducer through the chest wall via the left intercostal space in a symmetric position to the *right-sided* entry. The introducer will now be elevating the sternum

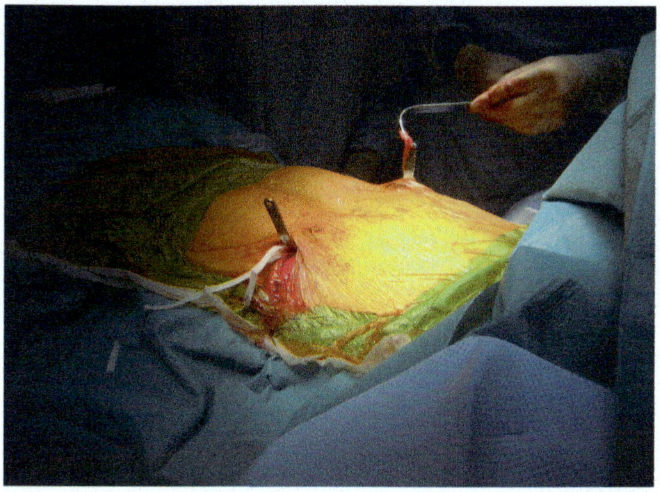

Fig. 9.10 Bend the Nuss bar to conform to the desired chest wall curvature

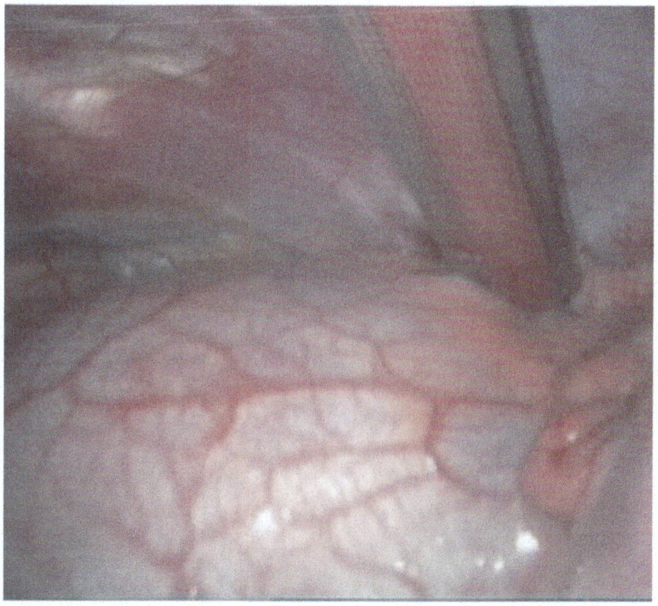

Fig. 9.11 Attach the Nuss bar end to end as close as possible to the introducer using umbilical tape, and slowly guide the bar through the tunnel with its concavity facing anteriorly by using the umbilical tape for traction

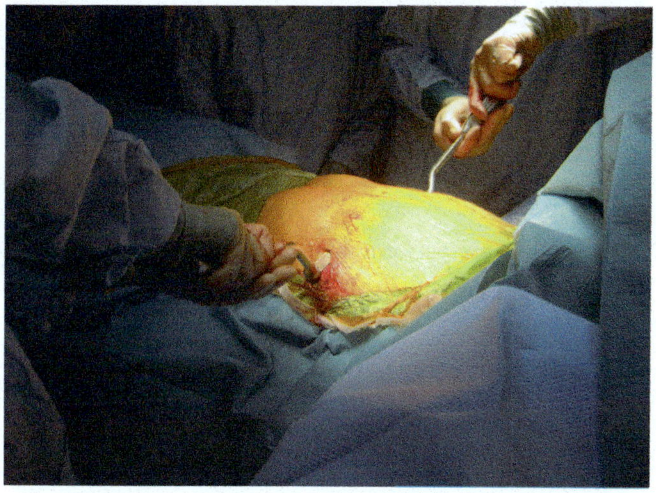

Fig. 9.12 When the Nuss bar is in position, use the rotational instrument (bar flipper) to turn the bar over by rotating 180° inferoposteriorly. Facilitate right lung reinflation by temporarily placing an intercostal drain via the thoracoscopy port site and connecting it to an underwater seal. Use positive pressure ventilation until the bubbling stops and tidal volume is restored

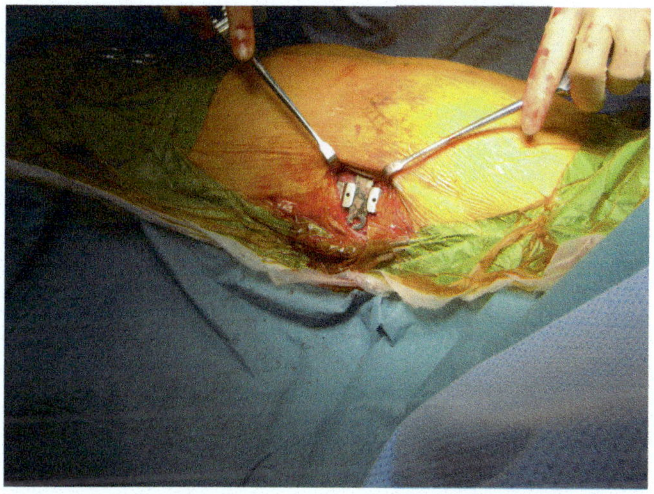

Fig. 9.13 Attach stabilisers to the Nuss bar, as well as to the underlying muscle, and secure them with 1–0 Vicryl sutures (Ethicon; Somerville, NJ, USA). Close the skin with subcuticular 4–0 polydioxanone sutures (PDS; Ethicon)

9.6 Alternatives

Wire tethering sutures may be used instead of stabilisers to secure the bar in position. Additional bars may be useful in cases of severe deformity. Donald Nuss [1] described the use of double bars in 27.7% and triple bars in 0.4% of his patients.

Some authors perform bilateral thoracoscopy to prevent inadvertent damage to the left lung.

9.7 Highlights and Pitfalls

- Doubling up the umbilical tape may avoid loss of the track.
- Placing a single 170° bend at the midpoint of the Nuss bar directed convex anteriorly enhances anterior displacement of the sternum.
- Using a 16 F intercostal drain and underwater seal to reinflate the right lung before closing the skin avoids routine postoperative insertion of an intercostal drain.

Reference

1. Nuss D, Kelly RE. Indications and technique of Nuss procedure for pectus excavatum. Thorac Surg Clin. 2010;20:583–97.

Suggested Reading

Hebra A. Minimally invasive repair of pectus excavatum. Semin Thorac Cardiovasc Surg. 2009;21:76–84.

Schaarschmidt K, Kolberg-Schwerdt A, Dimitrov G, Strauss J. Submuscular bar, multiple pericostal bar fixation, bilateral thoracoscopy: a modified Nuss repair in adolescents. J Pediatr Surg. 2002;39:1276–80.

10 Esophageal Atresia and Tracheoesophageal Fistula

Merrill McHoney, Fraser Munro, Jimmy Lam, and Gordon MacKinlay

Abstract

Thoracoscopic repair of esophageal atresia (OA) and tracheoesophageal fistula (TOF) is one of the most advanced and technically demanding minimally invasive surgery (MIS) operations in pediatric surgery. Since the first report by Rothenberg and Lobe in 1999 [1], there has been much investment in the development of this operation. There remains much debate about the benefits and risks of the thoracoscopic approach compared with those of open surgery. Nevertheless it is now performed by those with sufficiently advanced MIS skills with good outcomes.

Keywords

Oesophageal atresia • Tracheoesophageal fistula • Thoracoscopy • Neonates

10.1 General Information

Thoracoscopic repair of esophageal atresia (OA) and tracheoesophageal fistula (TOF) is one of the most advanced and technically demanding minimally invasive surgery (MIS) operations in pediatric surgery. Since the first report by Rothenberg and Lobe in 1999 [1], there has been much investment in the development of this operation. There remains much debate about the benefits and risks of the thoracoscopic approach compared with those of open surgery. Nevertheless it is now performed by those with sufficiently advanced MIS skills with good outcomes.

10.2 Working Instruments

- 3- (and 5-mm) Ports
- 3- or 5-mm 30-degree Telescope (short)
- 3-mm Needle holders (preferably left- and right-handed)
- 3-mm Hook diathermy
- or Bipolar forceps (e.g., LigaSure)
- 3-mm Johan forceps (×2)
- 3-mm Maryland forceps
- 3-mm Metzenbaum and hook scissors

M. McHoney (✉) • G. MacKinlay
Department of Paediatric Surgery,
Royal Hospital for Sick Children, Edinburgh, UK

F. Munro • J. Lam
Department of Paediatric Surgery,
Royal Hospital for Sick Children, Edinburgh, UK

M. McHoney et al. (eds.), *Color Atlas of Pediatric Anatomy, Laparoscopy, and Thoracoscopy*,
DOI 10.1007/978-3-662-53085-6_10

10.3 Positioning, Port Siting, and Ergonomic Considerations

The approach is through the right side of the chest with the patient in a semi-prone position, with the right side elevated 30–45°. This allows gravity to assist with lung retraction away from the posterior mediastinum. The right arm is positioned above the head with suitable padding.

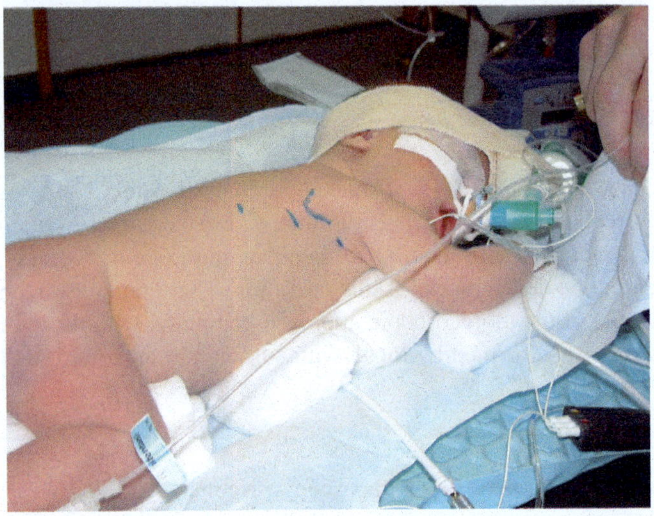

Fig. 10.1 Siting of ports. The scapula is outlined. The primary port site is shown at the interspace below the tip of the scapula

The primary port position is just below the tip of the scapula, in the fourth or fifth intercostal space in the midaxillary line (Fig. 10.1). The working ports are placed under vision in the anterior and posterior axillary lines. Lung collapse is achieved by low pressure CO_2 pneumothorax at 4–6 mmHg pressure. This takes a few minutes to achieve and can be aided by gentle compression of the lung with instruments.

10.4 Relevant Anatomy (Figs. 10.2 and 10.3)

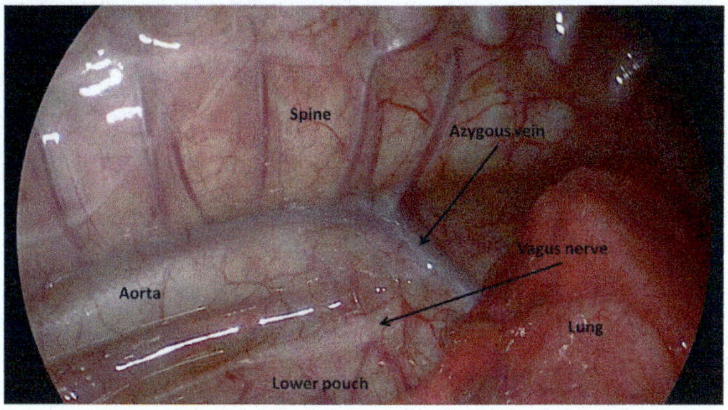

Fig. 10.2 Anatomy of the right posterior mediastinum as seen on entering. The azygos vein is seen crossing the upper mediastinum (over the upper aorta, which is just coming into view at this level) to enter the SVC. The vagus nerve is seen and is a good landmark to identify the lower pouch and eventually the fistula

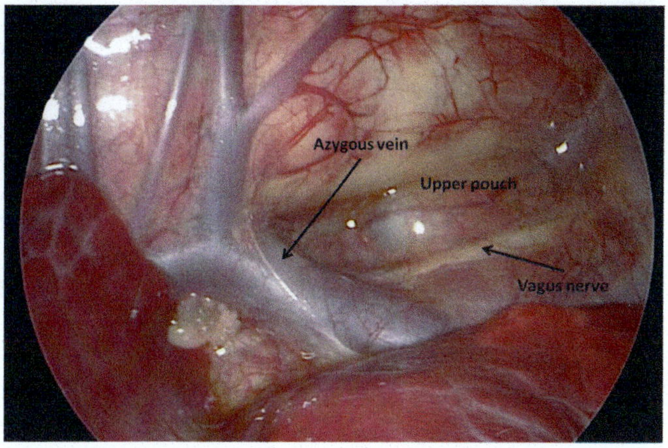

Fig. 10.3 View of the uppermost part of the mediastinum showing the upper pouch (made more visible by the presence of the Replogle tube). Again note the right vagus nerve lying between the upper esophagus and trachea. The azygos vein is crossing over the esophageal gap

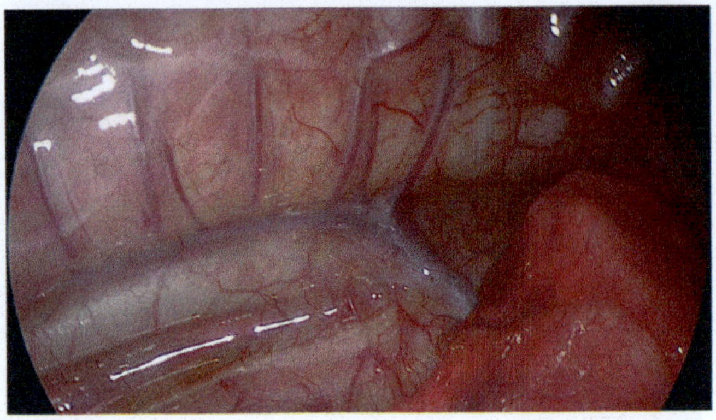

Fig. 10.4 Identification of the lower pouch and fistula. After entering the chest, identification of the azygos vein and the vagus nerve helps to orientate the anatomy and identify the esophagus (lower pouch)

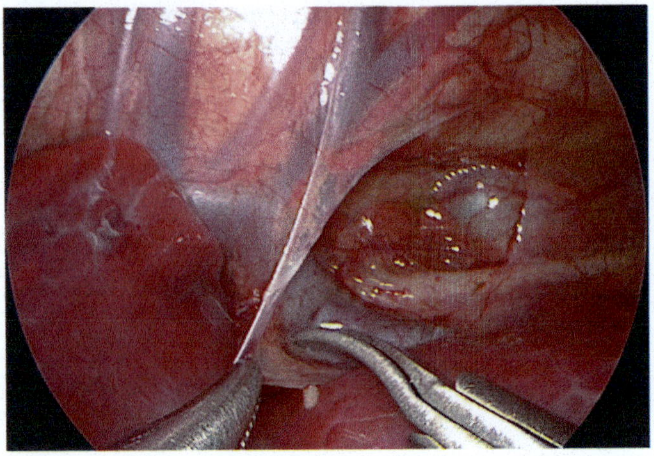

Fig. 10.5 Dissection of the azygos vein if needed. It may need to be divided if it obscures the approach to the lower pouch or fistula, but it is not absolutely necessary. The pleura is incised over the azygos vein to expose it's surface and then freed by blunt dissection until a sufficient length is achieved for safe coagulation and division. Note the upper pouch seen during this dissection

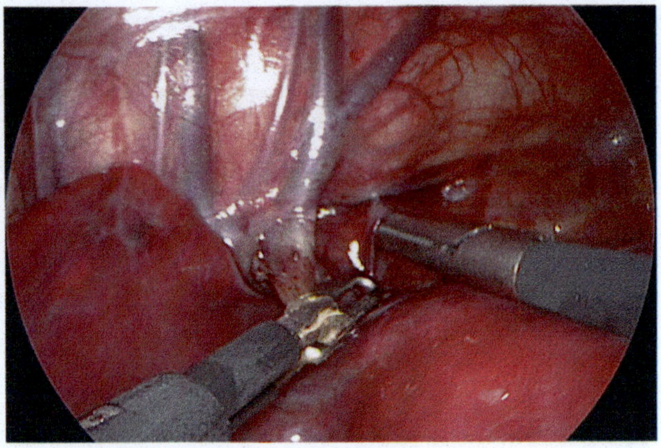

Fig. 10.6 Division of the azygos vein. The vein is then divided. Here it is cauterized with bipolar forceps and then divided with scissors. Other methods of division can be used, e.g., ligation or clips. This usually gives access to the lower pouch

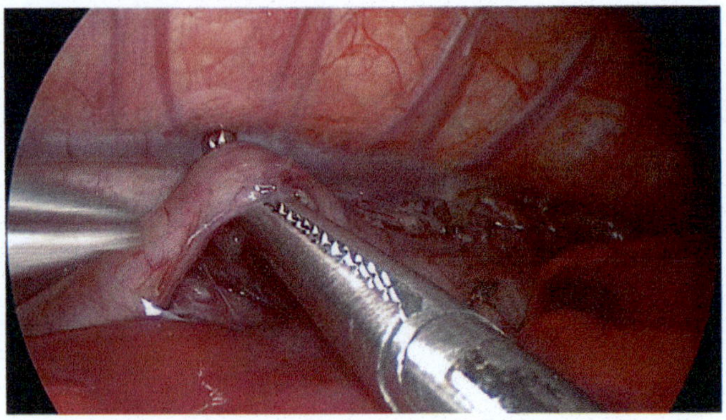

Fig. 10.7 Dissection of the lower pouch. The lower pouch is identified and dissected free mainly by blunt dissection. The entire circumference is then mobilized to achieve a sufficient length. Excessive mobilization is avoided to maintain its blood supply. Care should be taken to avoid the vagus nerves, particularly the left one on the deep surface of the fistula

10.5 Surgical Technique (Figs. 10.4, 10.5, 10.6, 10.7, 10.8, 10.9, 10.10, 10.11, 10.12, 10.13, 10.14, 10.15, 10.16, 10.17, 10.18, 10.19, 10.20, 10.21 and 10.22)

After the anastomosis is inspected, the right lung is visualized to inflate under positive pressure ventilation on exiting. A chest drain is left in the pleural space (through one of the port sites), close to but not in contact with the anastomosis. The chest drain can be removed if there is no drainage or leakage in the first few postoperative days. The use of a routine contrast examination of the anastomosis varies but can be done if reassurance is needed before oral feeding.

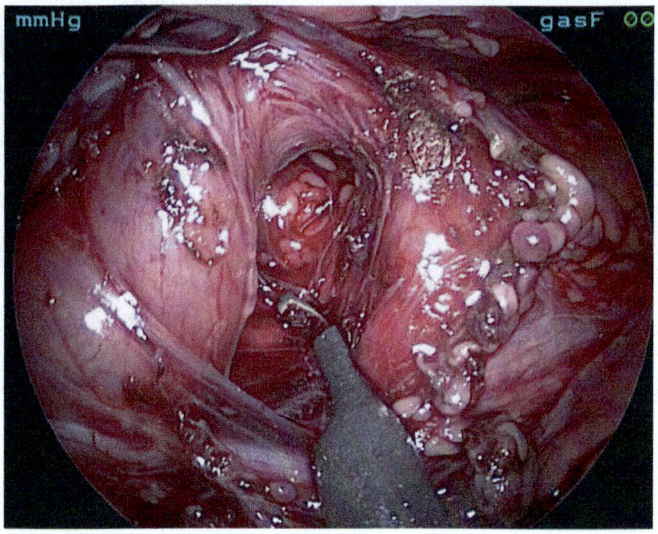

Fig. 10.8 Dissection of the lower pouch. The lower pouch and fistula may need to be further freed up using hook diathermy to small vascular bands, while preserving the vagus nerves and esophageal branches (seen in this figure)

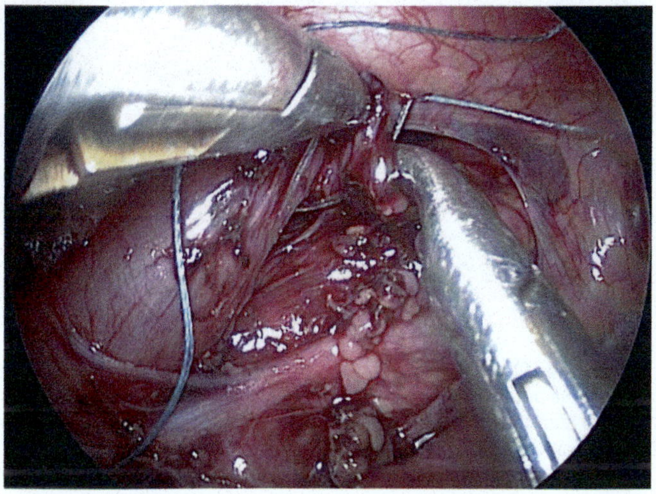

Fig. 10.9 Transfixion of the lower pouch fistula. On identification of the entry of the fistula into the trachea, temporary occlusion of the fistula is performed while ventilation confirms that the right lung insufflates. The lower pouch fistula is then transfixed using a 5/0 suture, close to the posterior wall of the trachea to avoid leaving a diverticulum. The needle should be passed through on the superficial side of the fistula with the intention of avoiding nonabsorbable suture material within the lumen

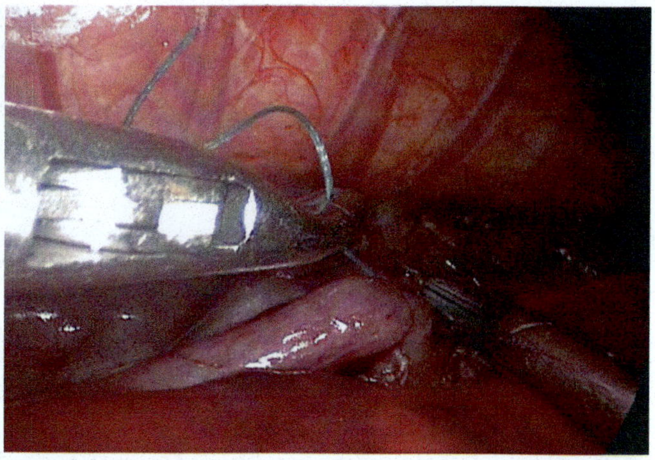

Fig. 10.10 Ligation of the upper pouch. The fistula is ligated after transfixion passing the suture material behind the fistula so that it is completely occluded. Ligation of the fistula improves ventilation. The lower pouch can be left in continuity until later in the procedure

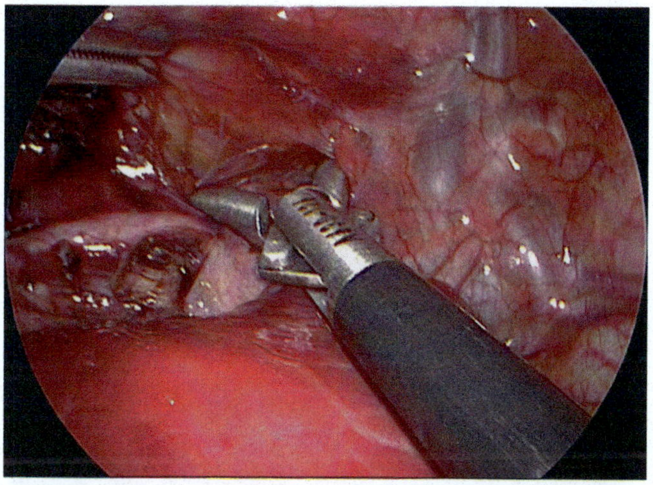

Fig. 10.11 Identification of the upper pouch. The upper pouch may be obvious without any maneuvres. If needed gentle manipulation of the Replogle tube by the anesthetist can help to identify it. The pleura overlying it is incised. The upper pouch is then grasped, and blunt dissection begins its mobilization

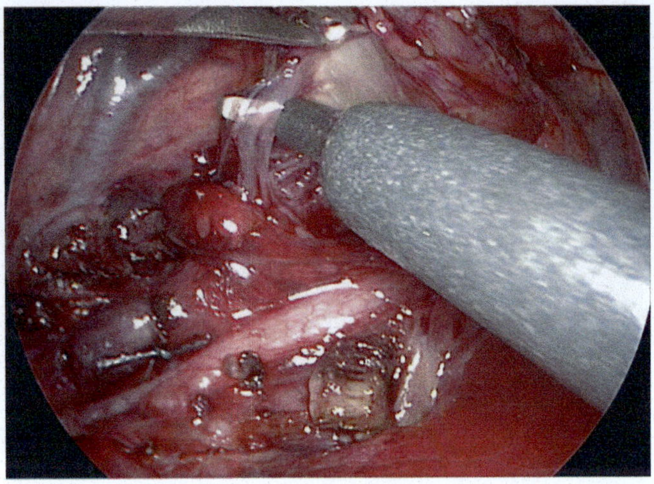

Fig. 10.12 Diathermy mobilization of the upper pouch. Adherent bands maybe encountered between the upper pouch and the trachea. These can be carefully divided by hook diathermy in the plane between these viscera. The "side on" view with the thoracoscope gives superior visualization for this phase of the procedure

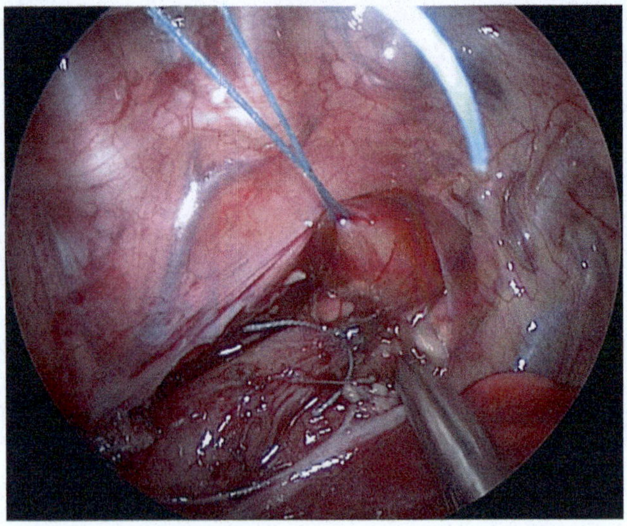

Fig. 10.13 Alternative mobilization of the upper pouch. The upper pouch can be mobilized using a suture passed through the tip of the upper pouch to control it without excessive handling. Note in this instance the almost overlapping upper and lower pouches

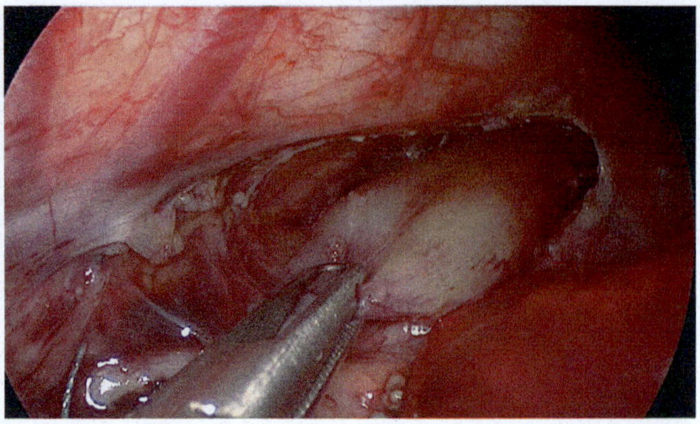

Fig. 10.14 Continued dissection of the upper pouch. Circumferential dissection of the upper pouch is then continued up into the upper mediastinum and neck, using a combination of blunt scissors and diathermy dissection as necessary. This allows sufficient length of the upper pouch to be established

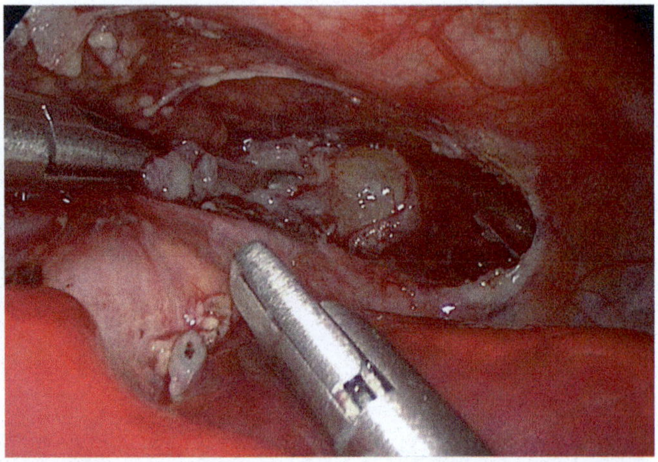

Fig. 10.15 Opening of upper pouch. The upper pouch is opened (as distally as possible) ensuring that the mucosa is seen and the Replogle tube is identified. A small attachment of the tip to the rest of the upper pouch can be left to assist with atraumatic manipulation. Hook scissors are best for opening the mucosa, which otherwise may push away

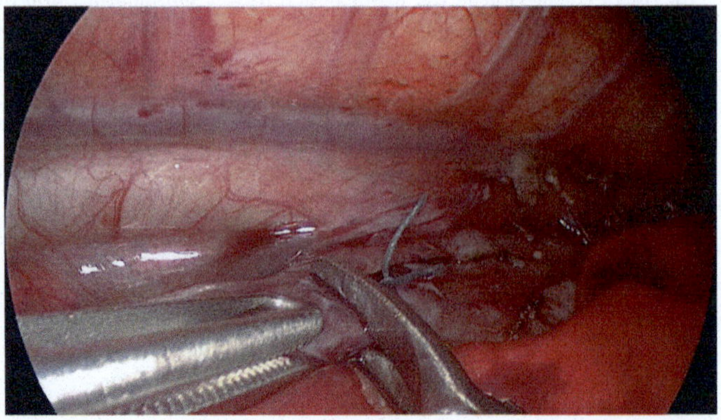

Fig. 10.16 Division of the lower pouch fistula. The fistula is then divided using scissors close to the posterior wall of the trachea where it was previously ligated

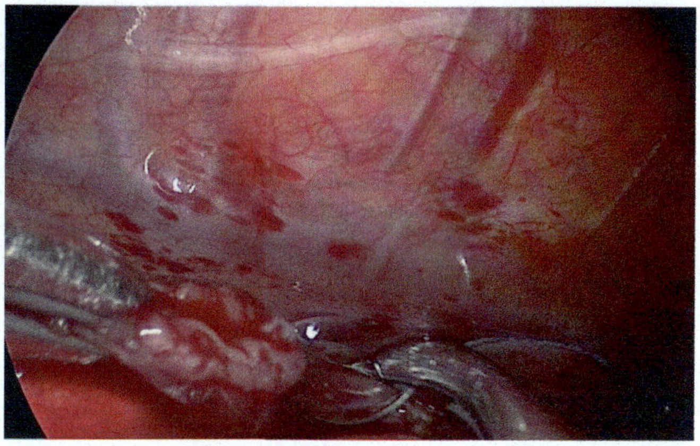

Fig. 10.17 Posterior wall anastomosis. The posterior wall anastomosis is performed with approximately four interrupted sutures of 5/0 Vicryl, ensuring that the mucosa is seen and included in the suture. Here the mucosa is seen as the first posterior wall suture is placed

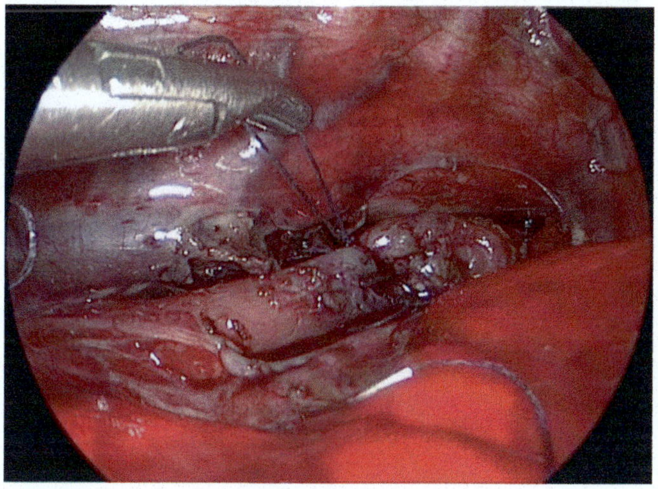

Fig. 10.18 Posterior wall completed. The four sutures in the posterior wall have been placed, with the lateral suture being used to aid mobilization, thus giving access to the anastomotic line to start the closure of the anterior wall

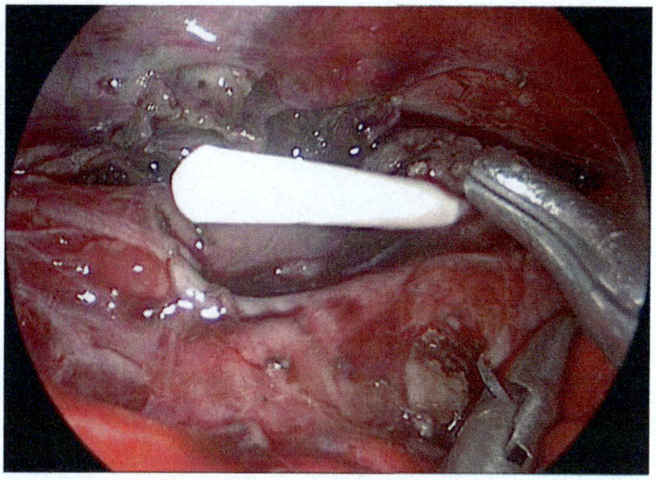

Fig. 10.19 Advancing of a transanastomotic tube. The tube is then advanced by the anesthetist and grasped and guided into the distal esophagus. It is then be advanced into the stomach

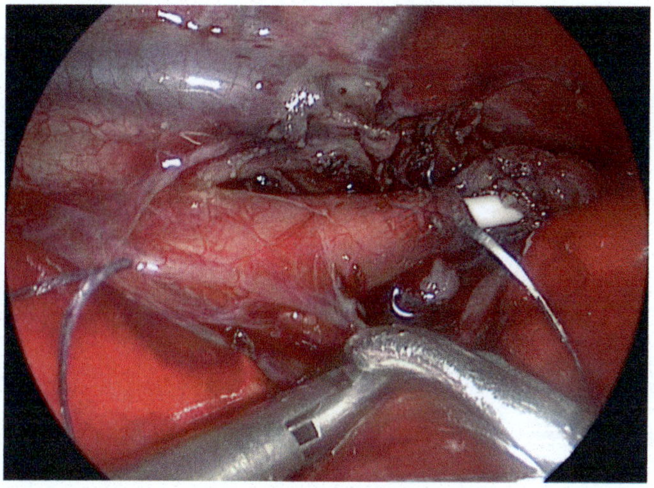

Fig. 10.20 Transanastonomic tube passed into lower pouch

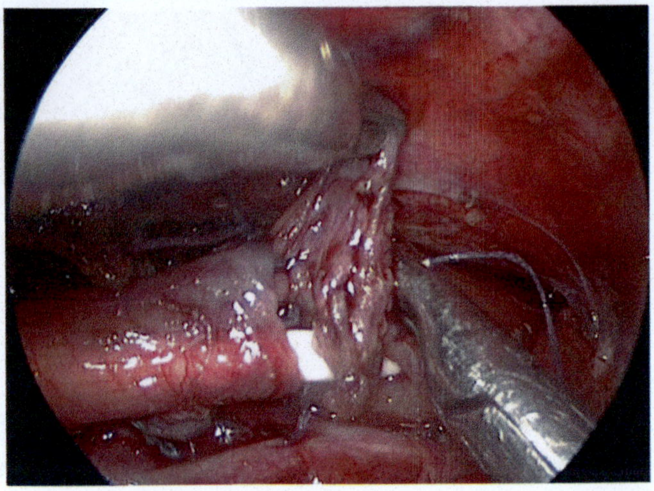

Fig. 10.21 Anterior wall anastomosis. The next stitches are placed in the anterior wall, again ensuring that the mucosa is seen and included in the sutures. Usually three or four sutures are needed to complete the anastomosis

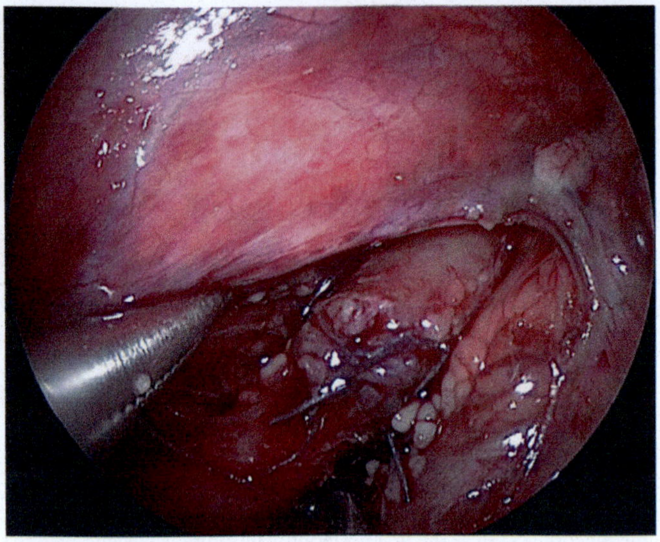

Fig. 10.22 Finished anastomosis. Once the anastomosis is complete, the sutures should be evenly spaced with no gaps. The transanastomotic tube should not be visible through the anastomotic line

10.6 Alternatives

- The division of the azygos vein can be performed with simple hook diathermy, but could also be performed with bipolar forceps. If needed one of the 3-mm ports could be upsized to 5 mm.
- The fistula may be dealt with as discussed (by suture transfixion), but other methods including ligation with clips (both titanium and plastic) have also been described.

10.7 Highlights and Pitfalls

- Close communication and cooperation between the surgeon and the anesthetist is needed for this procedure because surgical access by lung deflation and efficient ventilation by the anesthetist are in conflict with each other. Occasional manual ventilation (with frequent shallow tidal volumes) or other ventilation strategy (e.g., higher frequency ventilation) is sometimes needed to procure good respiratory gas exchange while maintaining exposure.
- Careful dissection of the upper pouch is needed to avoid breaching the posterior wall of the trachea. Finding the plane between the end of the pouch and the trachea is occasionally difficult. It may be necessary to find the correct plane higher up and then dissect down toward the tip. Primary suture closure is needed if a tracheal breach is caused.
- If the anastomosis is under some tension, the use of a tumbled square or sliding knot is useful. In this scenario two (or three) sutures are placed and tied widely apart without approximating the pouches. They are then sequentially and slowly slid together as the tension is spread across the sutures. Once the ends are in apposition, the rest of the posterior wall sutures can then be placed and tied.

Reference

1. Lobe TE, Rothenberg SS, Waldschmidt JE, Stroedter LU. Thoracoscopic repair of esophageal atresia in an infant: a surgical first. Ped Endosurg Innovat Tech. 1999;3:141–8.

Suggested Reading

Davenport M, Rothenberg SS, Crabbe DC, Wulkan ML. The great debate: open or thoracoscopic repair for oesophageal atresia or diaphragmatic hernia. J Pediatr Surg. 2015;50:240–6.

Holcomb III GW, Rothenberg SS, Bax KM, Martinez-Ferro M, Albanese CT, Ostlie DJ, van Der Zee DC, Yeung CK. Thoracoscopic repair of esophageal atresia and tracheoesophageal fistula: a multi-institutional analysis. Ann Surg. 2005;242:422–8.

MacKinlay GA. Esophageal atresia surgery in the 21st century. Semin Pediatr Surg. 2009;18:20–2.

Rothenberg SS. Thoracoscopic repair of esophageal atresia and tracheoesophageal fistula in neonates, first decade's experience. Dis Esophagus. 2013;26:359–64.

Szavay PO, Zundel S, Blumenstock G, Kirschner HJ, Luithle T, Girisch M, et al. Perioperative outcome of patients with esophageal atresia and tracheo-esophageal fistula undergoing open versus thoracoscopic surgery. J Laparoendosc Adv Surg Tech A. 2011;21:439–43.

11 Thoracoscopic Aortopexy

Joanna Stanwell and Edward Kiely

Abstract

When severe tracheomalacia is not controlled by conservative measures, minimally invasive aortopexy may be a treatment option. The procedure involves suturing the aortic arch to the posterior aspect of the sternum via a left-sided approach, with left lobe thymectomy to achieve adequate exposure. The risks of aortic injury, nerve damage, and haemomediastinum are minimised by performing the procedure under thoracoscopic guidance, though concerns remain about the rate of recurrence using this approach.

Keywords

Severe tracheomalacia • Minimally invasive aortopexy • Insufflation

11.1 General Information

When severe tracheomalacia is not controlled by conservative measures, minimally invasive aortopexy may be a treatment option. The procedure involves suturing the aortic arch to the posterior aspect of the sternum via a left-sided approach, with left lobe thymectomy to achieve adequate exposure. The risks of aortic injury, nerve damage, and haemomediastinum are minimised by performing the procedure under thoracoscopic guidance, though concerns remain about the rate of recurrence using this approach.

11.2 Working Instruments

- 5-mm camera port and 0° thoracoscope
- One 5-mm port, one 3-mm port
- Monopolar hook diathermy
- 14-gauge cannula × 3
- Prolene sutures

J. Stanwell
Department of Paediatric Surgery, University Hospital Southampton NHS Foundation Trust, Southampton, UK

E. Kiely (✉)
Department of Paediatric Surgery, Great Ormond Street Hospital for Children NHS Foundation Trust, London, UK

M. McHoney et al. (eds.), *Color Atlas of Pediatric Anatomy, Laparoscopy, and Thoracoscopy*,
DOI 10.1007/978-3-662-53085-6_11

11.3 Positioning, Port Siting, and Ergonomic Considerations

The patient is positioned as shown in Fig. 11.1, and three ports are inserted.

The surgeon and thoracoscopist stand to the left, with the scrub nurse to the right of the patient. Thoracic insufflation using a pressure of 8 mmHg and low flow of 1 L/min provides a good view of the left thoracic cavity (Fig. 11.2).

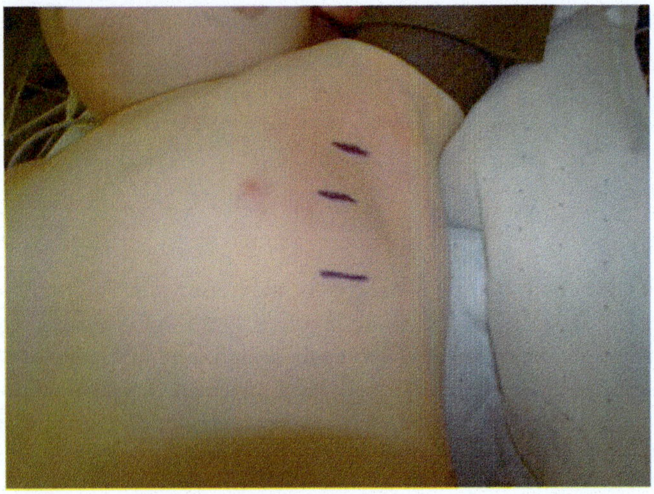

Fig. 11.1 The patient is positioned supine, with arms outstretched and secured. Under general anaesthesia, three ports are inserted in the 3rd, 4th, and 5th intercostal spaces on the *left side*

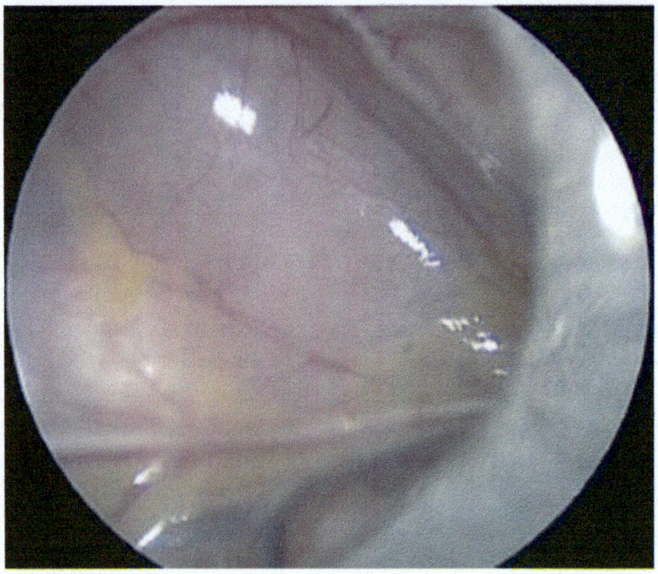

Fig. 11.2 With the left lung deflated, the superior mediastinum can clearly be seen thoracoscopically

11.4 Relevant Anatomy

Figures 11.3 and 11.4 show the anatomy encountered in the procedure.

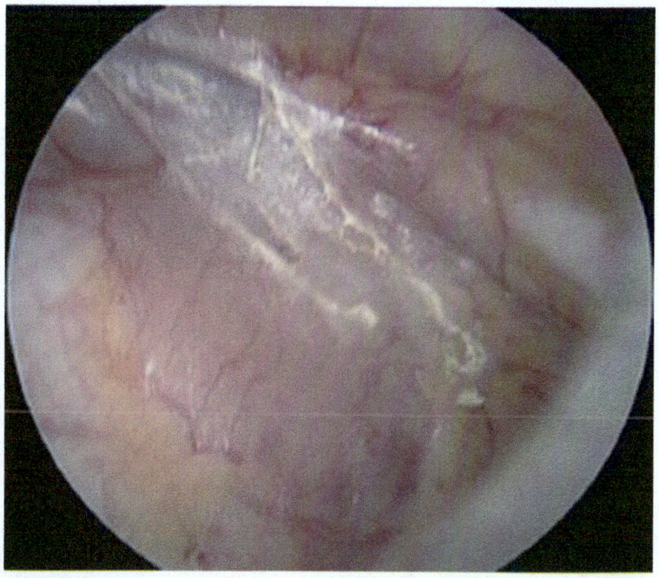

Fig. 11.3 The superior mediastinum, seen from the left thoracoscopic view, is bounded by the thoracic inlet superiorly, the oblique transverse thoracic plane inferiorly, the pleurae laterally, and the first four thoracic vertebral bodies posteriorly. Dissection is commenced onto the left lobe of the thymus using monopolar hook diathermy, avoiding the phrenic nerve. The left lobe of the thymus is completely freed with diathermy and removed piecemeal from the chest

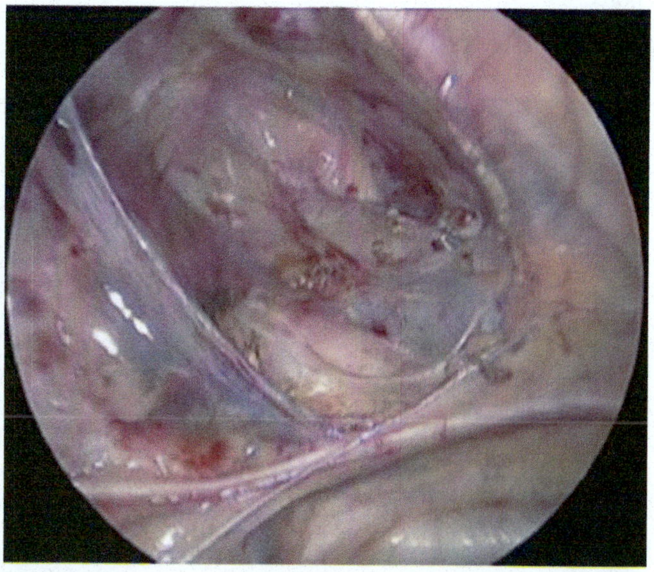

Fig. 11.4 The pericardium is identified and opened below its reflection on the ascending aorta. The aortic arch then can be seen clearly

11.5 Surgical Technique

Figures 11.5, 11.6, 11.7, 11.8, 11.9 and 11.10 illustrate the steps involved in thoracoscopic aortopexy. Reinflation of the left lung can be facilitated by opening the valve on one of the ports.

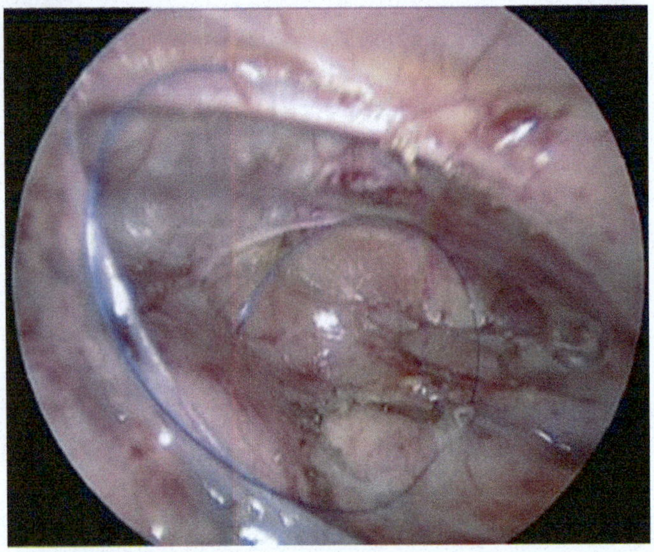

Fig. 11.5 Pass a 4/0 Prolene suture into the chest through an intercostal space. The needle is then passed through the pericardial reflection on the aorta, together with one or two bites of aortic adventitia. The suture is then divided and the needle removed through one of the ports

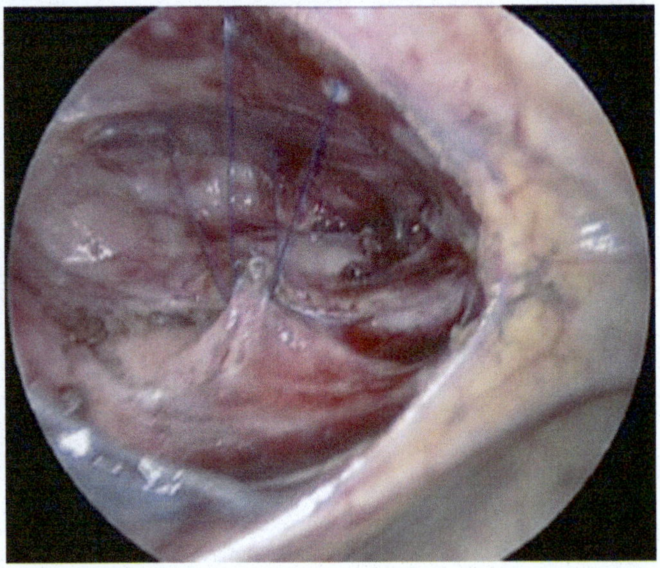

Fig. 11.7 Place the second Prolene suture in the same fashion via another 14-gauge cannula, and clip externally

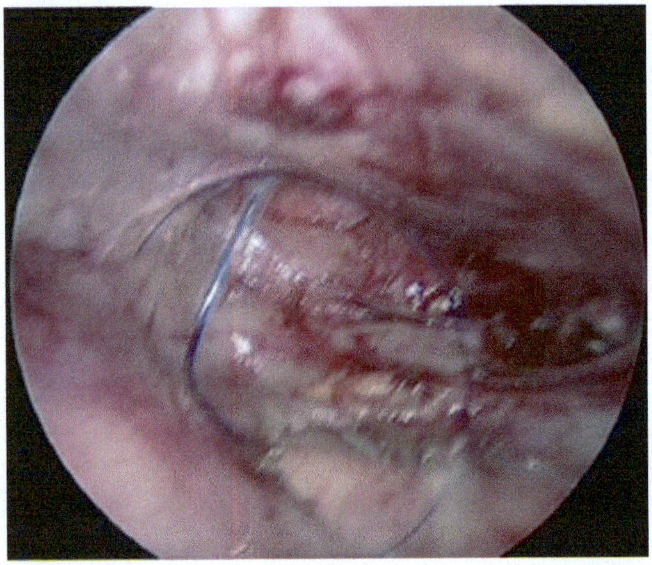

Fig. 11.6 Exteriorise the Prolene suture ends one at a time via a cannula inserted parasternally through a stab incision. The ends of the suture may be passed directly into the cannula or may be retrieved using a loop of suture material

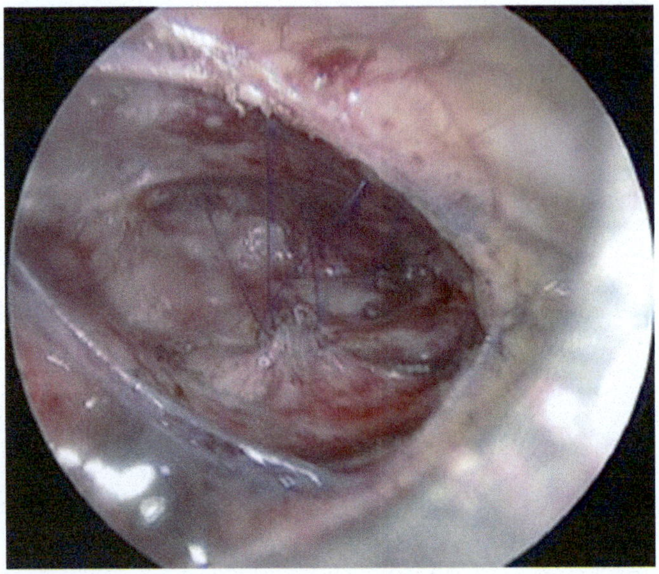

Fig. 11.8 Place the third Prolene suture and prepare for subcutaneous extracorporeal tying

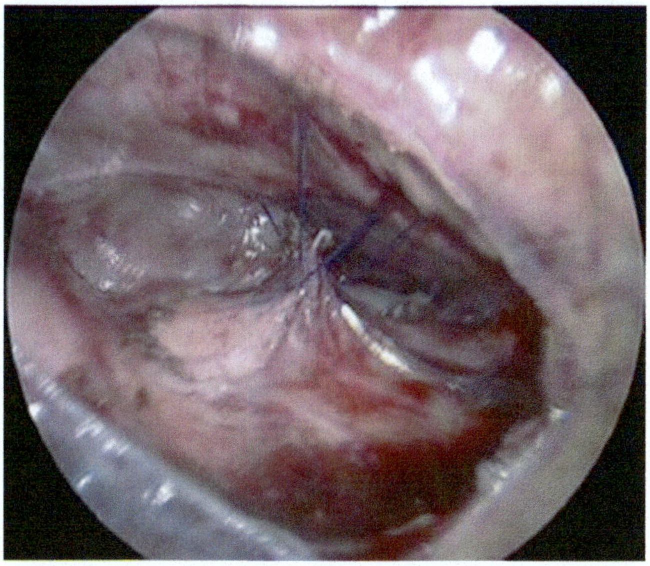

Fig. 11.9 The Prolene sutures are tied subcutaneously by the surgeon as the assistant firmly depresses the sternum. This manoeuvre may cause bradycardia, and the anaesthetist must briefly suspend ventilation whilst the surgeon promptly but securely ties the three Prolene sutures. Assess the effectiveness of the aortopexy thoracoscopically, and confirm haemostasis before removing the camera and desufflating the left thoracic cavity. Although the sutures are tied tightly, the aorta always lies suspended from the sternum once the assistant releases the compression

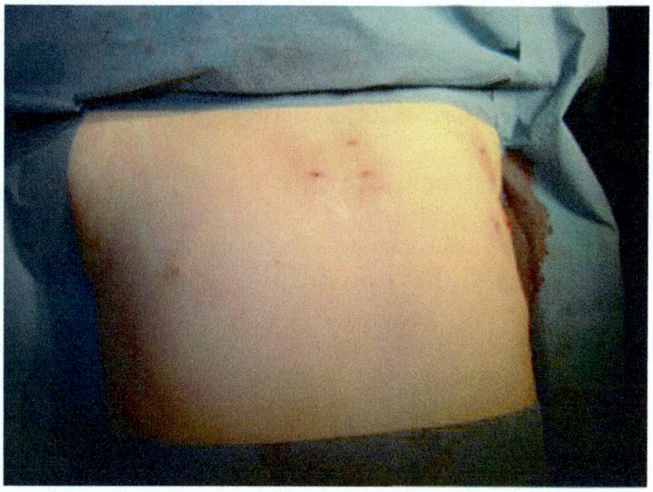

Fig. 11.10 Close the thoracoscopic port sites. Apply tissue glue to these sites and to the parasternal puncture sites, as illustrated

11.6 Alternatives

- Some authors advocate the use of simultaneous bronchoscopy to ensure that the aortopexy relieves tracheal occlusion.
- A right-sided approach has also been described.

11.7 Highlights and Pitfalls

- Aortopexy is successful in over 90 % of patients, but various small published case series have reported recurrence rates ranging from 1 to 13 %.
- Retrieving the Prolene suture ends via the 14-gauge cannula is made easier by removal of the cannula hub.
- Close communication with the anaesthetist is essential, particularly during compression of the sternum when tying the sutures.

Suggested Reading

Jennings RW, Hamilton TE, Smithers CJ, Ngerncham M, Feins N, Foker JE. Surgical approaches to aortopexy for severe tracheomalacia. J Pediatr Surg. 2014;49:66–70.

Kane T, Nadler E, Potoka D. Thoracoscopic aortopexy for vascular compression of the trachea: approach from the right. J Laparoendosc Adv Surg Tech A. 2008;18:313–6.

Thoracoscopic aortopexy for severe primary tracheomalacia. NICE interventional procedures guidance [IPG243]. National Institute for Health and Clinical Excellence. 2007. www.nice.org.uk/guidance/IPG243. Accessed 27 June 2014.

Torre M, Carlucci M, Speggiorin S, Elliott MJ. Aortopexy for the treatment of tracheomalacia in children: review of the literature. Ital J Pediatr. 2012;38:62.

Thoracoscopy for Congenital Lung Malformations

12

Fraser D. Munro

Abstract

Lobectomy for congenital pulmonary airway malformations (CPAMs) is one of the most demanding operations in pediatric MAS. Many patients are now diagnosed antenatally, and there is considerable controversy on the management of those cases that are asymptomatic at birth. Early resection and observation both have equally strong advocates. Large cyst lesions may be at higher risk for complications, including late bronchoalveolar carcinoma, and may not be distinguishable from type 1 pleuropulmonary blastoma on imaging alone; therefore the case for resection is stronger.

Keywords

Thoracaoscopy • Lobectomy • CPAM

12.1 General Information

Lobectomy for congenital pulmonary airway malformations (CPAMs) is one of the most demanding operations in pediatric MAS. Many patients are now diagnosed antenatally, and there is considerable controversy on the management of those cases that are asymptomatic at birth. Early resection and observation both have equally strong advocates. Large cyst lesions may be at higher risk for complications, including late bronchoalveolar carcinoma, and may not be distinguishable from type 1 pleuropulmonary blastoma on imaging alone; therefore the case for resection is stronger.

Most surgeons favor lobectomy rather than segmental or nonanatomic resections for CPAMs because there is less risk of leaving residual abnormality and of a troublesome postoperative air leak. CPAMs most commonly involve the lower lobes, with the middle and upper lobes only rarely affected. Lower lobe resections are illustrated, but the approach to the upper lobes is also discussed.

Some lesions have a systemic arterial supply, as is also the case for extralobar sequestrations. Good preoperative imaging (Fig. 12.1) is critical in demonstrating the anatomy, enabling planning, and avoiding unpleasant surprises!

F.D. Munro
Department of Paediatric Surgery,
Royal Hospital for Sick Children, Edinburgh, Scotland, UK

M. McHoney et al. (eds.), *Color Atlas of Pediatric Anatomy, Laparoscopy, and Thoracoscopy*,
DOI 10.1007/978-3-662-53085-6_12

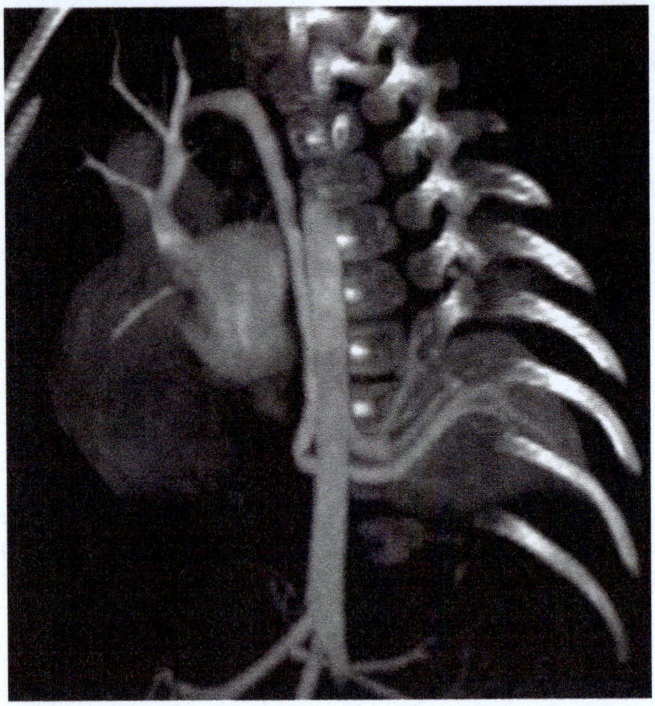

Fig. 12.1 CT reconstruction showing systemic supply to extralobar sequestration

12.1.1 Working Instruments

- 5-mm 30° Telescope
- Three or four 5-mm valved ports
- 5-mm Atraumatic graspers (e.g., a Johan grasper)
- 5-mm Curved dissector (e.g., Maryland or Kelly)
- 5-mm Right angle dissector (e.g., Mixter)
- Diathermy hook
- 5-mm Metzenbaum scissors
- 5-mm Needle holders
- 5-mm Hook scissors
- 5-mm Sucker/irrigator
- LigaSure 5-mm sealer (Covidien-Medtronic; Minneapolis, MN, USA)
- 5-mm Clip applier
- 12-mm Port (if stapler to be used)
- Linear stapler with thin/medium tissue staple loads

12.2 Position, Port Siting, and Ergonomic Considerations

Single lung ventilation is essential. This may be achieved by endobronchial intubation, a bronchus blocker, or a double-lumen tube, depending on patient size and anesthetic preference/expertise. The patient is positioned in the full lateral position as for a thoracotomy. A roll or bolster placed under the chest helps to open the upper rib spaces. There should be adequate support or fixation of the patient to the table to allow safe lateral tilting. If the operating screens can be positioned independently of the equipment stack, the stack is best placed at the foot of the table to allow unimpeded access to both sides of the chest.

The general approach is cross table, looking from the anterior aspect of the chest posteriorly up the major fissure, with the primary operating screen at the patient's back. A second screen to allow working from the posterior to the anterior chest may be useful in difficult cases. The first port is placed using a blunt "cut-down" technique in the fourth or fifth interspace between the mid-axillary line and the anterior axillary line (usually just below and a little anterior to the scapula tip). Once the position of the major fissure is confirmed, two operating ports are placed at the anterior axillary line, above and below the fissure. A fourth port can be placed in the posterior axillary line at the same level as the inferior operating port for a retracting instrument if necessary. The angles for the instruments and, in particular, the telescope are quite steep and so the table will need to be low for comfortable handling.

12.3 Relevant Anatomy

Sometimes the fissures may be incomplete, virtually to the extent of absence. In these cases it may not be possible to adequately demonstrate the anatomy thoracoscopically, and conversion to open thoracotomy may be advisable (Fig. 12.2). An understanding of the spatial relationships at the hilum on pulmonary artery, bronchi, and pulmonary veins on both sides is crucial as is an understanding of the normal segmental pulmonary arterial anatomy and variations.

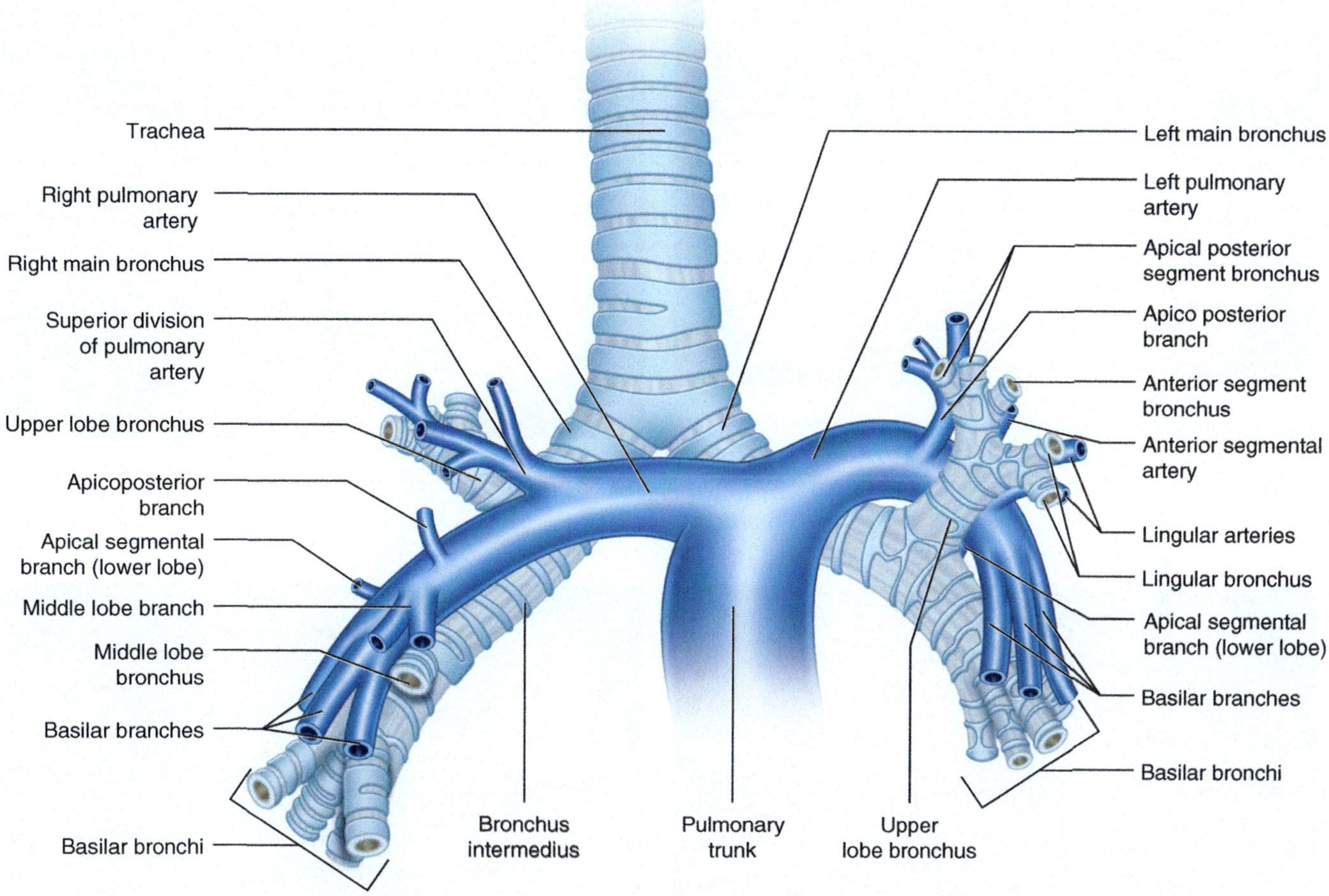

Fig. 12.2 Bronchial and pulmonary artery anatomy

12.4 Surgical Technique

If the oblique fissure is complete (Fig. 12.3), then the pulmonary artery and its branches are readily identified (Figs. 12.4 and 12.5).

When the fissure is incomplete, pulmonary parenchyma will need to be divided to expose the vessels. This can be accomplished with the diathermy hook, but the LigaSure or an equivalent sealer provides better hemostasis and pneumostasis (Figs. 12.6 and 12.7).

Dissection of the vessels is easiest if done right on the vessel wall. Clear identification and dissection of the segmental branches is essential before control and division (Figs. 12.8 and 12.9).

Care must be taken to identify the apical segmental branch to the lower lobe, which usually comes off proximal to the middle lobe and lingular vessels, which need to be preserved. Division of the segmental branches rather than the main trunk is probably safer. Vessels may be individually ligated prior to division, but this is difficult and time-consuming. Clips may also be used, but with multiple branches and further dissection of both the pulmonary vein and bronchus in a confined space there is a risk of accidental dislodgement (Fig. 12.10).

The LigaSure gives a secure seal, with the technique described by Rothenberg, of two seals made 3–5 mm apart with initial partial division of the vessel between them appearing the safest. With this technique, in the event of failure of the seal, control should be readily regained since vessel ends will not have retracted. In larger children, if vessels are beyond the size that can be sealed with the LigaSure a vascular stapler may be used (Figs. 12.11, 12.12, and 12.13).

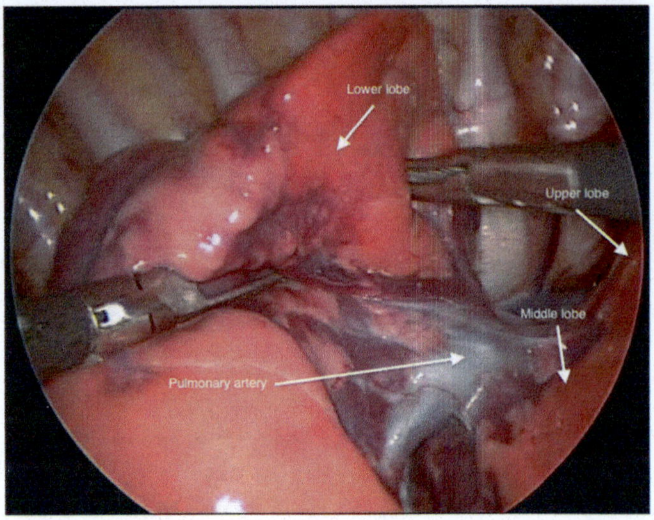

Fig. 12.3 Right oblique fissure

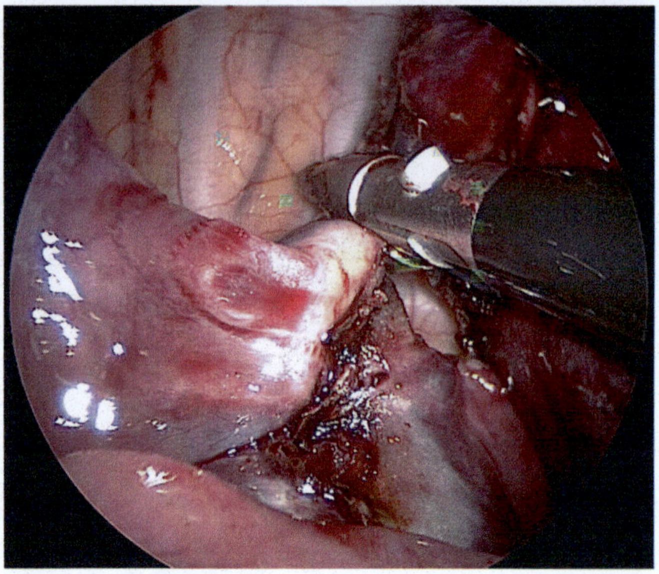

Fig. 12.5 LigaSure division of lung parenchyma

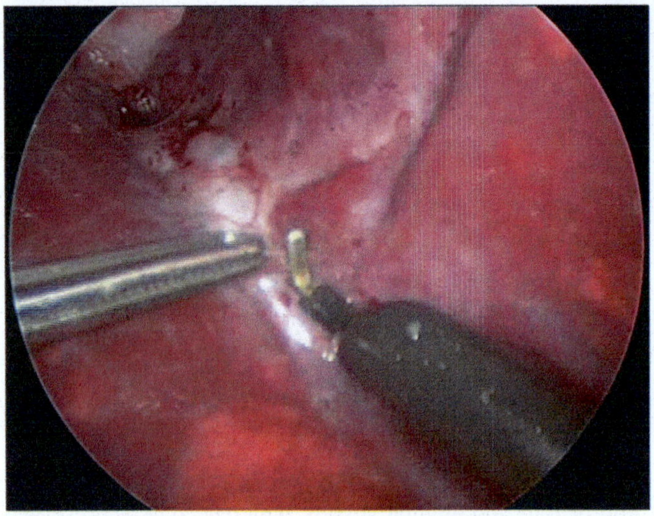

Fig. 12.4 Incomplete left oblique fissure

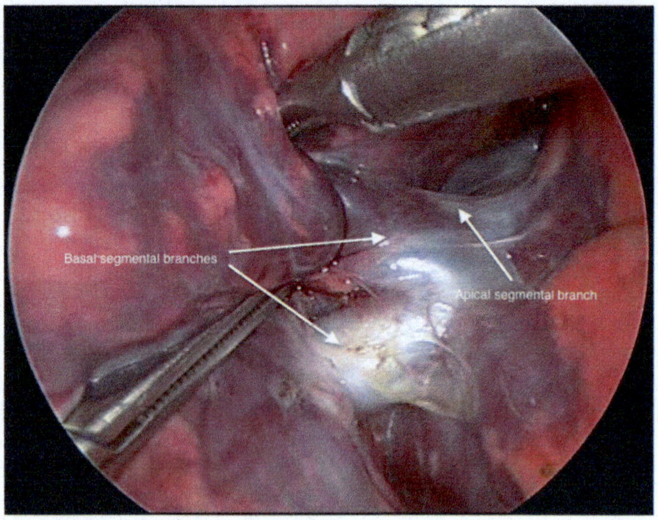

Fig. 12.6 Right lower lobe pulmonary artery branches

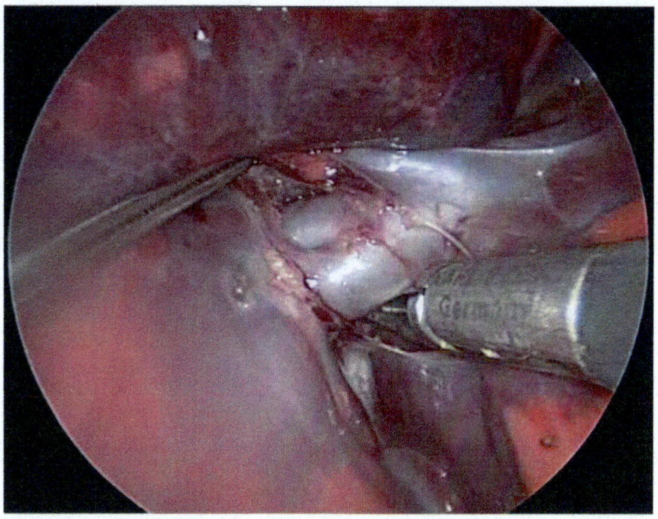

Fig. 12.7 Right lower lobe pulmonary artery basal branches

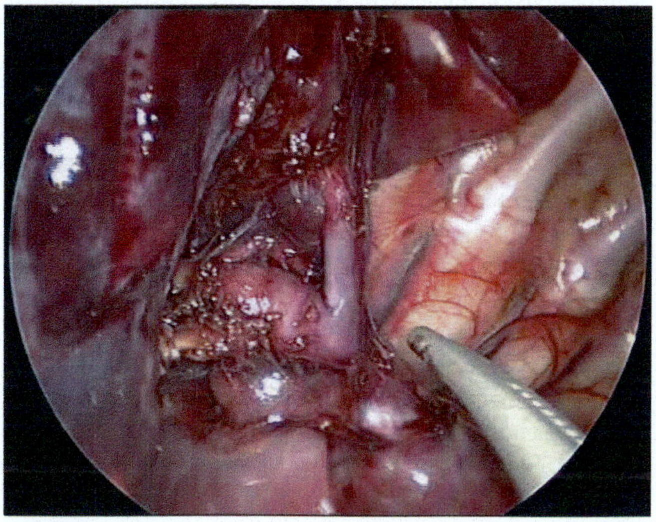

Fig. 12.8 Right lower lobe pulmonary artery apical branch

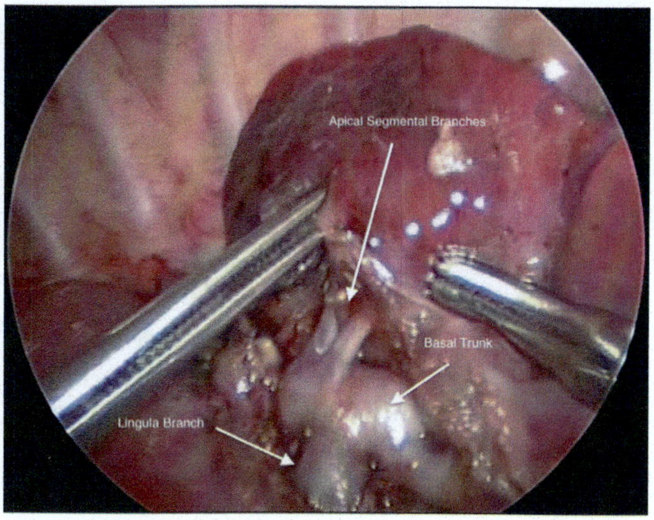

Fig. 12.9 Left lower lobe pulmonary artery branches

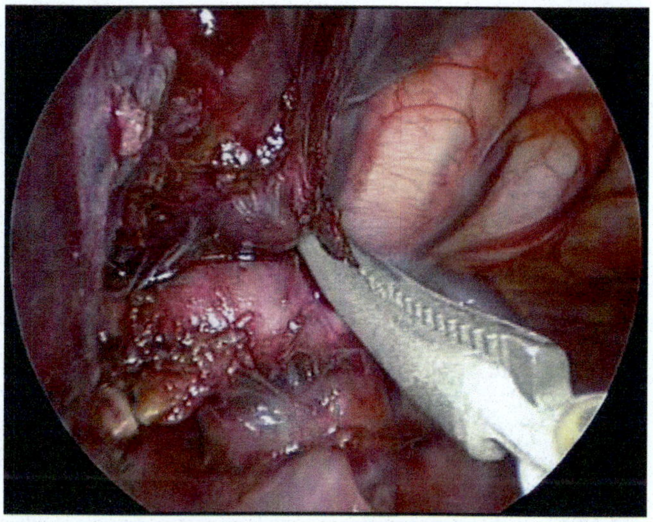

Fig. 12.10 LigaSure coagulation apical branch

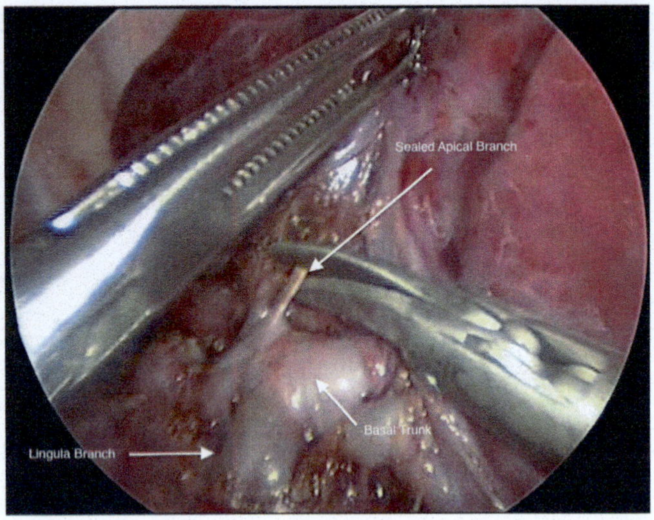

Fig. 12.11 Scissors division coagulated apical branch

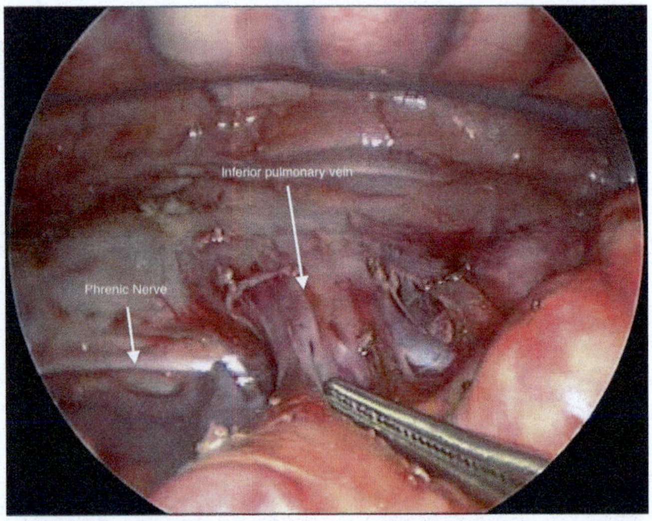

Fig. 12.12 Right inferior pulmonary vein from posterior aspect of hilum

Once the pulmonary artery branches are divided to the lobe to be resected, the pulmonary vein is then identified and dissected. For the lower lobes the inferior pulmonary vein can be approached from the posterior or anterior aspect of the hilum with appropriate retraction after division of the pleura of the pulmonary ligament. Occasionally moving the telescope to either the anterior or posterior inferior ports may provide a better view. Again dissecting the vein tributaries back into the lobe for a short distance allows control of the smaller vessels with a greater safety margin (Figs. 12.14 and 12.15).

After division of the pulmonary vein, the lobe should be left attached only by the bronchus (Fig. 12.16). In larger children, a stapler may be used to divide this, but in infants there is little space to deploy the instrument with adequate vision. In that instance we have divided the bronchus and then sutured the open end. Before stapling or dividing the lobar bronchus, it is critical to check that it is possible to inflate the remaining lobes with the bronchus occluded at the point chosen for division, either with an atraumatic grasper or with the stapler prior to firing. Occasionally occlusion of the bronchus, even when well clear of its origin, appears to compromise ventilation of the residual lobe(s). In that case dissection must be taken more distally, even dividing individual segmental branches. Once the bronchus has been closed, an "underwater test" (Fig. 12.17) will confirm this is airtight.

An upper lobectomy is less often required but is significantly more difficult to accomplish. On the left, the pulmonary artery lies behind the upper lobe bronchus and can be approached through the posterior part of the oblique fissure, identifying and dividing the segmental branches to the upper lobe. On the right, the pulmonary artery lies anterior to the lobar bronchus. It can be exposed in the anterior hilum by division of the superior pulmonary vein tributaries from the upper lobe and posterior reflection of the lobe (Figs. 12.18 and 12.19).

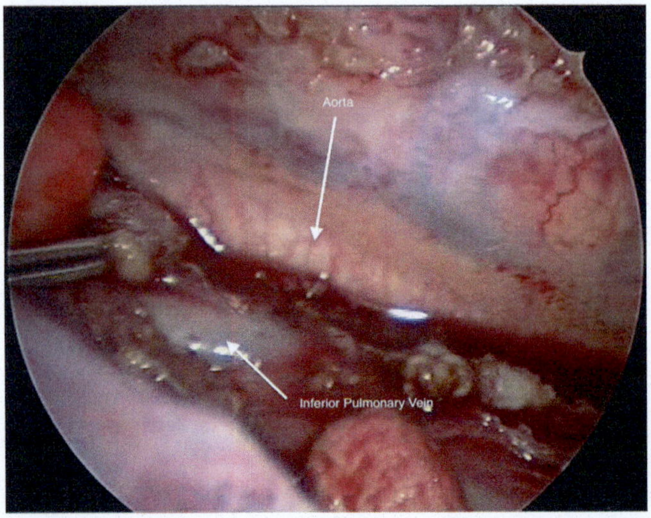

Fig. 12.13 Left inferior pulmonary vein from posterior aspect of hilum

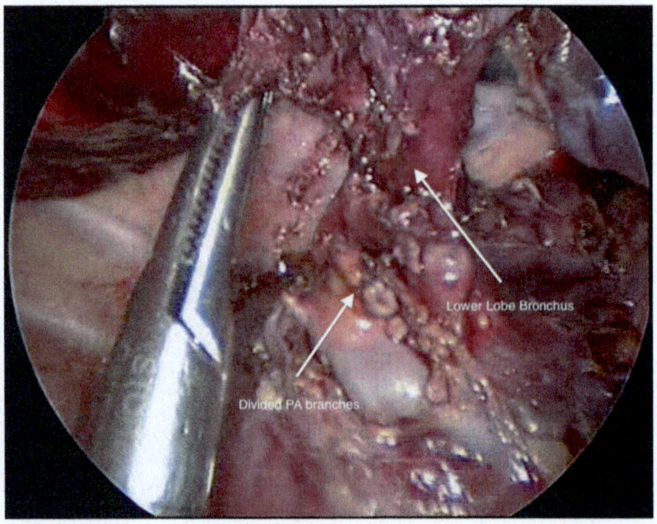

Fig. 12.14 Left lower lobe pedicled on bronchus

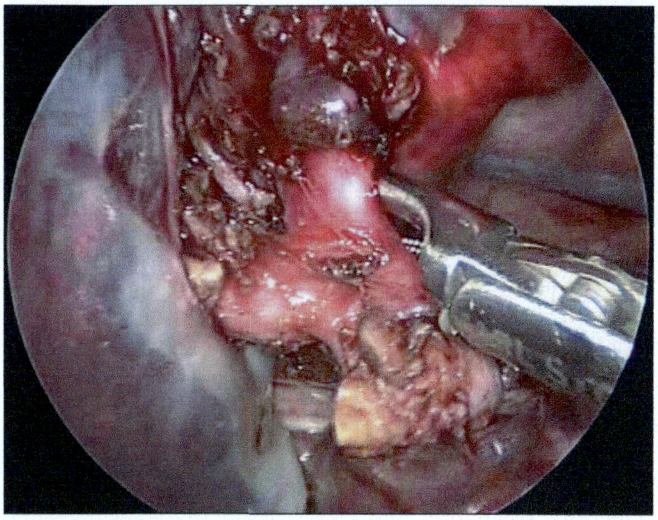

Fig. 12.15 Right lower lobe bronchus

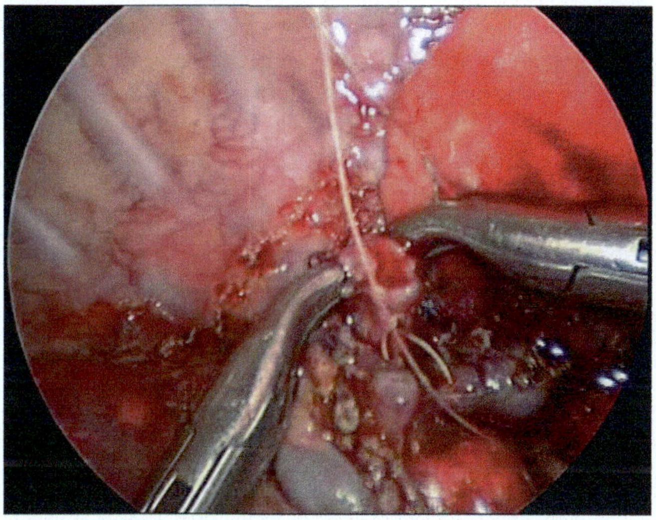

Fig. 12.16 Suturing left lower lobe bronchus

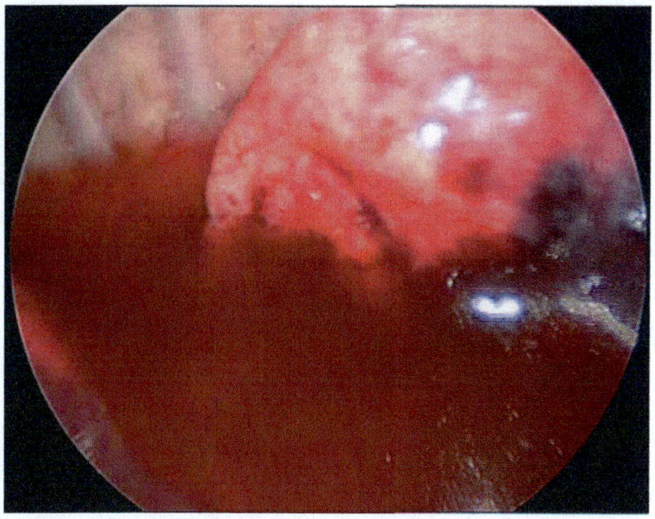

Fig. 12.17 Underwater test

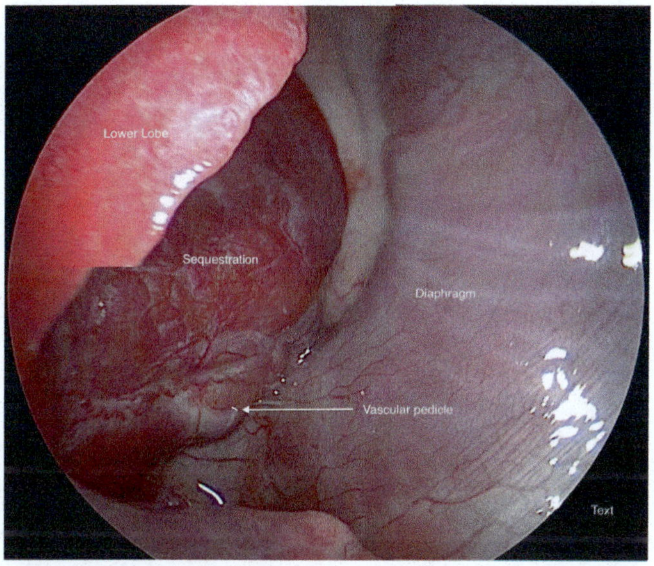

Fig. 12.18 Extralobar sequestration

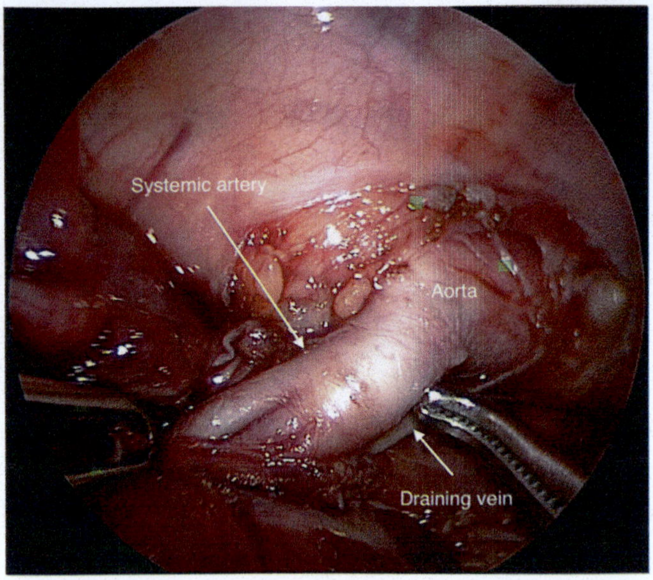

Fig. 12.19 Extralobar sequestration showing systemic vessels

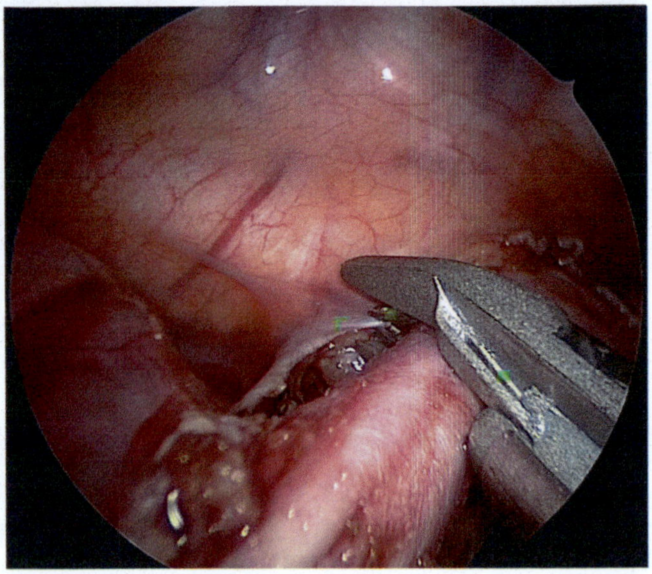

Fig. 12.20 Clipping systemic artery

In the case of a CPAM or an extralobar sequestration with a systemic arterial feeding vessel, this should be identified, controlled, and divided first. This may traverse the diaphragm but most frequently arises directly from the thoracic aorta with a short course into the malformation. The addition of a clip to coagulation with the LigaSure may provide additional security with proximal control (Fig. 12.20).

12.5 Alternatives

A 3-mm sealing device and a 5-mm linear stapler are now becoming available. These will allow downsizing to 3-mm instruments and ports in smaller children and infants. Other groups have demonstrated the safety of closure of the lobar bronchus with a simple 5-mm clip in patients less than 10 kg body weight as an alternative to suture closure. This may be both technically easier and quicker. In selected patients with a lesion clearly confined to a single broncho-pulmonary segment, a formal segmental resection may be possible.

12.6 Highlights and Pitfalls

One-lung ventilation with good collapse of the operated lung is key in enabling clear demonstration of the anatomy and safe dissection.

Do not staple or divide the bronchus without first checking that the residual lobe(s) can be inflated with the bronchus temporarily occluded. This is easier with a bronchus blocker than with endobronchial intubation, as one lung ventilation can be readily reestablished if further dissection is necessary.

Be prepared to convert to a thoracotomy if any concerns arise about defining the anatomy or the ability to safely control vessels.

Suggested Reading

Langston C. New concepts in the pathology of congenital lung malformations. Semin Pediatr Surg. 2003;12:17–37.

Puligandla PS, Laberge J-M. Congenital lung lesions. Clin Perinatol. 2012;39:331–7.

Priest JR, Williams GM, Hill DA, Dehner LP, Jaffe A. Pulmonary cysts in early childhood and the risk of malignancy. Pediatr Pulmonol. 2009;44:14–30.

Rothenburg SS, Middlesworth W, Kadennhe-Chisweshe A, Aspelund G, Kuenzler K, Cowles R, et al. Two decades of experience with thoracoscopic lobectomy in infants and children: standardising techniques for advanced thoracoscopic surgery. J Laparoendosc Adv Surg Tech A. 2015;25:1–6.

Thoracic Neuroblastoma

13

Michael Singh and Giampiero Soccorso

Abstract

Posterior mediastinal neurogenic tumors (neuroblastomas, ganglioneuroblastomas, and ganglioneuromas) represent 15–20 % of all neuroblastomas (apical 38 %, cervical 9.5 %, cervicothoracic 9.5 %, midthoracic 38 %, and thoracoabdominal 5 %). In small (<6 cm), mature, stage 1, mid-thoracic lesions, the thoracoscopic approach can result in adequate resection, low complication rate, shorter hospital stay, and long-term disease-free survival.

Keywords

Neuroblastoma • Thoracoscopic

13.1 General Information

Posterior mediastinal neurogenic tumors (neuroblastomas, ganglioneuroblastomas, and ganglioneuromas) represent 15–20 % of all neuroblastomas (apical 38 %, cervical 9.5 %, cervicothoracic 9.5 %, midthoracic 38 %, and thoracoabdominal 5 %). In small (<6 cm), mature, stage 1, mid-thoracic lesions, the thoracoscopic approach can result in adequate resection, low complication rate, shorter hospital stay, and long-term disease-free survival.

13.2 Working Instruments

- 5-mm Working ports, 10-mm port, and an endocatch bag for specimen extraction
- 5-mm, 0° laparoscope
- 5- mm Curved and straight graspers, monopolar hook diathermy
- 5-mm LigaSure sealing device
- 5-mm Endoclips

13.3 Positioning, Port Siting, and Ergonomic Considerations

For a midthoracic lesion, the patient is positioned semi-prone on a roll. The surgeon and assistant stand toward the patient's abdomen. The monitor is positioned opposite the surgeon. The ports should be positioned so that the triangulation occurs in relation to the tumor.

M. Singh (✉)
Department of Paediatric Surgery, Birmingham Children's Hospital, Birmingham, England

G. Soccorso
Birmingham Children's Hospital NHS Foundation Trust, Birmingham, UK

M. McHoney et al. (eds.), *Color Atlas of Pediatric Anatomy, Laparoscopy, and Thoracoscopy*,
DOI 10.1007/978-3-662-53085-6_13

13.4 Relevant Anatomy

The tumor usually lies along the posterior thoracic wall, lateral to the spine. On the left side it tends to be in close proximity to the thoracic aorta (Figs. 13.1 and 13.2). Its blood supply may come from the aorta and intercostal vessels. On the right side the related structures include the oesophagus, thoracic duct, pericardium, and superior mediastinum.

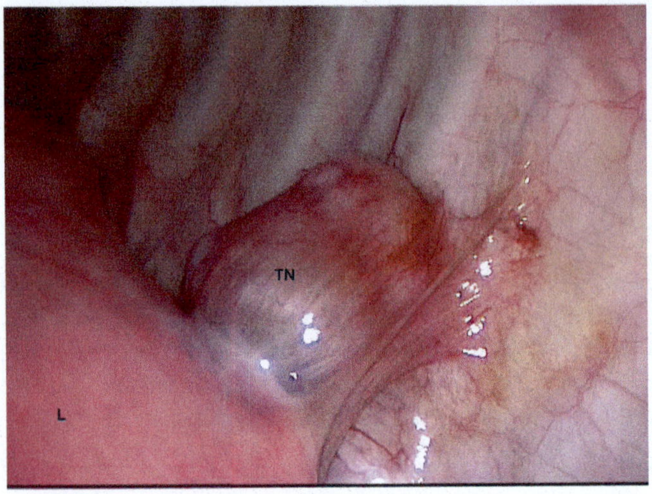

Fig. 13.1 The left-sided tumor is lateral to the spine and aorta. *L* lung, *TN* thoracic neuroblastoma

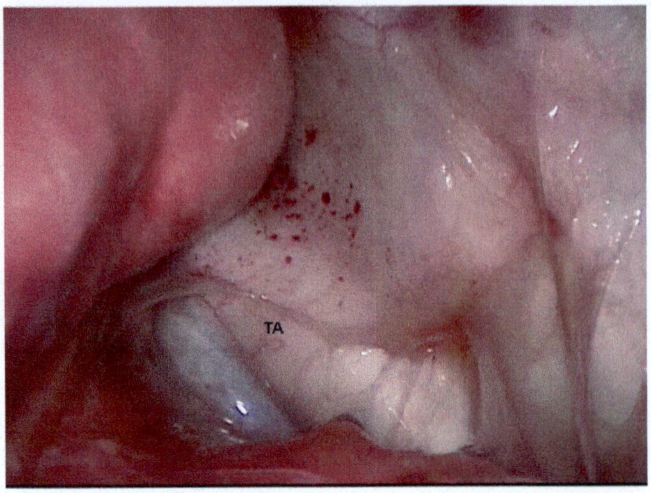

Fig. 13.2 The thoracic aorta (*TA*) is medial to the tumor

13.5 Surgical Technique

1. The first port (optical) is inserted in the anterior axillary line directly in line with the tumor.
2. A pneumothorax of 6 mmHg with flows of 1.5–2 L/min is created.
3. Two additional working ports are inserted in superior and inferior intercostal spaces to achieve triangulation.
4. Any adhesions between the lung and the tumor are divided with the monopolar hook diathermy (Fig. 13.3).
5. The pleura over the tumor is grasped and incised at the base with a monopolar hook diathermy (Fig. 13.4). The tumor is progressively dissected from the intercostal space with the hook diathermy. If the tumor is adjacent to the aorta, it can usually be easily separated with blunt dissection.
6. Any feeding vessels can be coagulated with the monopolar hook diathermy or LigaSure. Larger vessels can be coagulated with the LigaSure or clipped.
7. Once the tumor is excised, it can be removed by upgrading to a 10-mm port and placed in an endocatch bag.
8. The tumor bed can be coagulated with the monopolar hook diathermy (Fig. 13.5).
9. The pneumothorax can be evacuated with a no. 16 Fr chest drain and removed at the end of the procedure.
10. The wounds are closed.
11. A chest x-ray is done the following day, and the patient can be discharged once the pain is controlled.

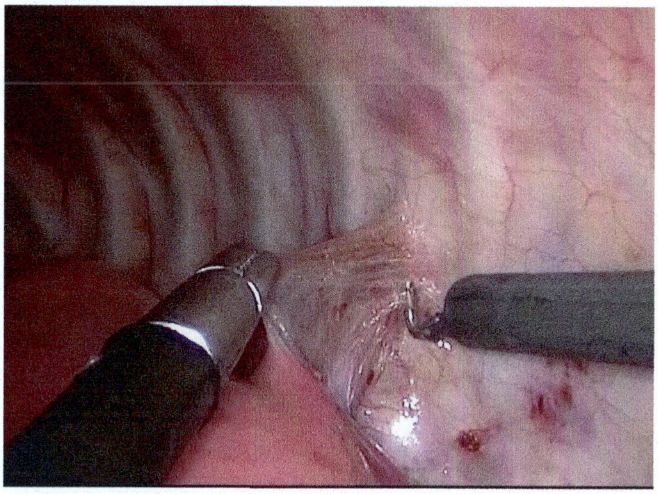

Fig. 13.3 Any adhesions between the tumor and the lung are divided with the monopolar hook diathermy

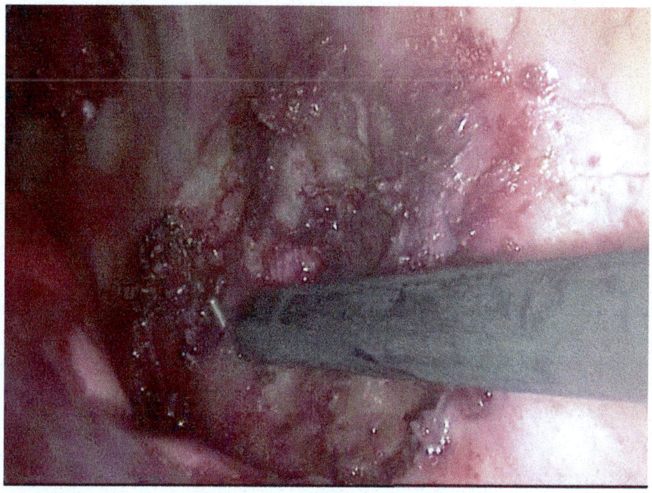

Fig. 13.5 The tumor bed can be coagulated with the monopolar hook diathermy

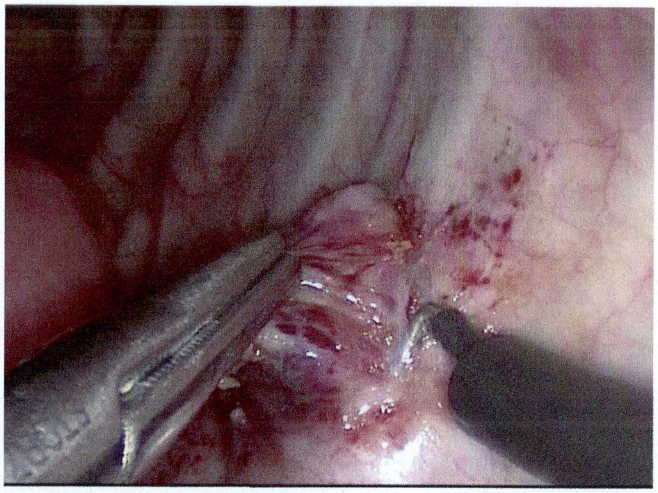

Fig. 13.4 The pleura at the base of the tumor is incised with the monopolar hook diathermy

13.6 Highlights and Pitfalls

- Careful and limited use of any electrosurgical devices will prevent any electrical or thermal conduction injuries to the surrounding structures.
- Large tumors with abdominal extension are better excised via a thoracoabdominal incision.

Suggested Reading

de Cou JM, Schlatter MG, Mitchell DS. Primary thoracoscopic gross total resection of neuroblastoma. J Laparoendosc Adv Surg Tech A. 2005;15:470–3.

Parikh DH, Crabbe D, Auldist A, et al. Paediatric thoracic surgery. Berlin: Springer; 2009.

Petty JK, Bensard DD, Partrick DA, et al. Resection of neurogenic tumours in children: is thoracoscopy superior to thoracotomy? J Am Coll Surg. 2006;203:577–81.

Thoracoscopic Placation for Eventration of the Diaphragm

14

Ashok Daya Ram and Michael Singh

Abstract

The major advantage of the thoracoscopic approach for the repair of eventration of the diaphragm is that it avoids the morbidity of a thoracotomy, including: pain, opioid analgesia, chest drainage, intensive care admission, a prolonged hospital stay, a large scar, and scoliosis. The indications for surgery include: respiratory distress, failure to wean from invasive ventilation and continuous positive airway pressure (CPAP), recurrent chest infections, and phrenic nerve palsy.

Keywords

Eventration • Diaphragm • Thoracoscopic

14.1 General Information

The major advantage of the thoracoscopic approach for the repair of eventration of the diaphragm is that it avoids the morbidity of a thoracotomy, including: pain, opioid analgesia, chest drainage, intensive care admission, a prolonged hospital stay, a large scar, and scoliosis. The indications for surgery include: respiratory distress, failure to wean from invasive ventilation and continuous positive airway pressure (CPAP), recurrent chest infections, and phrenic nerve palsy.

14.2 Working Instruments

A 5-mm, 0° scope is used in all cases, as it provides optimum visualisation. For infants weighing less than 7 kg, 3-mm working instruments are preferred; 5-mm instruments are used for larger children.

- 5-mm or 3-mm ports
- 5-mm 0° scope
- Straight and curved graspers
- Needle holder
- Scissors
- Knot pusher
- Sutures: 4-0 Ticron or 2-0 Ticron (ski needle)

14.3 Positioning, Port Siting, and Ergonomic Considerations

General anaesthesia with central endotracheal intubation is maintained. Infiltration with local anaesthetic prior to port insertion or paravertebral blocks can provide effective analgesia intraoperatively.

The patient is placed in the lateral decubitus position with the affected side up and a roll under the dependent axilla. The patient's head should be at the foot of the table, away from the anaesthetic machine. The monitor is placed over the patient's pelvis. The surgeon stands at the foot of the table.

A.D. Ram • M. Singh (✉)
Department of Paediatric Surgery,
Birmingham Children's Hospital, Birmingham, UK

M. McHoney et al. (eds.), *Color Atlas of Pediatric Anatomy, Laparoscopy, and Thoracoscopy*,
DOI 10.1007/978-3-662-53085-6_14

14.4 Relevant Anatomy

Figures 14.1 and 14.2 show an example of the affected anatomy.

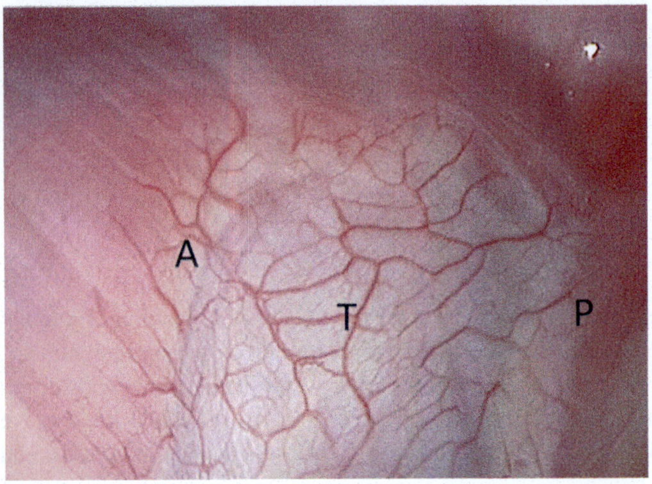

Fig. 14.1 Right-sided eventration, showing the anterior musculotendinous rim (*A*), central diaphragm (*T*), and posterior musculotendinous rim (*P*)

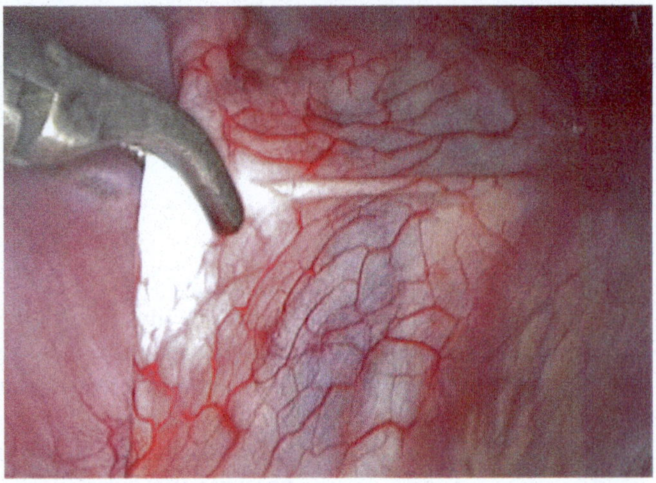

Fig. 14.2 The lax diaphragm is easily picked up

14.5 Surgical Technique

1. The first port is placed just anterior to the inferior angle of the scapula (Fig. 14.3). A pneumothorax of 5–6 mmHg with flows of 1.5–2 L/min is maintained.
2. Two working ports are inserted under vision between the anterior and midaxillary line and posteriorly to achieve effective triangulation (Fig. 14.4).
3. The plication starts laterally. After the needle is placed at the junction of the muscular rim and the tendinous part, it is then passed through the central tendinous part of the diaphragm and the opposite musculotendinous portion (Fig. 14.5). The knot can then be tied extracorporeally or intracorporeally.
4. Plication is continued medially with further interrupted sutures (Fig. 14.6).
5. The pneumothorax is evacuated by inserting a 16 Fr nasogastric tube via the 5-mm port and immersing the other end in a dish of saline. The anaesthetist is then asked to manually ventilate the patient to re-expand the lung. The nasogastric tube is removed and the wounds closed. It is not necessary to insert a chest drain.
6. Postoperatively, oral analgesia is prescribed. A chest x-ray is performed the following day. If there is adequate lung re-expansion, the patient can be discharged.

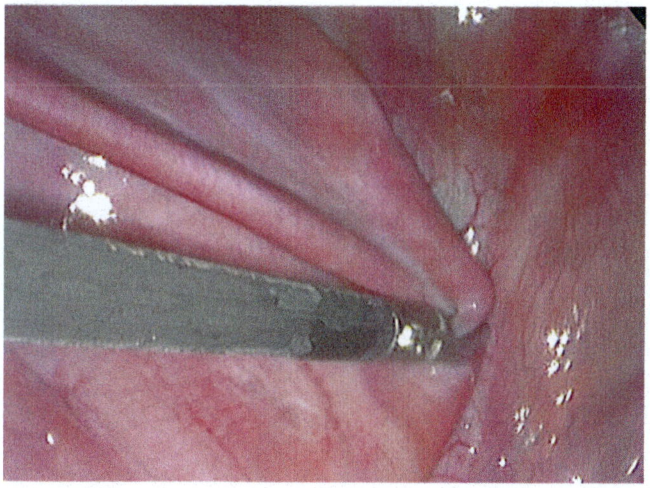

Fig. 14.3 The first port is placed just anterior to the inferior angle of the scapula

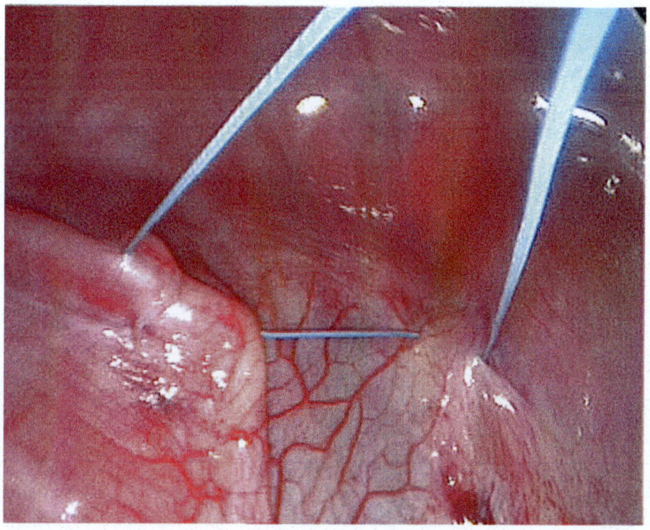

Fig. 14.4 Two working ports are inserted under vision between the anterior and midaxillary line and posteriorly to achieve effective triangulation

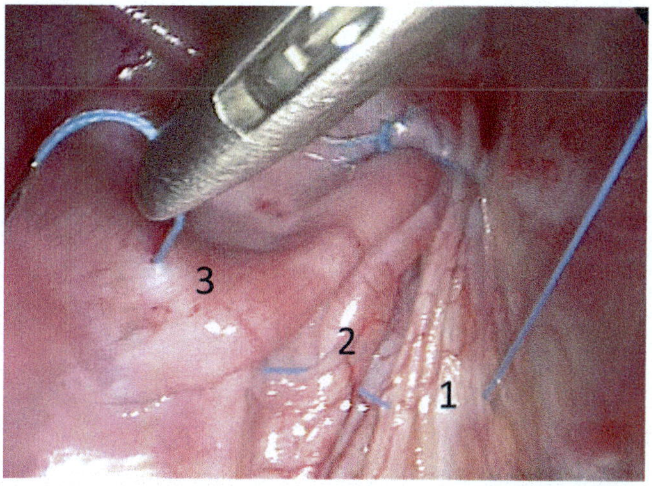

Fig. 14.5 The plication begins by placing the needle at the junction of the muscular rim and the tendinous part (*1*). The needle is then passed through the central tendinous part of the diaphragm (*2*) and the opposite musculotendinous portion of the diaphragm (*3*)

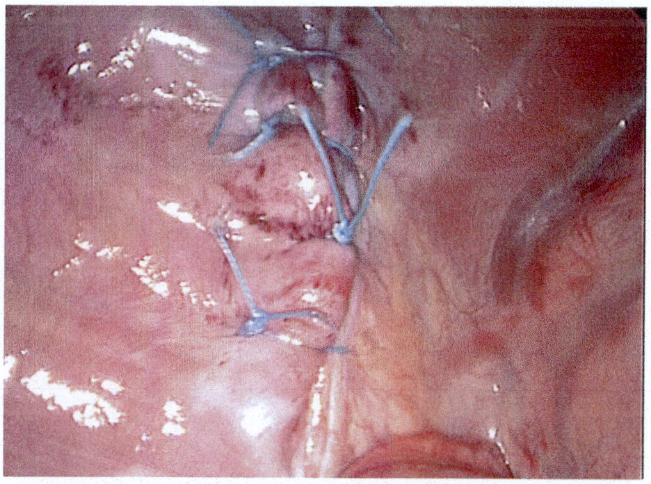

Fig. 14.6 Plication is continued medially with further interrupted sutures

14.6 Alternatives

Laparoscopic repair is more complicated, owing to the presence of the abdominal viscera, and it has a higher recurrence rate.

14.7 Highlights and Pitfalls

- The first port should be inserted cautiously, as the diaphragm can be quite elevated, presenting a risk of injury to the abdominal viscera underneath. A head-up position is helpful during this stage, as it allows the diaphragm to fall inferiorly.
- To perform a secure repair, the surgeon should use the suturing technique that he or she is most comfortable with.
- Caution should be exercised medially, as the inferior vena cava and aorta are in close proximity.

Suggested Reading

Becmeur F, Talon I, Schaarschmidt K, Phillippe P, Moog R, Kauffmann I, et al. Thoracoscopic diaphragmatic eventration repair in children: about 10 cases. J Pediatr Surg. 2005;40:1712–5.

Crabbe DCG. Diaphragmatic eventration and phrenic palsy. In: Parikh DH, Crabbe DCG, Auldist AW, Rothenberg SS, editors. Pediatric thoracic surgery. London: Springer; 2009. p. 501–8.

Laparoscopic Repair of Morgagni Hernia

15

Michael Singh

Abstract

A Morgagni (anterior) diaphragmatic hernia is usually diagnosed on a chest x-ray in a child with recurrent respiratory symptoms. The Morgagni hernia lends itself easily to a laparoscopic repair.

Keywords

Morgagni (anterior) diaphragmatic hernia • Laparoscopic repair

15.1 General Information

A Morgagni (anterior) diaphragmatic hernia is usually diagnosed on a chest x-ray in a child with recurrent respiratory symptoms. The Morgagni hernia lends itself easily to a laparoscopic repair.

15.2 Working Instruments

- 3-mm or 5-mm straight and curved graspers
- Hook diathermy
- Needle holder
- Laparoscopic suture retriever
- 2/0 Nonabsorbable sutures
- 5-mm umbilical port and two 3-mm or 5-mm working ports
- 0° or 30°, 5-mm laparoscope

15.3 Positioning, Port Siting, and Ergonomic Considerations

The patient is positioned supine towards the foot of the table. The legs can be placed in the frog-leg position or hung over the edge at the knees. The monitor is placed over the patient's lower chest. The surgeon stands at the foot of the table.

M. Singh
Department of Paediatric Surgery, Birmingham Children's Hospital, Birmingham, UK

M. McHoney et al. (eds.), *Color Atlas of Pediatric Anatomy, Laparoscopy, and Thoracoscopy*,
DOI 10.1007/978-3-662-53085-6_15

15.4 Relevant Anatomy

The hernia is in the midline and anterior. The transverse colon may be within the sac and can be easily removed with traction. There is normally a well-developed posterior rim but no anterior rim. The pericardium and pleura can be seen just outside the fundus of the sac (Fig. 15.1).

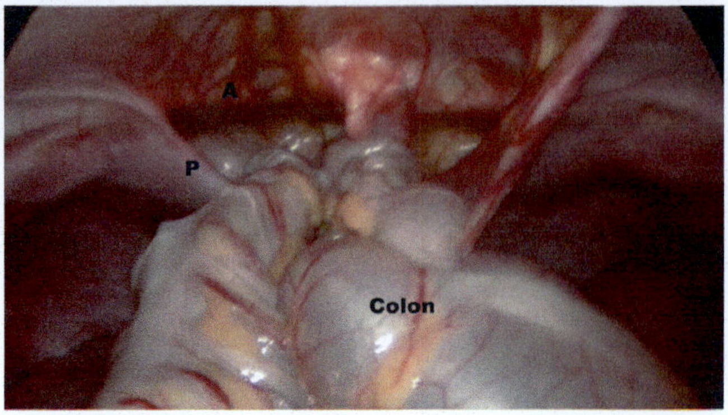

Fig. 15.1 The transverse colon may be seen within the sac of the Morgagni hernia. There is normally a well-developed posterior rim (*P*) and no anterior rim (*A*). The pericardium and pleura can be seen just outside the fundus of the sac

15.5 Surgical Technique

1. The first 5-mm port is inserted with an open technique via the umbilicus. A pneumoperitoneum of 8–10 mmHg with flows of 1.5–2 L/min is established. Two additional working ports are inserted superior to the umbilicus and at the right and left midclavicular lines, to produce effective triangulation. The hernia contents are reduced.
2. The posterior rim of the defect is lifted to the anterior abdominal wall and the right and left limits of the defect are marked externally on the skin. The sac of the hernia is incised at the posterior rim. The sac is always dissected and removed to prevent recurrence (Fig. 15.2).

   In most cases the posterior rim can reach the anterior abdominal wall without tension, and a primary repair can be performed. If there is significant tension, then a Gore-Tex sheet should be used to bridge the gap.
3. A 2-mm stab incision is made in the skin over the midpoint of the defect. A 2/0 nonabsorbable suture is passed via the stab incision, through the anterior abdominal wall (Fig. 15.3).
4. The needle is then grasped with a laparoscopic needle holder and passed through the midpoint of the posterior rim (Fig. 15.4).
5. A laparoscopic suture retriever is passed through the same stab incision but through a different path through the anterior abdominal wall into the peritoneal cavity. The suture retriever is used to grasp the previously inserted suture and pull it externally via the stab incision (Fig. 15.5).
6. The suture is placed in a clip and not tied until all the sutures have been placed (Fig. 15.6).
7. One or two additional sutures are placed on either side via stab incisions. Once all the sutures have been placed, they can be progressively tied (Fig. 15.7). The knots are buried in the subcutaneous tissues and the umbilicus and skin closed. Figure 15.6 shows two sutures tied and buried.
8. Postoperative opioid analgesia is not required, and the patient is allowed to eat once awake. A chest x-ray is done the following day and the patient is discharged.

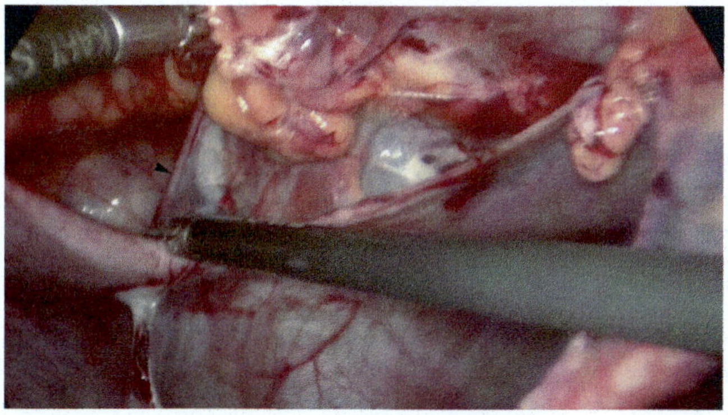

Fig. 15.2 The sac of the hernia is incised at the posterior rim (*arrowhead*) with the monopolar hook diathermy or scissor

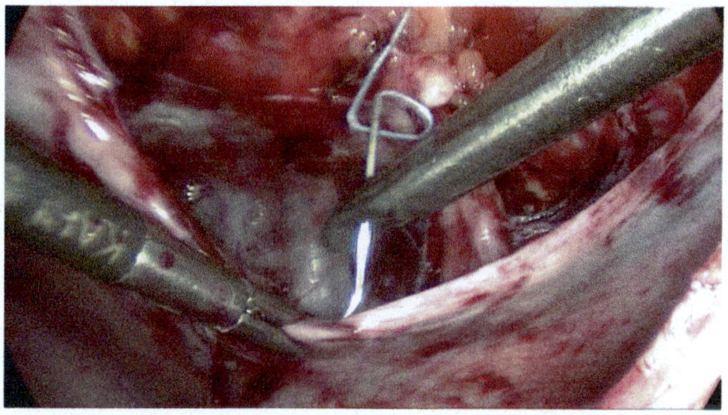

Fig. 15.4 The needle is then grasped with a laparoscopic needle holder and passed through the midpoint of the posterior rim

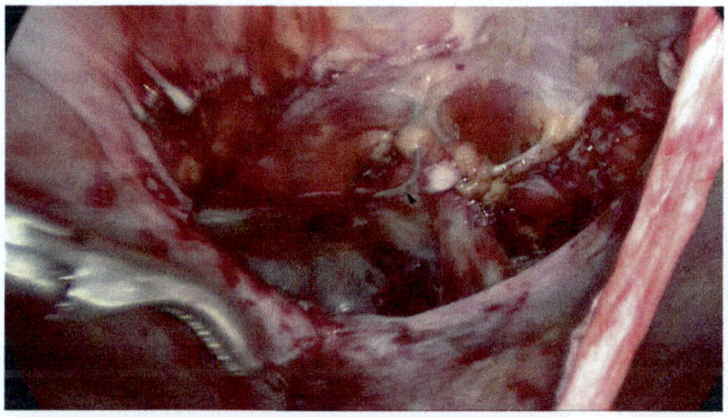

Fig. 15.3 A 2-mm stab incision is made in the skin over the midpoint of the defect and a 2/0 nonabsorbable suture is passed via the incision, through the anterior abdominal wall (*arrow*)

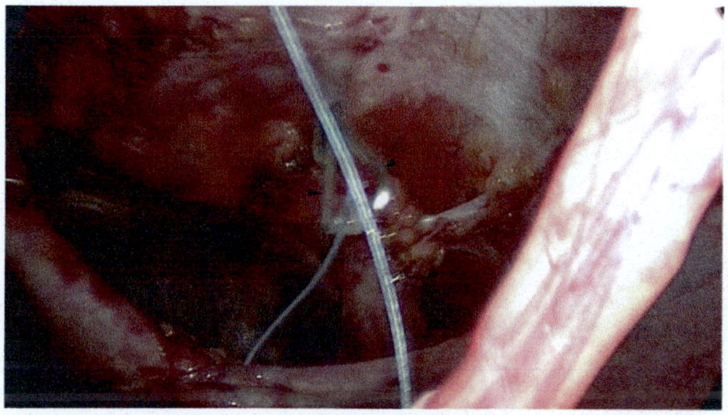

Fig. 15.5 A laparoscopic suture retriever (*arrowheads*) is passed through the same stab incision but through a different path into the peritoneal cavity. The suture retriever is used to grasp the previously inserted suture and pull it externally via the stab incision

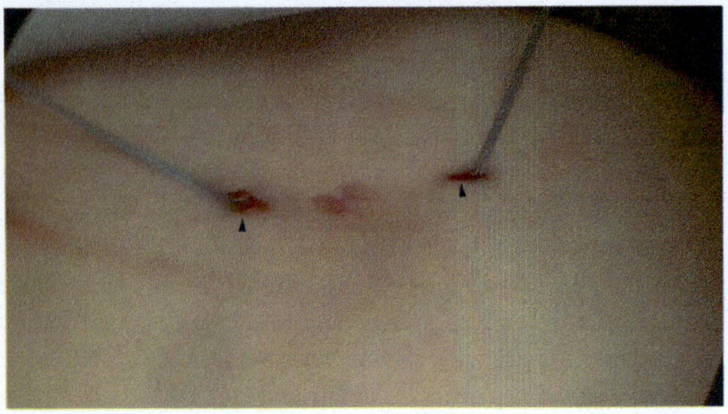

Fig. 15.6 The suture (*arrowheads*) is placed in a clip and not tied until all the sutures have been placed

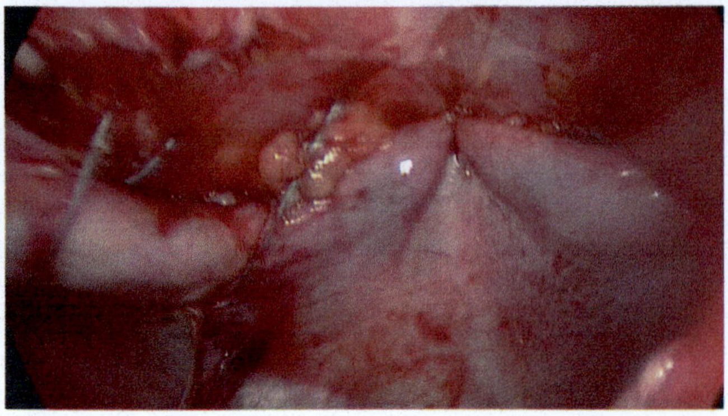

Fig. 15.7 Once all the sutures have been placed, they can be progressively tied

15.6 Alternatives

None.

15.7 Highlights and Pitfalls

- Always remove the sac to prevent recurrence.
- Do not attempt a primary repair if the posterior rim will not appose without tension. In this circumstance, use a Gore-Tex sheet to repair the defect.

Suggested Reading

Bax KN, Georgeson KE, Rothenberg SS, Valla J-S, Yeung CK, editors. Endoscopic surgery in infants and children. New York: Springer; 2008.

Primary Button Gastrostomy

16

Clarie Clark

Abstract

The laparoscopic approach offers several potential advantages; including good visualisation, no need for further surgical procedures and cosmesis.

Keywords

Button Gastrostomy, Laparoscopy

16.1 General Information

The laparoscopic approach offers several potential advantages:

- Good visualisation of the stomach, to avoid colonic spearing and enable correct placement within the gastric body.
- Avoidance of the need for a secondary procedure associated with traditional gastrostomy tubes, eliminating the serious complication of gastric separation and peritonitis.
- Early cosmetic advantage associated with the button rather than the larger percutaneous endoscopic gastrostomy (PEG) tube.

16.2 Working Instruments

- 5-mm Hassan port
- 30° telescope
- Atraumatic grasper (3 or 5 mm, depending on size of child)
- AMT PEG Assist™ Initial Placement Gastrostomy Kit (Applied Medical Technology; Brecksville, OH, USA)

Or

- Cook endovascular dilators, sizes 6 Fr to 18 Fr (Cook Medical, Bloomington, IN, USA)
- 18G needle
- Floppy-tip guide wire, diameter 0.035 or 0.038 in.
- 11 blade scalpel

Also

- 1–0 or 0 vicryl or polydioxanone suture (PDS) on a MO45 round-body needle
- Gastrostomy button (MIC-KEY low profile size 12–14) (Kimberley-Clark; Roswell, GA, USA)

C. Clark
Royal Hospital for Sick Children, Edinburgh, Scotland, UK

M. McHoney et al. (eds.), *Color Atlas of Pediatric Anatomy, Laparoscopy, and Thoracoscopy*,
DOI 10.1007/978-3-662-53085-6_16

16.3 Positioning

Patient is placed supine on the operating table. The surgeon stands at the end of the table for small infants, or to the right of larger children. The camera person is on the left. The patient requires a nasogastric (NG) tube.

Before preparation of the abdomen, the costal margin on both sides should be marked out.

16.4 Surgical Technique

1. Infraumbilical 5-mm port is inserted using the Hassan cut down technique.
2. Pneumoperitoneum is established, with typical pressure settings 5–10 mmHg, depending on the size and weight of the child.
3. The stomach is identified.
4. A small, 2-mm incision is made through the abdominal wall under direct vision with the 11 blade scalpel over the gastric area where you wish the button to sit, ensuring that there is enough space between the costal margin and the button for comfort when the pneumoperitoneum is released. Insert the atraumatic grasper directly though the incision without port placement (Fig. 16.1).
5. Identify the pylorus, antrum, and body of the stomach and decide on the best siting of the gastrostomy. Grasp this area of the stomach and bring it up to the anterior abdominal wall (Figs. 16.2 and 16.3). *Problem shooting: If the stomach does not easily reach the anterior abdominal wall, reduce the pneumoperitoneum pressure.*
6. Using the MO45 round-bodied needle, place the 1–0 or 0 vicryl or PDS stay sutures through the anterior abdominal wall and stomach, ensuring good bites of the stomach. Make sure these bites are placed near the atraumatic grasper externally, as they will be used to secure the button at the end of the procedure. Leave the stay sutures long and place on a clip. As shown in Figs. 16.4, 16.5, and 16.6, this is done under direct vision. *Top Tip: Ensure the needle is at a right angle to the skin, to allow easy suturing of the stomach wall.*
7. For correct placement of the guide wire, the stomach is inflated via the NG tube and then punctured with the 18G needle under direct vision. The stomach should deflate through the needle to ensure good intragastric needle placement. Place the guide wire into the stomach and remove the needle (Fig. 16.7). *Top Tip: To ensure that the guide wire does not fall out, an assistant should hold onto the wire externally at the entry point at all times.*
8. Dilation of the stomach now occurs over the guide wire (under vision), starting with size 6 Fr and graduating up to size 18 Fr (Figs. 16.8 and 16.9). Hold the stay sutures so the stomach is up against the anterior abdominal wall.
9. Remove the dilators but keep the guide wire in place. Place the gastrostomy measuring device over the guide wire into the stomach and blow up the balloon to measure the size of MIC-KEY button required (*e.g.*, 14 Fr, 1.5 cm MIC-KEY). Deflate the balloon and remove the measuring device, leaving the guide wire in the stomach.
10. To place the MIC-KEY button, place the 8 Fr dilator through the MIC-KEY button and then place it over the guide wire and into the stomach under direct vision. Blow up the balloon. To ensure correct intragastric placement, insufflate the stomach via the NG tube again (Fig. 16.10) and connect the MIC-KEY external tubing to deflate the stomach via the button (Fig. 16.11). If the stomach does not deflate easily, the button is not in the correct place.
11. Once correct placement is confirmed, tie the stay sutures over the button flanges (Fig. 16.12).
12. Release the pneumoperitoneum. Suture closed the infraumbilical port site.

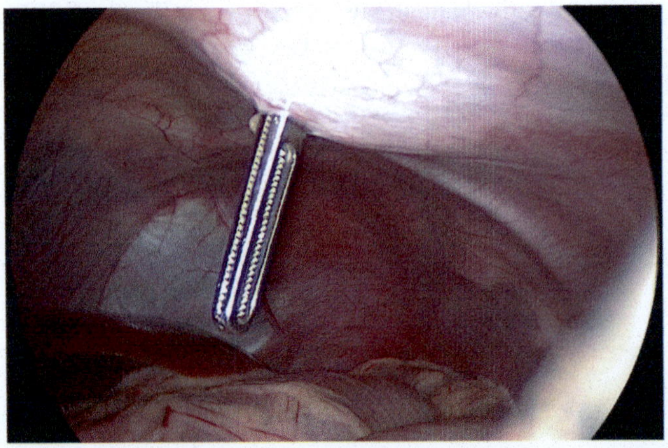

Fig. 16.1 The atraumatic grasper is inserted directly though the incision

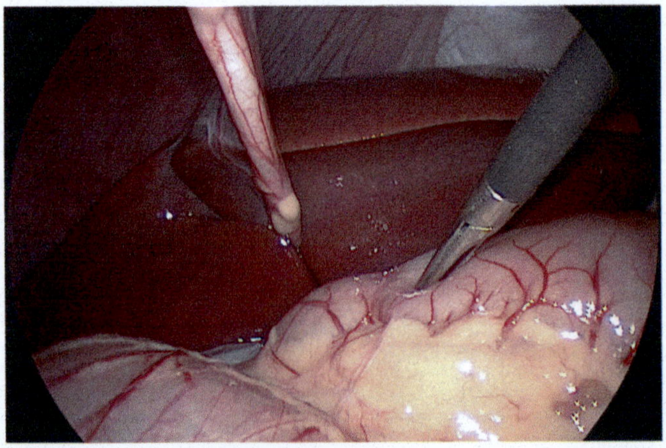

Fig. 16.2 The best site for the gastrostomy is chosen

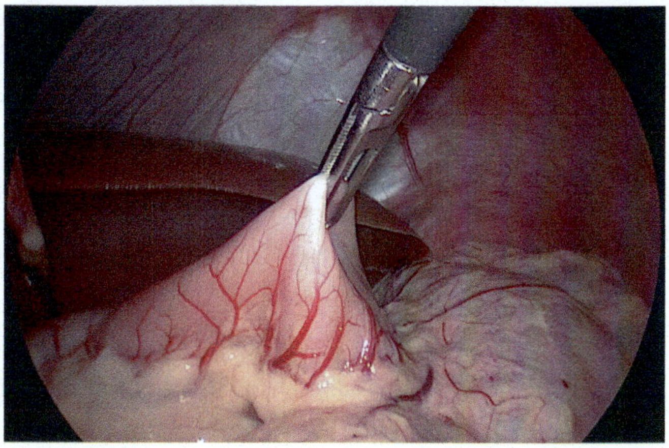

Fig. 16.3 This area is grasped and brought up to the anterior abdominal wall

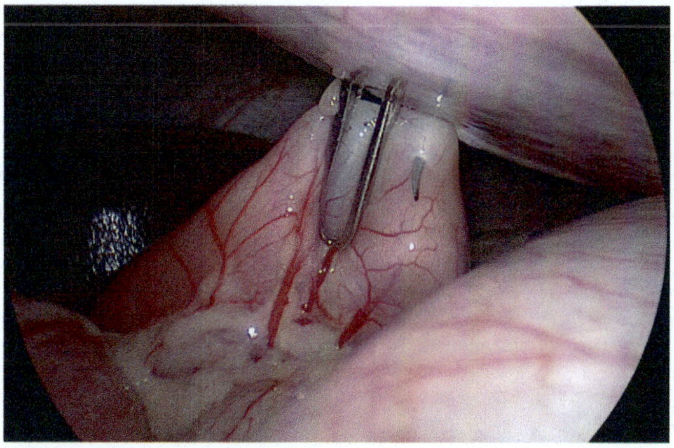

Fig. 16.4 Place the stay sutures through the anterior abdominal wall and stomach, ensuring good bites of the stomach. The needle should be at a right angle to the skin

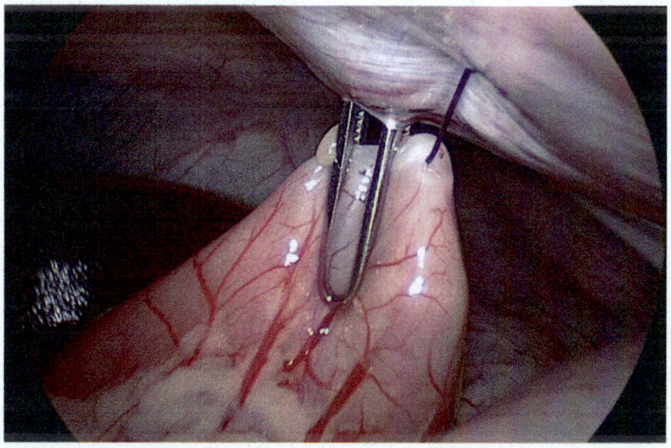

Fig. 16.5 Make sure these bites are placed near the atraumatic grasper

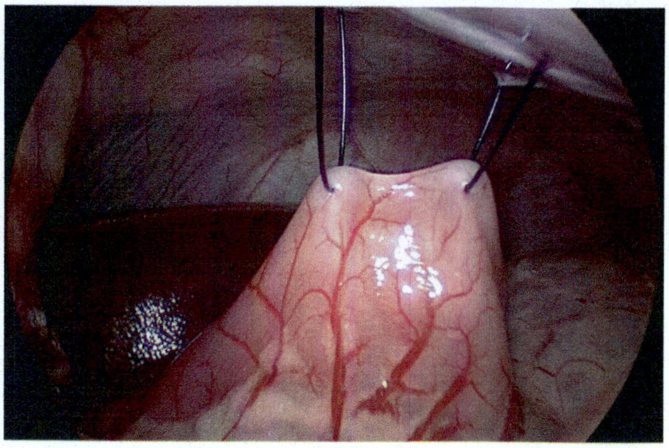

Fig. 16.6 Leave the stay sutures long

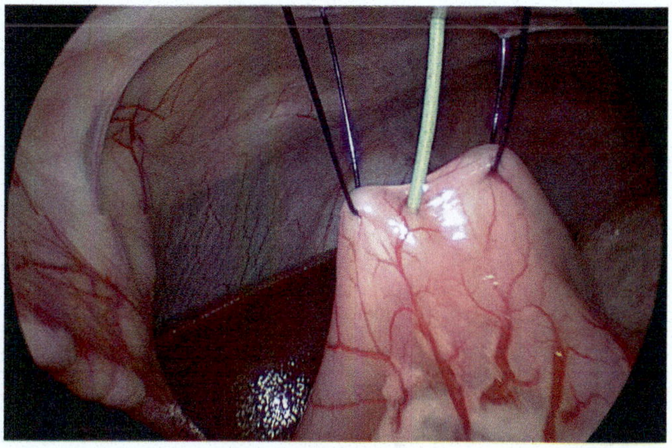

Fig. 16.7 The guide wire is placed into the stomach

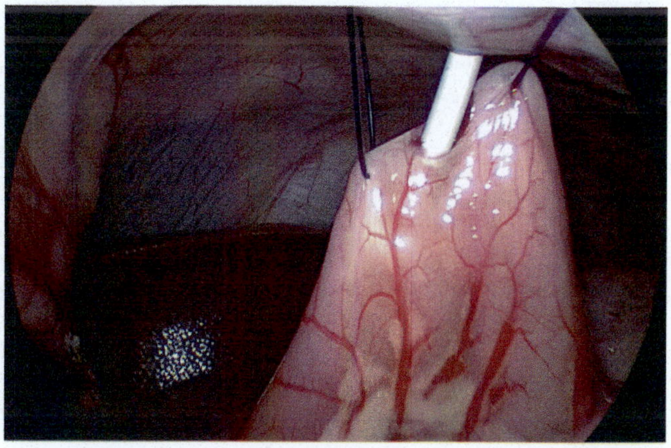

Fig. 16.8 Dilators of increasing size are now used

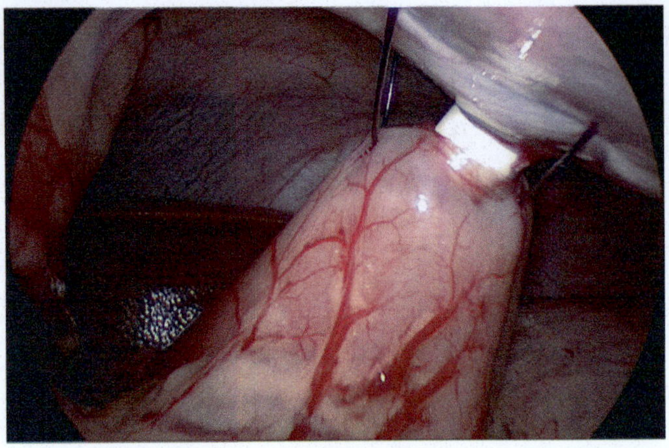

Fig. 16.9 During dilation, the stomach is held against the anterior abdominal wall using the stay sutures

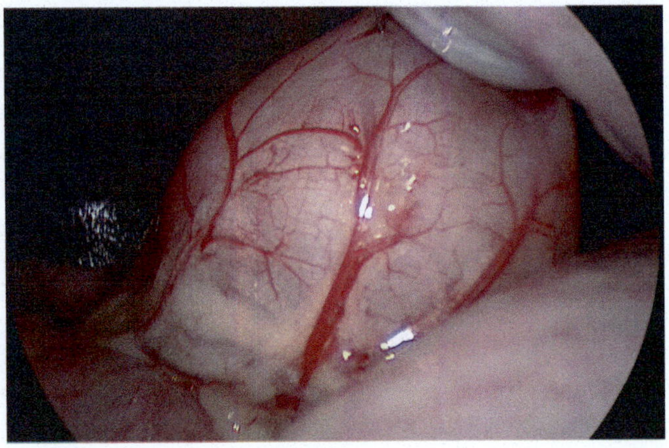

Fig. 16.10 The stomach is inflated to ensure correct intragastric placement of the MIC-KEY button

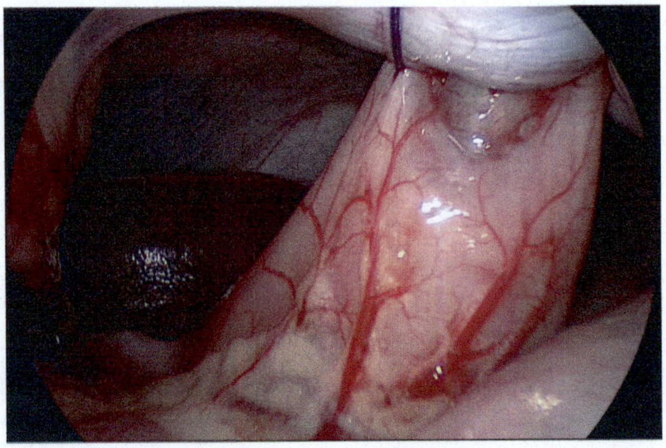

Fig. 16.11 If the stomach is easily deflated via the button, the placement is correct

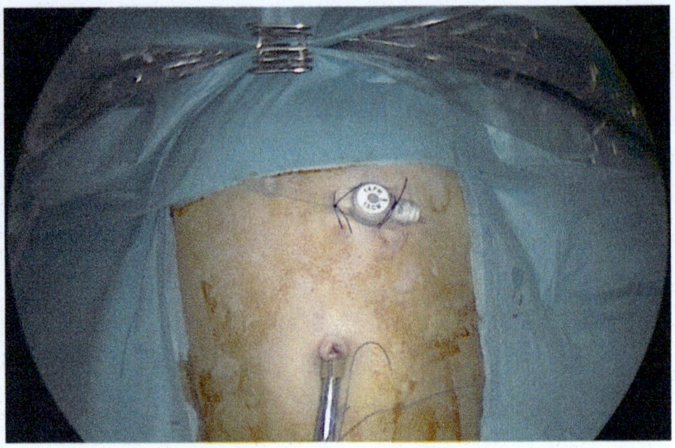

Fig. 16.12 When correct placement is confirmed, tie the stay sutures over the button flanges

13. Leave the sutures for 5 days. The button must not be twisted during this time, so that good tract formation can occur.
14. Remove the NG tube and place the button on free drainage overnight. The patient receives nothing by mouth overnight, with essential medication given only through the gastrostomy until the morning. Allow graduated use of the gastrostomy as required.

16.5 Highlights and Pitfalls

- Always ensure correct intragastric placement of the needle and button by insufflating and deflating the stomach as described above. If there is no deflation of the stomach, reposition the needle and button until it occurs.
- Ensure that the guide wire is always in the stomach throughout the procedure. If it comes out, replace it under direct vision.
- Beware of overly vigorous insertion of the guide wire, which may cause gastric perforation in unintended places.
- This technique can only be used if the stay suture needle can be passed through the anterior abdominal wall easily. In obese patients, a different primary button kit (Gastrointestinal Anchor Set with Saf-T-Pexy T-Fasteners; Kimberly-Clark, Roswell, GA, USA) can be used, via upper gastrointestinal (UGI) endoscopy rather than laparoscopy.

17 Nissen Fundoplication

Merrill McHoney

17.1 General Information

The laparoscopic approach to fundoplication may offer advantages over the open approach. It offers good access to the hiatus, oesophagus and stomach, while minimising the surgical wound. There is less postoperative need for use of high dependency units, less respiratory depression due to pain, less postoperative adhesions, better cosmesis and at least equal efficacy. There is less morbidity and equal efficacy with the laparoscopic approach.

17.2 Working Instruments

- 5 mm Hasson port
- 30° telescope (length proportionate to patient size)
- 3 or 5 mm instruments depending on patient size
- needle holders
- hook diathermy
- scissors
- Maryland and/or Yohan forceps
- Natheson's liver retractor (size depending on patient weight) with table attachment
- Ultrasonic scalpel or ligasure (optional for dissection and dealing with short gastric vessels)

17.3 Positioning, Port Siting and Ergonomic Considerations

The patient is positioned at the end of the table in frog legged position (the operating surgeon will be between the legs during the procedure). Patients with severe contractures secondary to neuro-muscular disorders may have to be placed with legs together, which are then placed on one side of the patient. Primary port is placed in the umbilical fold (the upper or lower fold may be chosen depending on patient size and body habitus to maximise ergonomics). Two further working ports are placed in the left and right upper abdomen under vision. The port in the right side of the patient may need to be slightly long to be beyond the falciform ligament, thereby avoiding catching it on introducing instruments. Slight head up positioning may be useful in allowing the intestines to fall away from the operating field.

A Natheson's retractor is placed through an epigastric incision under vision. The size is chosen appropriate to patient size. It is positioned to allow the left lobe of the liver to be lifted out of the operative field of the oesophagus.

M. McHoney, MD, PhD
Royal Hospital for Sick Children,
9 Sciennes Road, Edinburgh EH9 1LF, UK

M. McHoney et al. (eds.), *Color Atlas of Pediatric Anatomy, Laparoscopy, and Thoracoscopy*,
DOI 10.1007/978-3-662-53085-6_17

17.4 Relevant Anatomy

1. View of the oesophageal hiatus

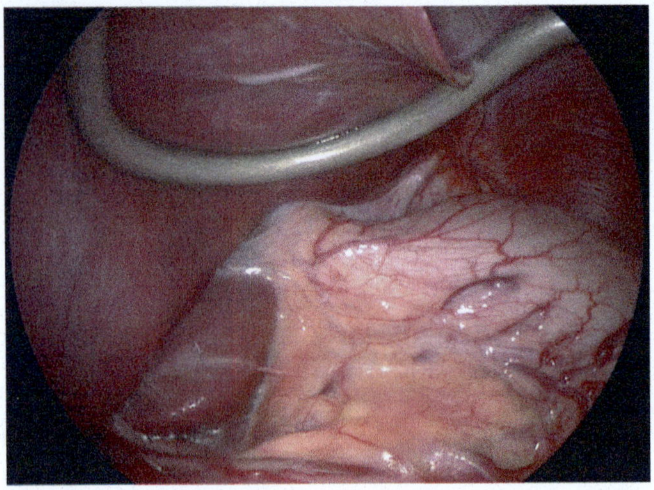

Fig. 17.1 This view shows the initial view seen in-situ. The zona pellucida of the lesser omentum overlying the caudate lobe of the liver is seen. The left lobe of the liver is seen (beneath the Natheson's liver retractor), ending in the left triangular ligament. The oesophagus and stomach is seen with the gastro-epiloic vessels along the lesser curvature

2. View of the anatomy of the hiatus

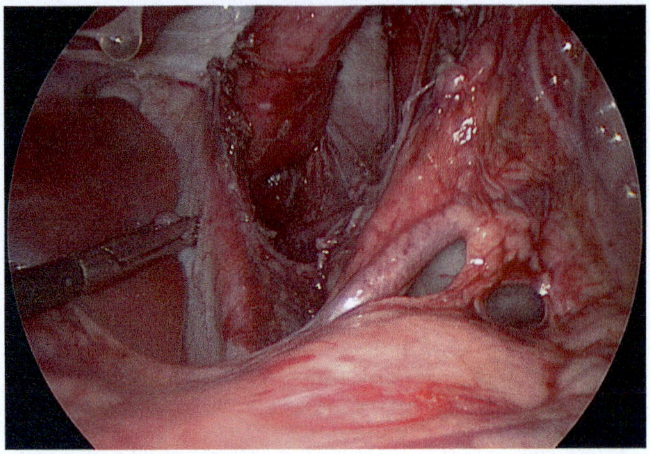

Fig. 17.2 The zona pellucuda has been opened and the right (and left) crus of the diaphragm is seen, creating the oesophageal hiatus. The posterior vagus can be seen on the oesophagus

3. Another view of anatomy of the hiatus

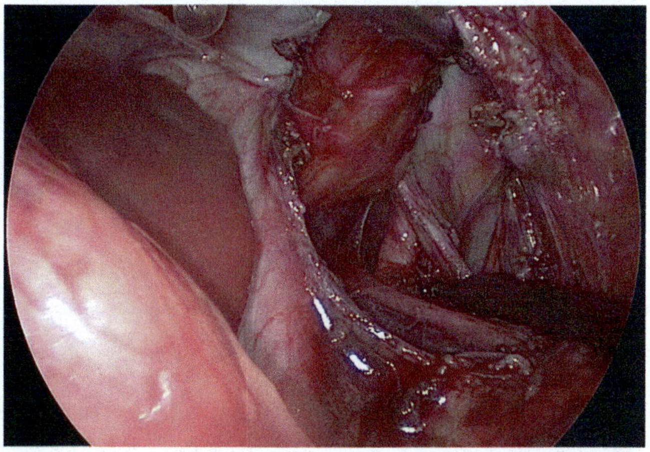

Fig. 17.3 In this figure the left crus is more visible. The posterior vagus is again seen applied to the back of the oesophagus and an oesophageal vessel is also seen going through the hiatus

17.5 Surgical Technique

4. Exposure of the oesophagus

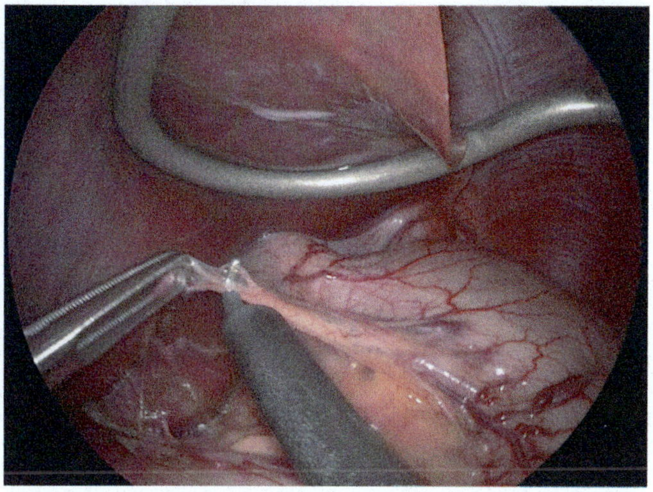

Fig. 17.4 An initial incision is made in the zona pellucida of the lesser omentum overlying the caudate lobe of the liver. Hook diathermy dissection is useful, and care is taken when approaching a small vessel bundle in the otherwise relatively avascular structure. The dissection is continued caudally to expose the oesophagus

5. Exposure of the oesophagus GOJ and crus

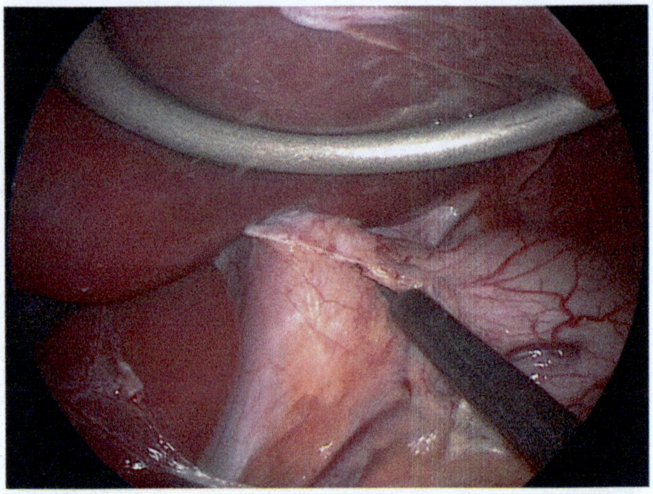

Fig. 17.5 The lower oesophagus and GOJ are now seen, and further exposed by dividing the phreno-oesophageal ligament and other attachments of the oesophagus to the diaphragm, some of which may represent inflammatory reaction to oesophagitis. Blunt and hook/bipolar dissection is used. The anterior and posterior vagi are identified and if possible preserved on the oesophageal wall

6. Mobilisation of the lower oesophagus to create intra-abdominal length

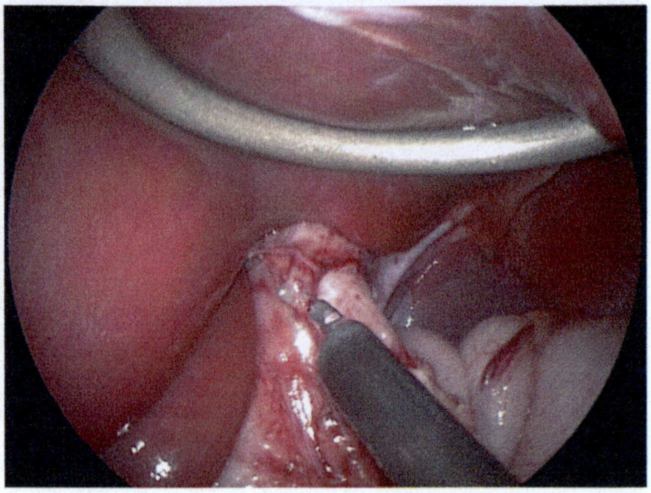

Fig. 17.6 Further adhesions of the lower oesophagus to the diaphragm and abdominal wall may need to be divided. Care is taken to avoid the pleura in the lower part of the chest. The aorta is also in close proximity behind the crura

7. Creation of window behind oesophagus for fundal wrap

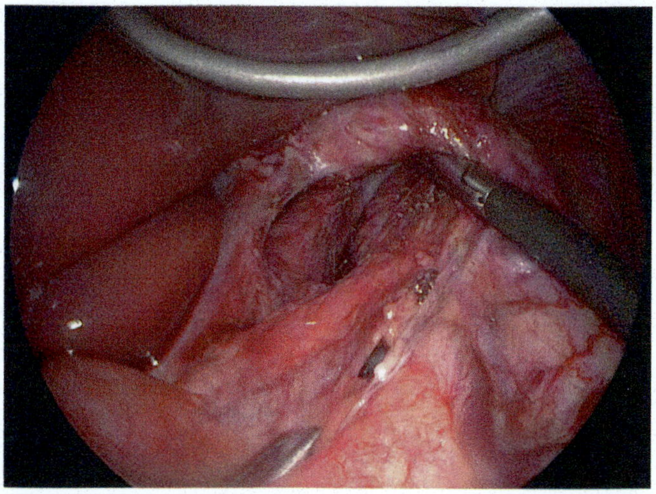

Fig. 17.7 The oesophagus is lifted forward and the space behind it developed by dividing the tissue behind using a combination of blunt (mostly) and diathermy dissection of the mainly loose areolar tissue behind it. Be aware and cautious of the vessels encountered towards the back of the space. The posterior vagus nerve should be sought and kept intact along the posterior wall of the oesophagus

8. Completed posterior window and crural exposure

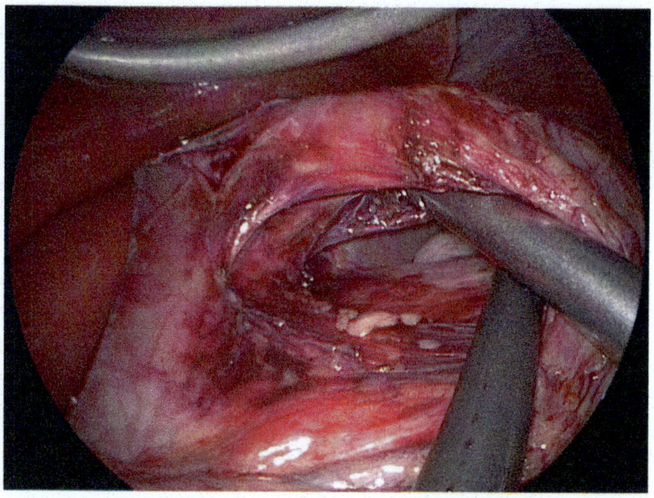

Fig. 17.8 The window has been created and developed. Note the posterior vagus nerve. The fundus is visible through the window and can then be grasped and mobilisation of the stomach begun

9. Mobilisation of the fundus

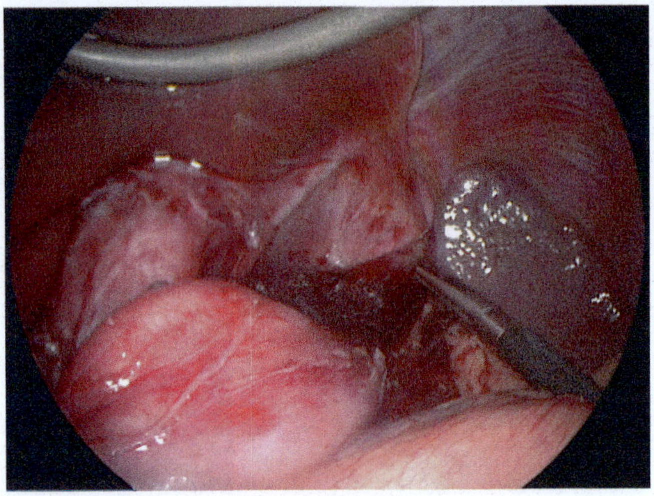

Fig. 17.9 Mobilisation of the fundus is then performed by dividing the adhesion between it and the abdominal wall and spleen. The loose adhesions can be seen here and are divided using hook diathermy. Very small vessels (vasa brevis) between fundus and spleen can be encountered and divided using hook diathermy. The short gastrics need not be routinely divided if not needed to create a loose wrap. If needing dividing a harmonic scalpel or ligasure bipolar is useful

10. Repairing the crura

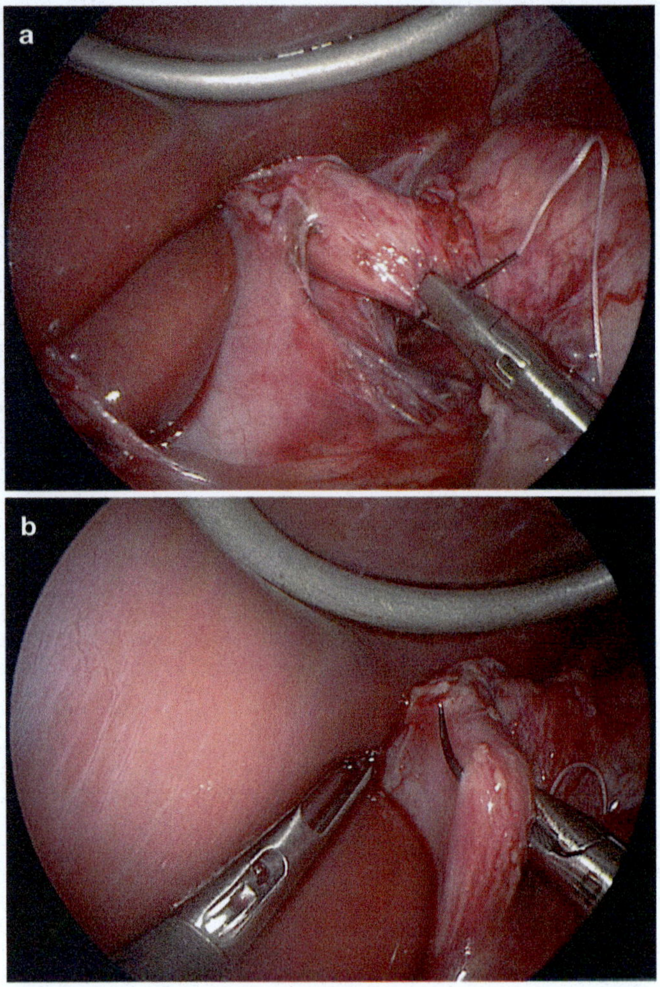

Fig. 17.10 (**a**) A non absorbable suture (on a ski needle) is introduced into the abdomen. The crura is approximated using one or two stitches to close the oesophageal hiatus, without closing too tightly (which can cause dysphagia). The ski needle is shown in this figure to be entering the left crus. (**b**) The ski needle is seen here exiting the right crus. The suture is tied using intracorporeal sutures. One or two are usually needed

11. Closed oesophageal hiatus

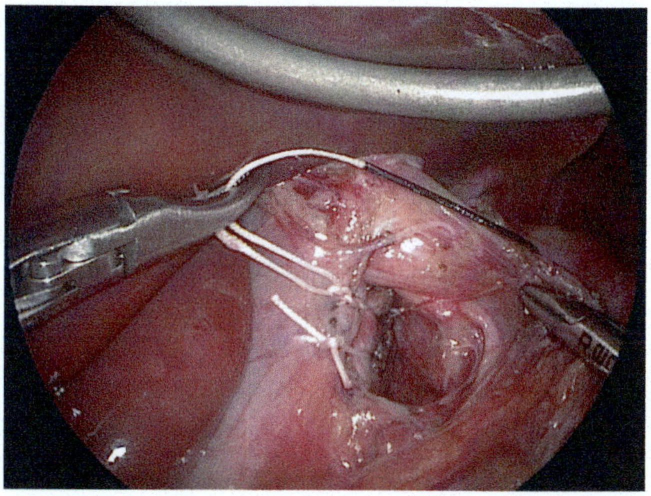

Fig. 17.11 Closed oesophageal hiatus is seen after sufficient interrupted sutures are placed. An adequate space can be judged by gently introducing a blunt instrument into the space (akin placing a little finger) to test tightness. The anterior and posterior vagi can be seen on the relevant oesophageal walls

12. Creation of loose wrap behind oesophagus

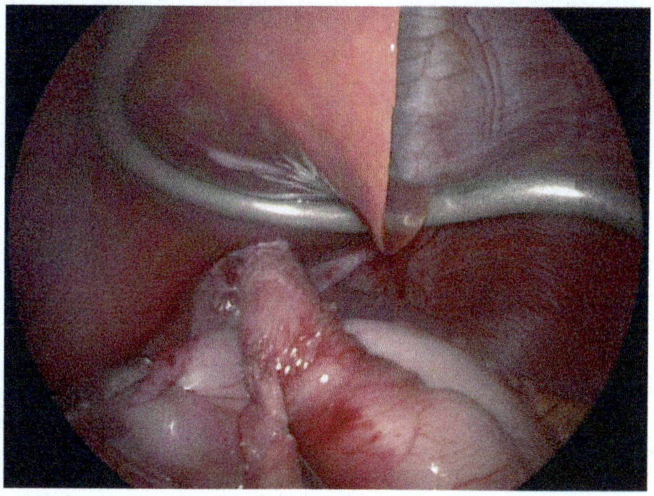

Fig. 17.12 The mobilised fundus is then passed through the window posterior to the oesophagus and it is ensured that it can be loosely wrapped around the oesophagus without tension. If released it should not shoot back across

13. Suturing of the wrap.

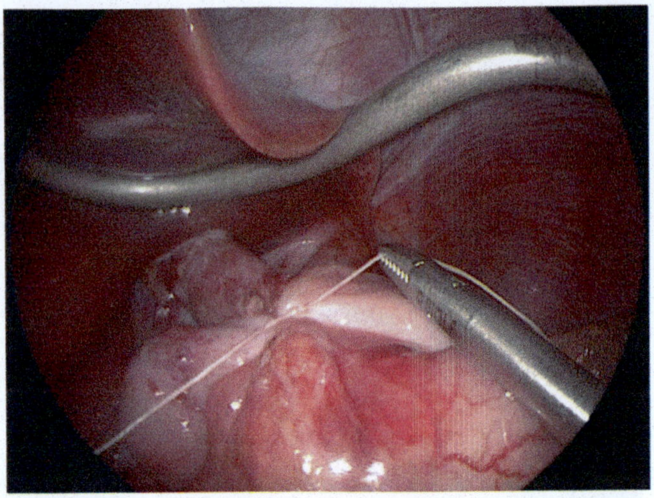

Fig. 17.13 The wrap is then created by suturing the fundus anteriorly over the oesophagus. The most superior (first) suture can incorporate the oesophageal wall between the fundus. Usually three sutures are needed to form the row of sutures

14. Final fundal suture

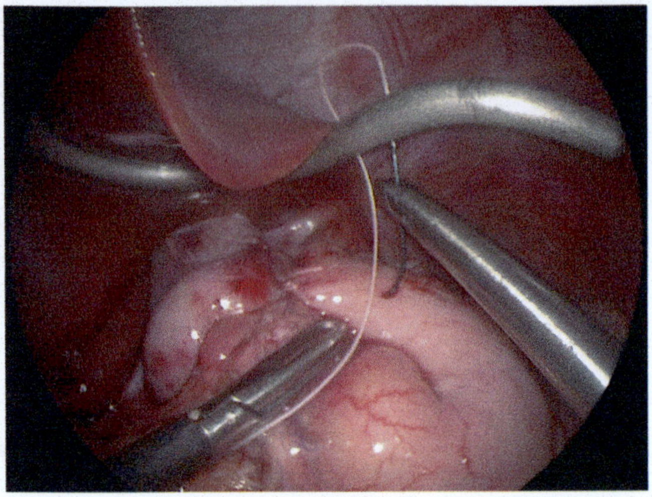

Fig. 17.14 The last of the three sutures is being placed to complete the wrap. The natural line of the fundus can be seen. It allows for a natural wrap

15. Completed fundoplication

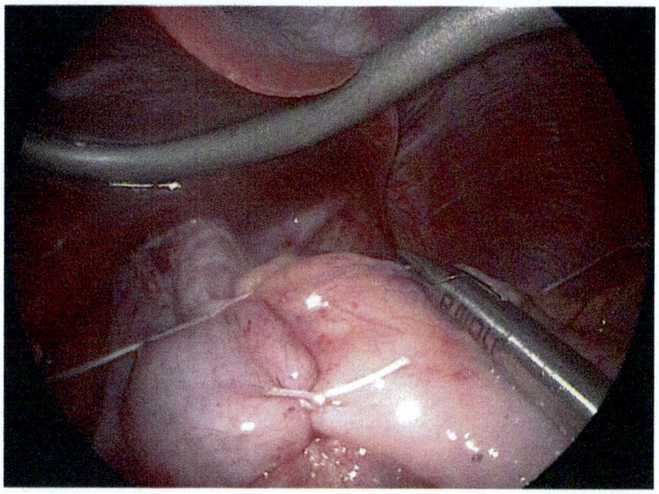

Fig. 17.15 The completed fundoplication is seen here

The Nathenson's retractor is first removed under vision, followed by the other working ports, and closed under vision. The abdomen is then desufflated and port closed.

17.6 Alternatives

Minimal mobilisation of the oesophagus has been highlighted by some as a technique to adopt to minimise the risk of recurrence due to wrap herniation. This should not be at the expense of identifying the appropriate anatomy to allow an accurate construction of wrap and repair of hiatus.

17.7 Highlights and Pitfalls

- Dissection in the lower thorax is kept to a minimal and is carefully visualised to avoid incising the pleura and creating a capnothorax. This is usually asymptomatic and sometimes only seen when a chest X-ray is performed for another reason. Most can be managed conservatively as the CO_2 is quickly absorbed. Symptoms may call for a chest drain to be inserted temporarily.
- Positioning in the neurologically impaired or those with musculoskeletal deformities may be difficult, while trying to maintain ergonomic positioning of ports.
- The use of a wide bore nasogastric tube is not mandatory to prevent a tight wrap. Ensuring the wrap is loose, and not too tight a crural wrap is sufficient.
- Care is needed when suturing the hiatus to prevent trauma to the liver with the needle, and bleeding.

Further Reading

1. Dingemann J, Ure BM. Systematic review of level 1 evidence for laparoscopic pediatric surgery: do our procedures comply with the requirements of evidence-based medicine? Eur J Pediatr Surg. 2013;23(6):474–9.
2. Siddiqui MR, Abdulaal Y, Nisar A, Ali H, Hasan F. A meta-analysis of outcomes after open and laparoscopic Nissen's fundoplication for gastro-oesophageal reflux disease in children. Pediatr Surg Int. 2011;27(4):359–66.
3. St Peter SD, Barnhart DC, Ostlie DJ, Tsao K, Leys CM, Sharp SW, et al. Minimal vs extensive esophageal mobilization during laparoscopic fundoplication: a prospective randomized trial. J Pediatr Surg. 2011;46(1):163–8.
4. McHoney M, Wade AM, Eaton S, Howard RF, Kiely EM, Drake DP et al. Clinical outcome of a randomized controlled blinded trial of open versus laparoscopic Nissen fundoplication in infants and children. Ann Surg 2011;254(2):209–16.

Laparoscopic Thal Fundoplication

18

Khaled Ashour, Alex C.H. Lee, and Hugh W. Grant

Abstract

The Thal fundoplication is a 180–270° anterior wrap. The basic dissection is similar to that of any other fundoplication, except that it does not require a posterior window or posterior mobilisation of the oesophagus and requires less dissection of the crus. It achieves the same surgical objectives as other types of fundoplication, i.e., ensuring an intra-abdominal oesophagus, creating an acute angle at the gastro-oesophageal junction, and creating a high-pressure zone around the lower oesophagus [1]. Thal fundoplication has an equivalent success rate in neurologically normal children but has a higher failure rate in neurologically impaired children when compared with the Nissen fundoplication. It also results in less postoperative dysphagia [2].

Keywords

Laparoscopic • Thal • Fundoplication

18.1 General Information

The Thal fundoplication is a 180–270° anterior wrap. The basic dissection is similar to that of any other fundoplication, except that it does not require a posterior window or posterior mobilisation of the oesophagus and requires less dissection of the crus. It achieves the same surgical objectives as other types of fundoplication, i.e., ensuring an intra-abdominal oesophagus, creating an acute angle at the gastro-oesophageal junction, and creating a high-pressure zone around the lower oesophagus [1]. Thal fundoplication has an equivalent success rate in neurologically normal children but has a higher failure rate in neurologically impaired children when compared with the Nissen fundoplication. It also results in less postoperative dysphagia [2].

K. Ashour • H.W. Grant (✉)
Department of Paediatric Surgery, John Radcliffe Hospital, Oxford University, Oxford, UK

A.C.H. Lee
Department of Paediatric Surgery, Oxford University Hospitals NHS Trust, Oxford Children's Hospital, John Radcliffe Hospital, Oxford, UK

M. McHoney et al. (eds.), *Color Atlas of Pediatric Anatomy, Laparoscopy, and Thoracoscopy*, DOI 10.1007/978-3-662-53085-6_18

18.2 Working Instruments

- For children <7 kg:
  - A 5-mm 30° short scope placed via the umbilicus
  - One working 3-mm port in the right upper quadrant
  - One working 5-mm port in the left upper quadrant (with reducer for 3-mm instruments)
  - Nathanson retractor placed via the epigastrium (Fig. 18.1).

- For children >7 kg:
  - A 5- or 10-mm 30° scope and 5-mm ports should be used
  - Hook (monopolar) diathermy usually is sufficient for dissection, but an ultrasonic dissector (Thunderbeat [Olympus, Southborough, MA] or Harmonic [Ethicon, Somerville, NJ]) or LigaSure (bipolar; Covidien, Mansfield, MA) may be used, although they usually are not required because division of the short gastric vessels seldom is necessary for a Thal fundoplication.

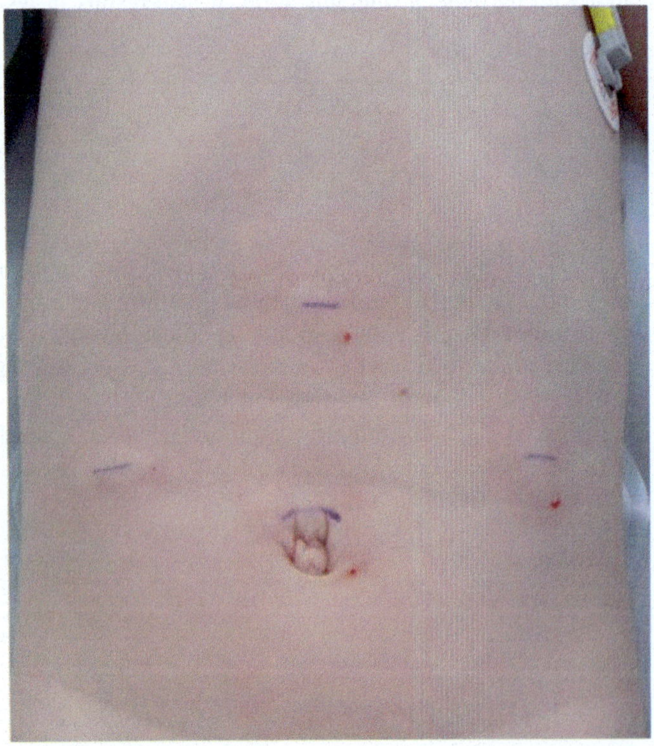

Fig. 18.1 Port sites

18.3 Positioning, Port Siting, and Ergonomic Considerations

If the child is <7 kg, he/she should be positioned supine with the head up and the legs abducted at the foot of the operating table. If the patient is >7 kg, the Lloyd-Davies position with the hips abducted and minimal flexure of the hip and knees is helpful (Fig. 18.2). The surgeon should stand between the patient's legs with the monitor over the patient's head for the best ergonomics.

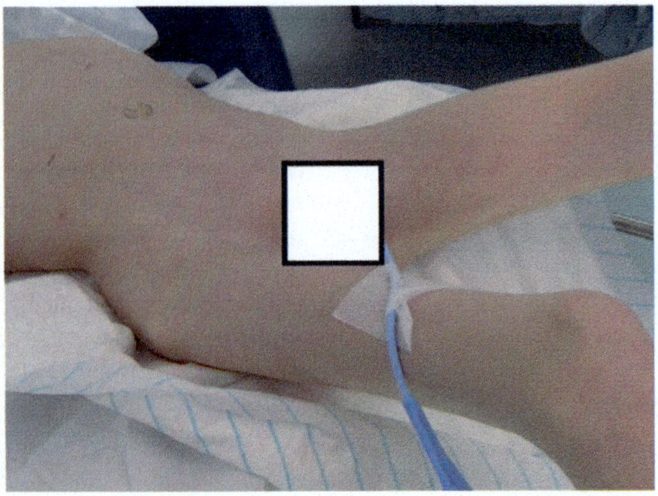

Fig. 18.2 Patient position

18.4 Surgical Technique

18.4.1 Entry

Empty the bladder with a catheter; use an umbilical or supraumbilical "Hasson" entry. Subsequent ports should be placed under laparoscopic vision in the right upper quadrant and left upper quadrant (for maximum triangulation), and a Nathanson retractor should be inserted at the level of (or just below) the edge of the liver, through the linea alba in the epigastrium.

18.4.2 Retraction

Good liver retraction is crucial; the Nathanson is a very useful retractor because it lifts the liver up and away from the operative field and is one less instrument to get tangled with the surgeon's instruments (Fig. 18.3).

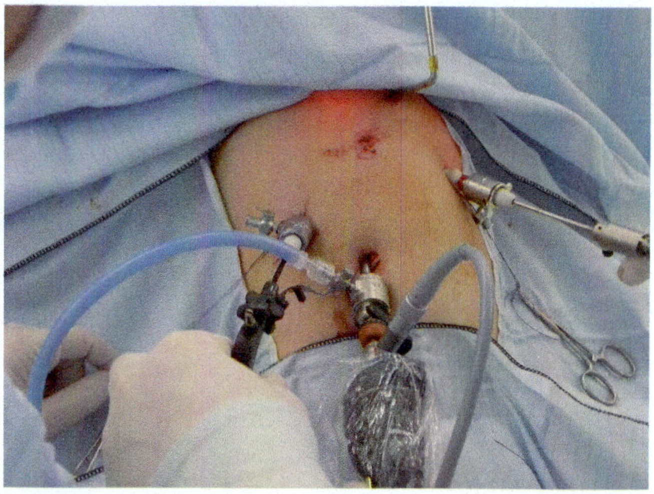

Fig. 18.3 Ergonomics

18.4.3 Exposure

Start the dissection over the caudate lobe by dividing the zona pellucida of the lesser omentum (Fig. 18.4). This opens a window that leads to the gastro-oesophageal junction (Fig. 18.5). Dissection should proceed superiorly towards the oesophageal hiatus. Accessory vessels from the left gastric vessels to the left lobe of the liver are common, and these may be ligated or diathermised as required.

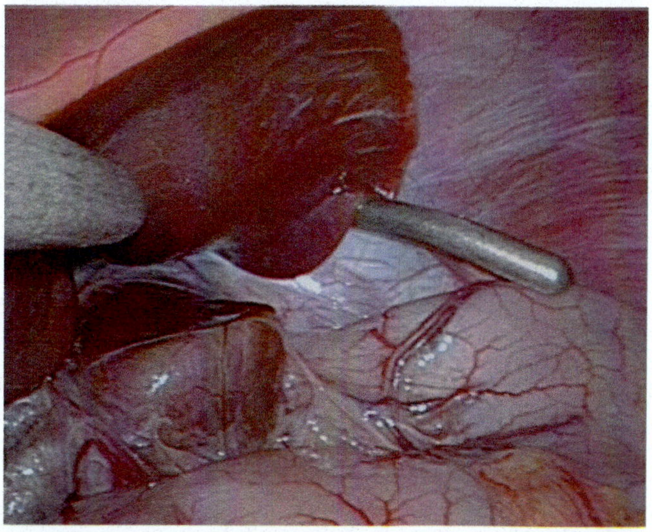

Fig. 18.4 Exposure of the operative field

18.4.4 Dissection

Identify and mobilise the phreno-oesophageal ligament; it usually is apparent in children as there is little fat. Occasionally in overweight children, it is difficult to identify this location, so look for lymph nodes that often lie over the gastro-oesophageal junction and are a useful landmark.

It is easier to start the dissection of the phreno-oesophageal ligament at the "9 o'clock" position (when looking towards the oesophageal hiatus). A superficial incision should be made vertically between the right leaflet of the right crus and the oesophagus (Fig. 18.6), and the oesophagus should be pushed to the patient's left. This usually opens the window and exposes the intra-abdominal oesophagus and the oesophageal hiatus. If the area is inflamed, some minor bleeding may be present.

Minimum dissection of the crus is important to minimise the risk of inducing a hiatus hernia or causing recurrence of a hiatus hernia [3]. The oesophagus should not be "stripped." In a Thal fundoplication the crural fibres have to be sufficiently dissected on the anterior aspect of the oesophagus to enable the surgeon to place secure sutures on the anterior part of the intra-abdominal oesophagus (Fig. 18.7).

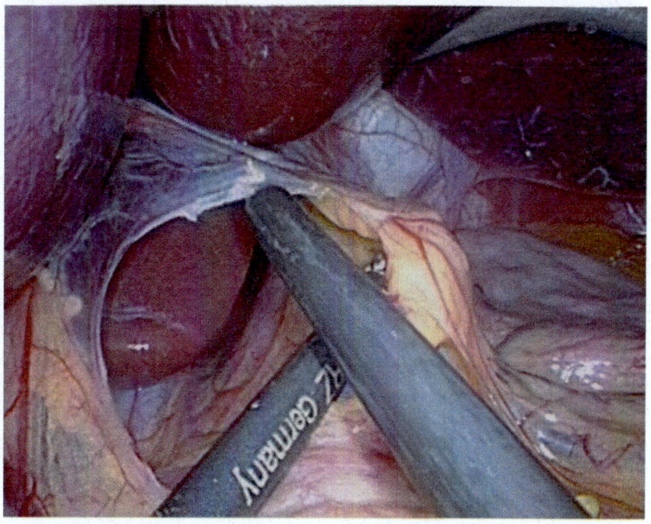

Fig. 18.5 Division of the zona pellucida of the lesser omentum

It is important in a Thal fundoplication to mobilise the greater curve of the stomach as far as the short gastric vessels, because the fundus is folded across the anterior aspect of the oesophagus in a Thal fundoplication. If the fundus is not mobilised sufficiently, there is too much tension to make a Thal fundoplication feasible.

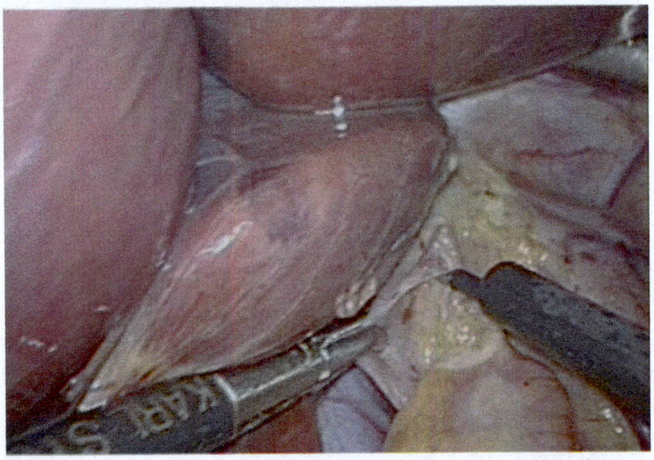

Fig. 18.6 Dissection of phreno-oesophageal ligament at the 9 o'clock position

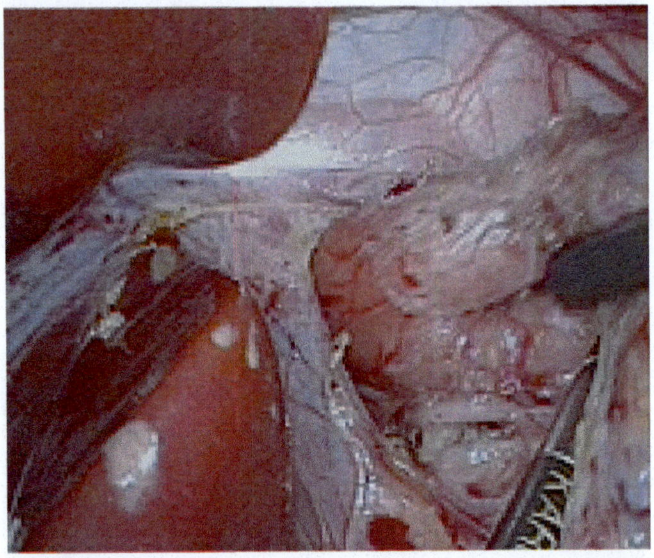

Fig. 18.7 Mobilising the oesophagus (minimal dissection)

18.4.5 Placing the Sutures

Interrupted 2–0 polyester (nonabsorbable and easy to tie) on a "ski" needle works well; intra- or extracorporeal suturing may be used.

Thal fundoplication originally was described as a 180° anterior wrap [1]; however, a 270° wrap is preferable. The sutures in the oesophagus are placed in an inverted "U" starting at the gastro-oesophageal junction and running up the left side of the patient's oesophagus, across the top of the intra-abdominal oesophagus, and down the right-hand side. The "mirror image" sutures are placed in the greater curve of the stomach starting at the gastro-oesophageal junction and extending laterally along the crest of the greater curve (Fig. 18.8). Approximately seven to nine sutures are required. The first suture should be placed at the gastro-oesophageal junction between the greater curve of the stomach and the left side of the descending oesophagus (Fig. 18.9). Sequential stitches are placed as described earlier (Fig. 18.10). The surgeon continues the fundoplication by suturing the top of the fundus to the top of the intra-abdominal oesophagus (Fig. 18.11). The wrap is completed by suturing the lateral aspect of the fundus to the right side of the descending oesophagus. At this point, the posterior vagus usually is seen (indicating that the wrap is around the back of the oesophagus). Once completed, the Thal fundoplication looks a like a pancake across the front of the oesophagus (Fig. 18.12).

There is no need for a wide-bore nasogastric tube as a partial wrap is not likely to be too tight.

18.4.6 Closure

The Nathanson retractor should be removed under laparoscopic vision, the ports removed under vision, additional local anaesthetic instilled around the port-sites, and the port sites closed.

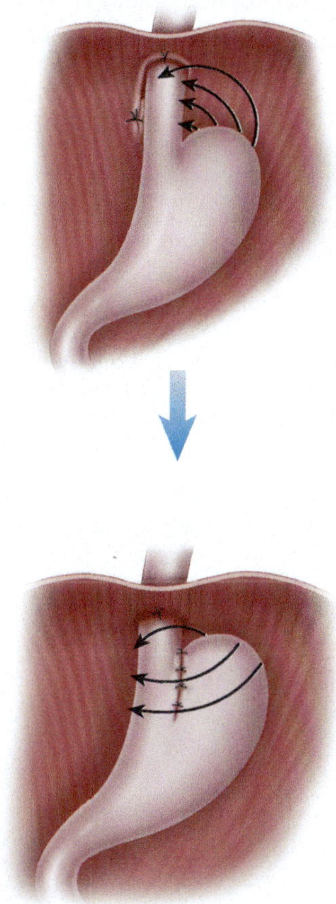

Fig. 18.8 Thal fundoplication

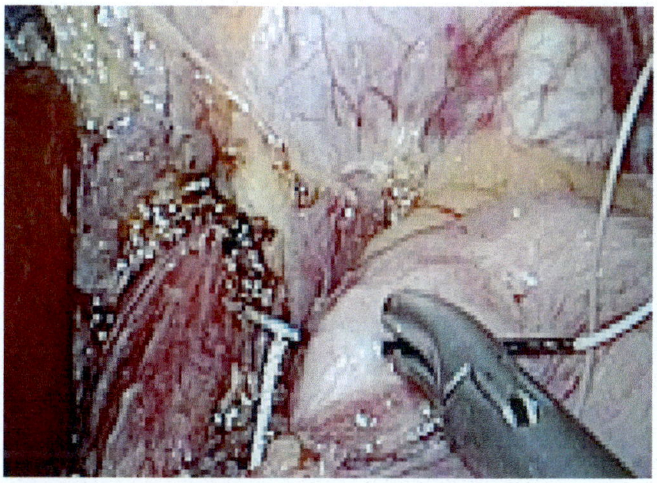

Fig. 18.9 Starting the Thal fundoplication: suturing the lower part of the fundus to the lower part of the oesophagus

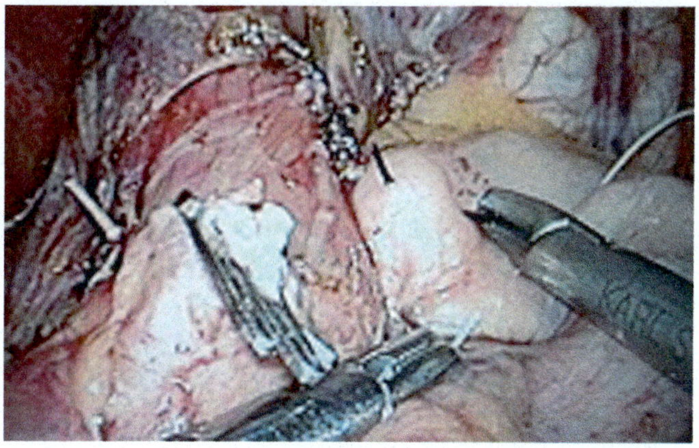

Fig. 18.10 Suturing the fundus to the left side of the oesophagus

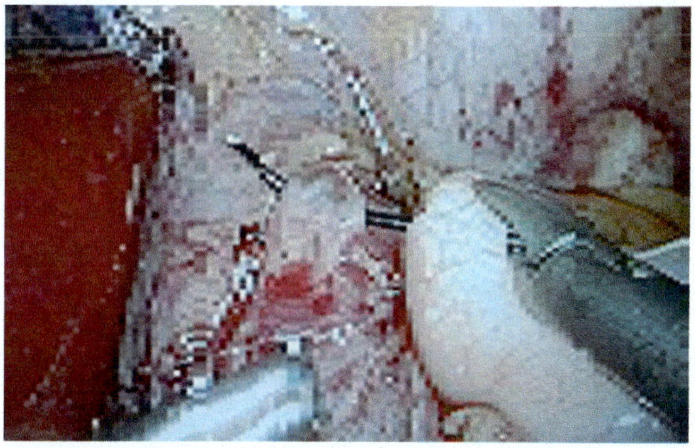

Fig. 18.11 Suturing the top of the fundus to the top of the intra-abdominal oesophagus

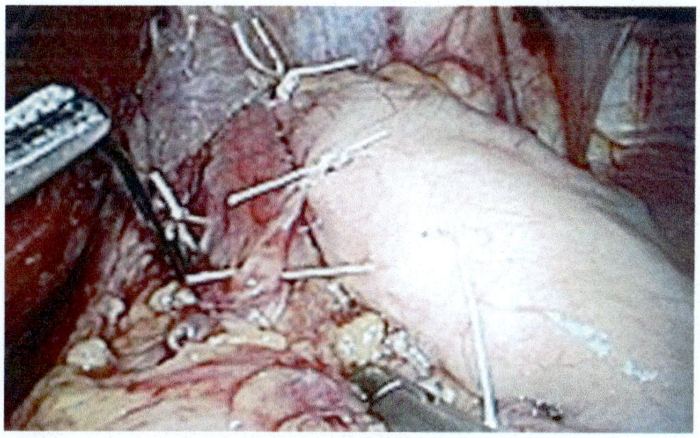

Fig. 18.12 Completed Thal fundoplication

18.5 Highlights and Pitfalls

- If the surgeon is right-handed, position the main working port high in the patient's left upper quadrant, just below the costal margin, with the (inner) tip lying at the level of the transverse colon and greater curve of stomach. This allows good triangulation, and instruments can be inserted and removed easily without too much movement of the camera, saving time and making the surgery easier.
- Make sure the (inner) tip of the right upper quadrant port is to the (patient's) left of the falciform ligament. This keeps the instruments from snaring in the falciform ligament each time they are inserted.
- Do not start dissection until there is good exposure and liver retraction. Spend an extra few minutes optimising the access to the oesophagus; this saves time and effort during the operation.
- If access and visibility deteriorate during the operation, check for leaks, pressure settings, patient position, and muscle paralysis. Discourage the anaesthetist from using nitrous oxide.

References

1. Ashcraft KW, Holder TM, Amoury RA, McGill CW, Holder TM. Thal fundoplication: a simple and safe operative treatment for gastroesophageal reflux. J Pediatr Surg. 1978;13(6D):643–7.
2. Kubiak R, Andrews J, Grant H. Long-term outcome of laparoscopic Nissen fundoplication versus laparoscopic Thal fundoplication: comparison of long-term outcomes. Ann Surg. 2011;253(1):44–7.
3. St. Peter SD, Barnhart DC, Ostlie DJ. Minimal vs extensive esophageal mobilisation during laparoscopic fundoplication: a prospective randomised trial. J Pediatr Surg. 2011;46(1):163–8.

Pyloric Stenosis

19

Merrill McHoney

Abstract

The laparoscopic approach for pyloromyotomy is gaining in popularity, mainly as a result of its cosmetic advantage, ease of operation, and good outcome [1, 2]. It requires very basic, inexpensive instrumentation, which probably helps make it cost-effective compared with open surgery [3]. Whatever the reasons for the increasing popularity of laparoscopic pyloromyotomy, the approach may be embarked upon safely by those with suitable skills and resources, and is taught easily.

Keywords

Pyloric stenosis • Ramstedt's operation • Pyloromyotomy • Laparoscopy

19.1 General Information

The laparoscopic approach for pyloromyotomy is gaining in popularity, mainly as a result of its cosmetic advantage, ease of operation, and good outcome [1, 2]. It requires very basic, inexpensive instrumentation, which probably helps make it cost-effective compared with open surgery [3]. Whatever the reasons for the increasing popularity of laparoscopic pyloromyotomy, the approach may be embarked upon safely by those with suitable skills and resources, and is taught easily.

19.2 Working Instruments

- 5-mm port
- 30° telescope
- 3-mm blade for pyloromyotomy (e.g., Conmed Linvatec [Largo, FL] or Smith and Nephew [London, UK] disposable 3-mm menisectomy knife)
- Two 3-mm Johan forceps
- 3-mm pyloromyotomy (Tan) spreader (optional)

M. McHoney, MD, PhD
Department of Paediatric Surgery,
Royal Hospital for Sick Children,
9 Sciennes Road, Edinburgh EH9 1LF, UK

M. McHoney et al. (eds.), *Color Atlas of Pediatric Anatomy, Laparoscopy, and Thoracoscopy*,
DOI 10.1007/978-3-662-53085-6_19

19.3 Positioning, Port Siting, and Ergonomic Considerations

The infant (with nasogastric tube in situ) is placed in the supine position at the foot of the operating table (Fig. 19.1a) or across the table for good ergonomics. Usually, no other specific positioning is required. Occasionally, the pylorus is tucked under the liver, and a small amount of head up-tilt will encourage the rest of the intestines to fall away from the operative field, facilitating access to the pylorus.

A 5-mm supraumbilical incision is made through layers into the abdominal cavity. A 5-mm primary port is inserted and secured (Fig. 19.1b, c). (The supraumbilical incision is preferable to the infraumbilical one, as it may be extended to convert to an open supraumbilical approach if necessary.) Pneumoperitoneum is established to 6 mmHg pressure, which may be increased to 10 mmHg if required and tolerated. The initial flow rate is set between 0.5 and 1 L/min.

Two other working instruments are placed in the right and left sides of the upper abdomen. It is worth noting that the best ergonomics are achieved with the surgeon's left-handed instrument (grasper) placed laterally in line with the duodenum, whereas the surgeon's right-handed instrument is best placed medially almost vertically over the pylorus (Fig. 19.2).

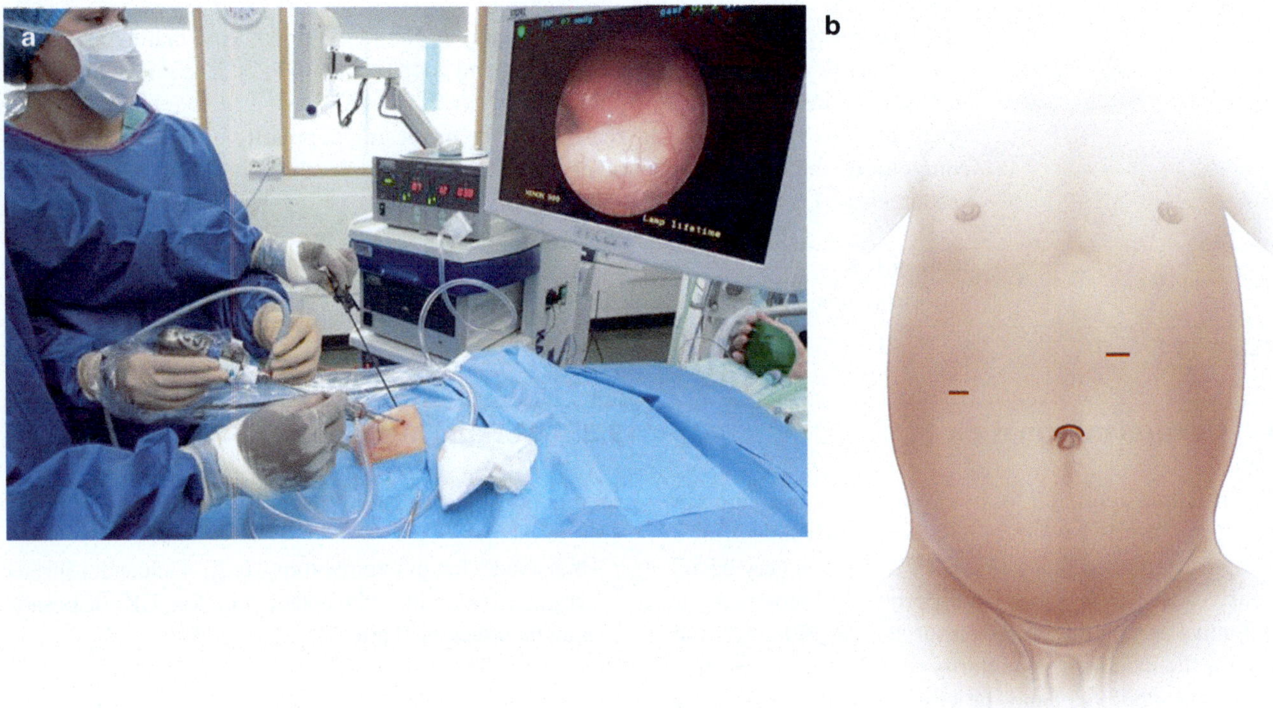

Fig. 19.1 Patient position (**a**) and ports (**b**). Patient positioning at the end of bed and the positioning of instruments. The screen is placed above the patient for good visualisation

19.4 Relevant Anatomy

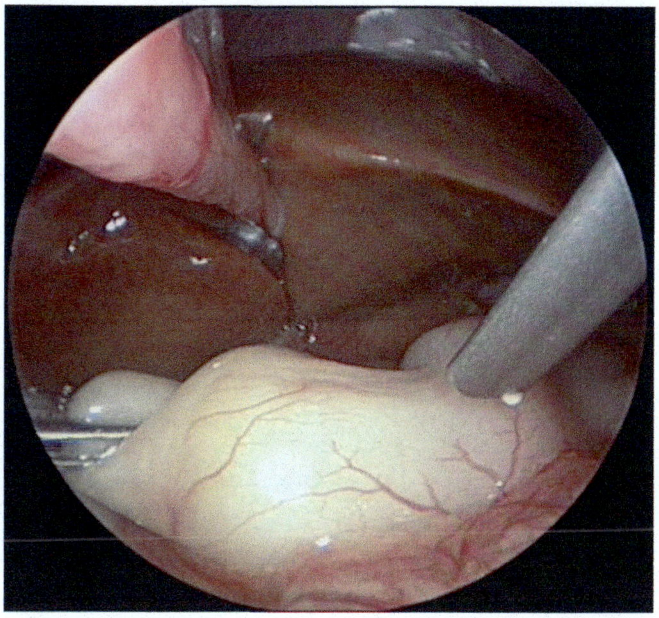

Fig. 19.2 Anatomy of the pyloric tumour. The falciform ligament is seen in the left-hand side running towards the liver, beneath which lies the pylorus. The prepyloric vein of Mayo may be identified. The intended incision is sited on a relatively avascular line along the pylorus

19.5 Surgical Technique

Two additional working instruments are passed directly into the abdomen (without ports) via stab incisions using a #11 blade and operation procedes as in the following figures (Fig. 19.3, 19.4, 19.5, 19.6, 19.7, 19.8, and 19.9).

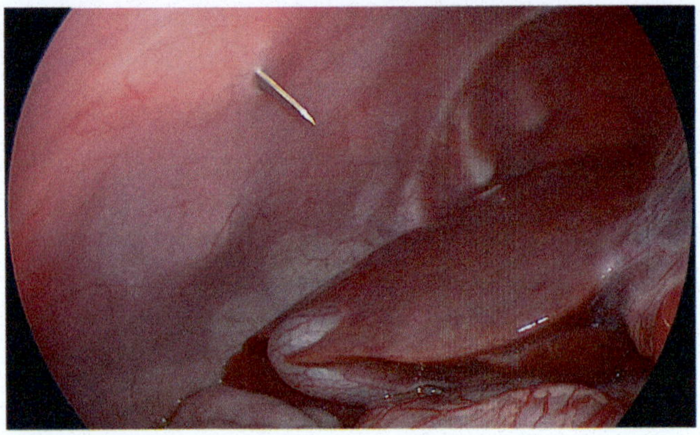

Fig. 19.3 Stab incision for first grasper. The first incision is used to place a grasper is in the patient's right upper quadrant (to enter below the liver edge) so that the duodenum can be grasped easily and stabilised

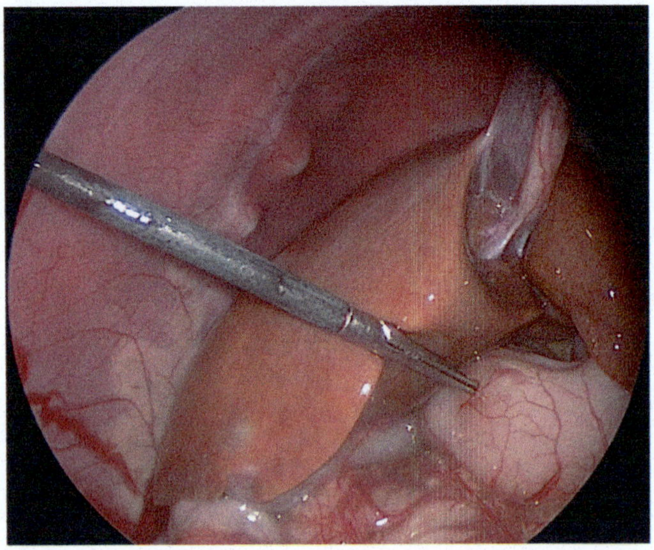

Fig. 19.4 Grasper inserted in the right upper quadrant. The grasper is inserted directly through the abdominal wall, below the liver, and extending below the falciform ligament to the pylorus

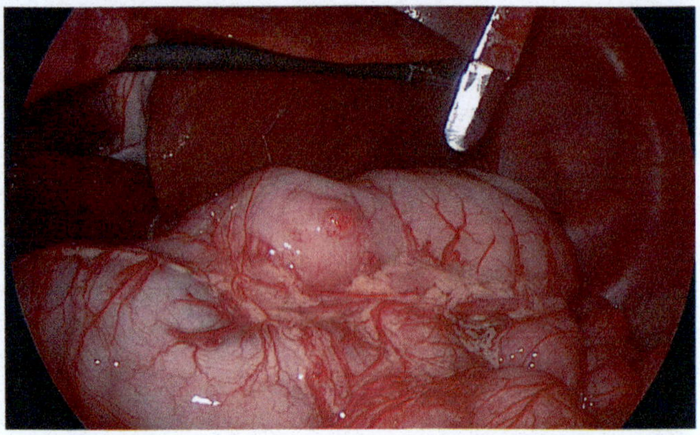

Fig. 19.5 Position of the pyloromyotomy blade. The second instrument (initially, the pyloromyotomy blade) is placed in the epigastrium below the liver and almost vertically over the pylorus. Initially, this is a blade for incising the pylorus and later a spreader to open the pylorus

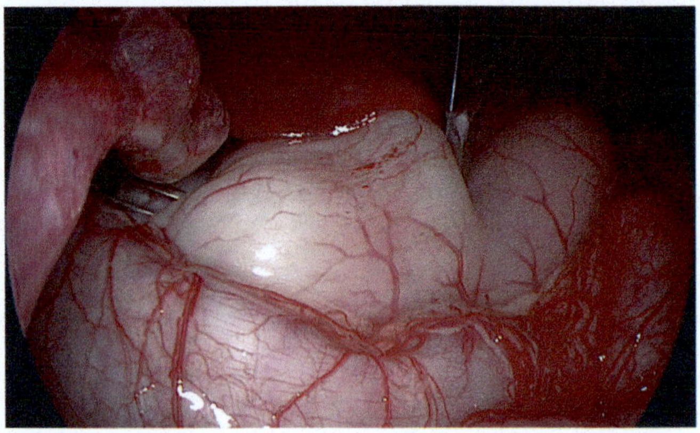

Fig. 19.6 Extent of hypertrophied pyloric muscle. The image shows the extent of pyloric muscle thickening by palpation between instruments. A relatively avascular plane may be seen in the middle of the pylorus

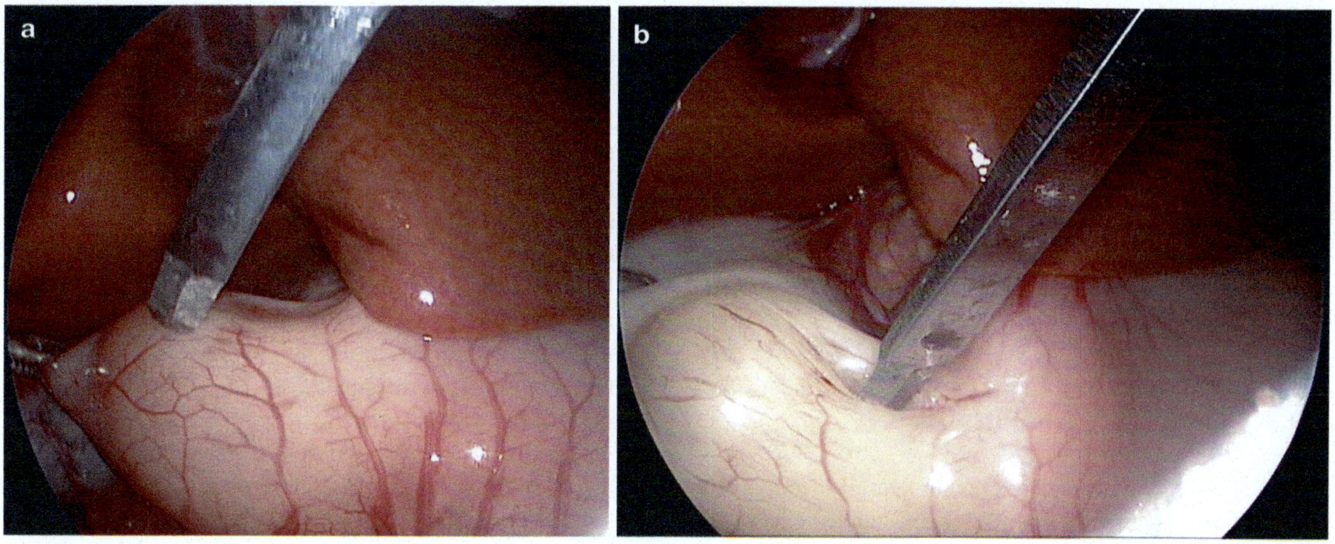

Fig. 19.7 Incision on the pylorus. The duodenum is grasped with the grasper to stabilise the pylorus. (**a**) The incision in the pylorus is initially made in the middle of the pylorus (thickest area of hypertrophy and least likely to perforate) and is extended superficially along the length of the pyloric thickening. (**b**) A gentle twist of the blade helps indicate a sufficient incision into which the spreader can be inserted

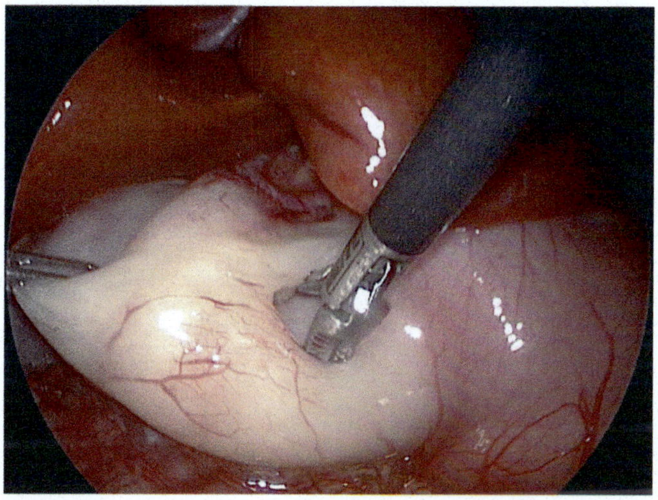

Fig. 19.8 Spreading of the pyloric muscle. A spreader introduced via the right-handed working incision into the middle of the tumour, with careful but sustained spread, is used to deepen and spread the myotomy

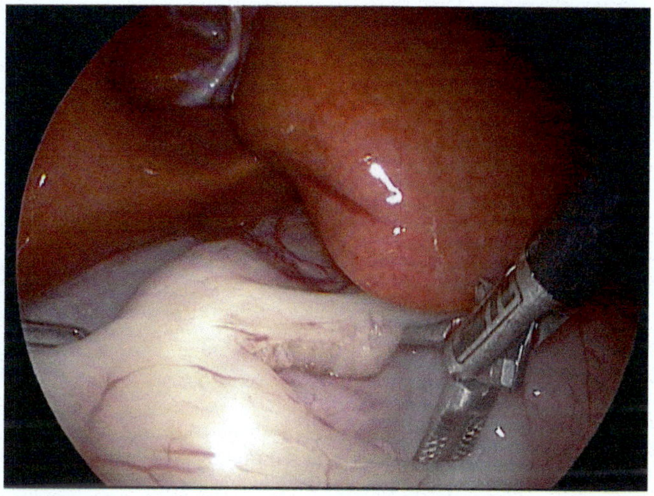

Fig. 19.9 Completing the myotomy. The myotomy is completed by spreading along the length of the pylorus with pouting of the gastric mucosa throughout the length of the incision. Any obvious leaks of air or bile should by looked for at this point. Adequacy may be checked by grasping the sides of the tumour and ensuring independent movement

Minor oozing is common and not a problem. It is not mandatory to check for leaks if the mucosa has been visualised clearly, with no obvious leaks. If desired, 20–40 mL of air may be injected via the nasogastric tube to distend the stomach. This air then is coaxed into the pylorus with pressure from the instruments, and any leak should be obvious. Leaks may be repaired primarily (via an open or laparoscopic approach, depending on operator's preference or experience), with or without an omental patch. The wounds are closed with absorbable sutures to the muscle and a subcuticular skin stitch or skin glue. The wound used for the working instruments should be closed before desufflation of the abdomen to help prevent omentum prolapse during closure. The nasogastric tube may be removed at the end of the operation. Feeds may be started and graduated according to local policy. We feed patients after 4–6 h and graduate to full feedings in three or four increments. Infants may be discharged when on full feedings.

19.6 Alternatives

- Although a supraumbilical approach is described and preferable, an infraumbilical cut-down may be used.
- A bowel-holding (Johan) forceps may be used in place of the pyloric spreader.
- If no suitable 3- (or 5-) mm blade is available, a 3-mm hook diathermy may be used to perform the initial myotomy. However, in the author's opinion, this requires somewhat more experience and is less controlled than the blade.

19.7 Highlights and Pitfalls

- Place the left-handed instruments laterally (to avoid the falciform ligament) but the right-handed instrument medially (almost vertically over the pylorus). Placing the right-hand instrument too laterally makes performance of the myotomy very unergomonic and may cause difficulty.
- It is safest to begin the myotomy with the blade in the middle of the tumour (where it is thickest); this helps avoid perforation. The myotomy incision may be deeper in the middle but more superficial in the lateral ends of the tumour.
- Initial firm and sustained spread in the middle of the pylorus can complete the myotomy in one swift movement with ease and efficacy.
- Omental prolapse through the accessory site may be minimised by closing these wounds under laparoscopic vision with the abdomen still inflated. This keeps the abdominal wall away from the roaming omentum until securely closed.
- The cosmetic outcome is excellent (Fig. 19.10).

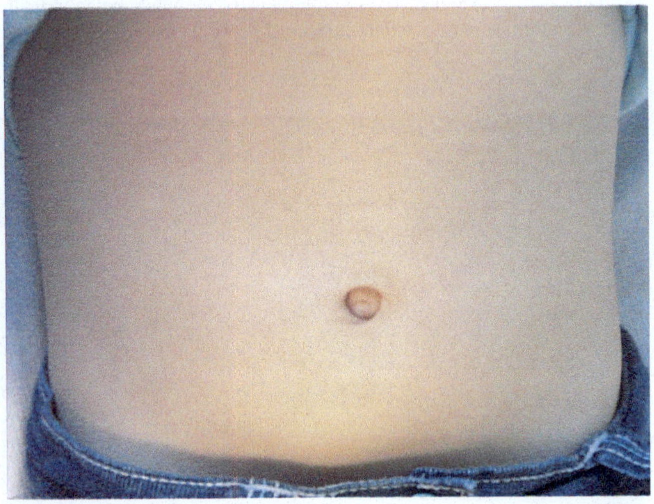

Fig. 19.10 Cosmetic outcome. The cosmetic outcome is excellent, with imperceptible scars. This image shows the cosmetic appearance 3 years after laparoscopic pyloromyotomy in one patient

References

1. Hall NJ, Pacilli M, Eaton S, Reblock K, Gaines BA, Pastor A, et al. Recovery after open versus laparoscopic pyloromyotomy for pyloric stenosis: a double-blind multicentre randomised controlled trial. Lancet. 2009;373:390–8.
2. Sola JE, Neville HL. Laparoscopic vs open pyloromyotomy: a systematic review and meta-analysis. J Pediatr Surg. 2009;44:1631–7.
3. Carrington EV, Hall NJ, Pacilli M, Drake DP, Curry JI, Kiely EM, et al. Cost effectiveness of laparoscopic versus open pyloromyotomy. J Surg Res. 2012;178:315–20.

Duodenal Atresia Repair

20

Aimee Gibson and Nada Sudhakaran

Abstract

Laparoscopic duodenal atresia repair (duodenoduodenostomy) was initially described at the beginning of the twenty-first century; some centres abandoned the laparoscopic approach due to high anastomotic leak rates [1]. One particular centre [1] reported an anastomotic leak rate of just under 30 %, in their initial early series before abandoning the procedure for some time. After modifying their technique from interrupted to continuous suturing, they revisited the procedure in a new cohort of patients and, with this, had no complications. As a result, they have been performing and teaching the procedure ever since. Others have also reported similar results [1]. They have themselves suggested that laparoscopic duodeno-duodenostomy should be restricted to paediatric centres with extensive laparoscopic experience.

Keywords

Laparoscopic duodenal atresia repair • Duodenoduodenostomy

20.1 General Information

Laparoscopic duodenal atresia repair (duodenoduodenostomy) was initially described at the beginning of the twenty-first century; some centres abandoned the laparoscopic approach due to high anastomotic leak rates [1]. One particular centre [1] reported an anastomotic leak rate of just under 30 %, in their initial early series before abandoning the procedure for some time. After modifying their technique from interrupted to continuous suturing, they revisited the procedure in a new cohort of patients and, with this, had no complications. As a result, they have been performing and teaching the procedure ever since. Others have also reported similar results [1]. They have themselves suggested that laparoscopic duodenoduodenostomy should be restricted to paediatric centres with extensive laparoscopic experience.

Advantages of the laparoscopic approach include faster recovery and earlier resumption of oral feeding, leading ultimately to earlier discharge.

A. Gibson • N. Sudhakaran, MD (✉)
Paediatric Surgery, Gold Coast University Hospital, Queensland, Australia

M. McHoney et al. (eds.), *Color Atlas of Pediatric Anatomy, Laparoscopy, and Thoracoscopy*,
DOI 10.1007/978-3-662-53085-6_20

20.2 Relevant Anatomy

There are three categorised types of duodenal atresia. Type 1 involves either a diaphragm or web that includes submucosa and mucosa. Type 1a is termed the "windsock" deformity, where the diaphragm has ballooned distally. 1b involves a membrane without ballooning, whereas 1c involves a web between the duodenal segments. Type 2 atresias have a dilated proximal segment, with collapsed distal segment connected by a fibrous cord. Type 3 atresias have no connection between proximal and distal segments. Most atresias occur at the level of D2 (Fig. 20.1).

More than 50 % of duodenal atresias are associated with other congenital anomalies, and approximately 30 % are associated with trisomy 21. Other associations include cardiac anomalies and other gastrointestinal abnormalities, the most important of which to recognise is malrotation.

Diagnosis may be made antenatally, with findings of a double bubble sign. Most were detected within the seventh and eighth months of pregnancy.

Although the duodenum has numerous close anatomical relations, those most important in laparoscopic duodenoduodenostomy include:

1. The falciform ligament: containing the left umbilical vein, it should not be transected but carefully secured superiorly to retract the liver.
2. The right lobe of the liver: in infants, the liver is quite large with respect to the abdominal cavity size and hangs over the duodenum.
3. The transverse colon: also overlying the duodenum, it must be gently peeled away from the duodenum to get exposure.
4. The pancreas: locating the pancreas helps identify the proximal and distal parts of the duodenum in duodenal atresia as it generally separates the two. In some cases, an annular pancreas may be identified

20.3 Working Instruments

- 3 mm hasson port
- Either 30° or 0° laparoscope
- 3 mm needle holders
- 3 mm scissors
- 3 mm suture cutting scissors
- 3 mm Maryland forceps
- 2×3 mm soft bowel grasping forceps
- 3 mm monopolar hook
- 3 mm Reddick Olsen grasper
- 3 mm bipolar scissors/grasper (optional)

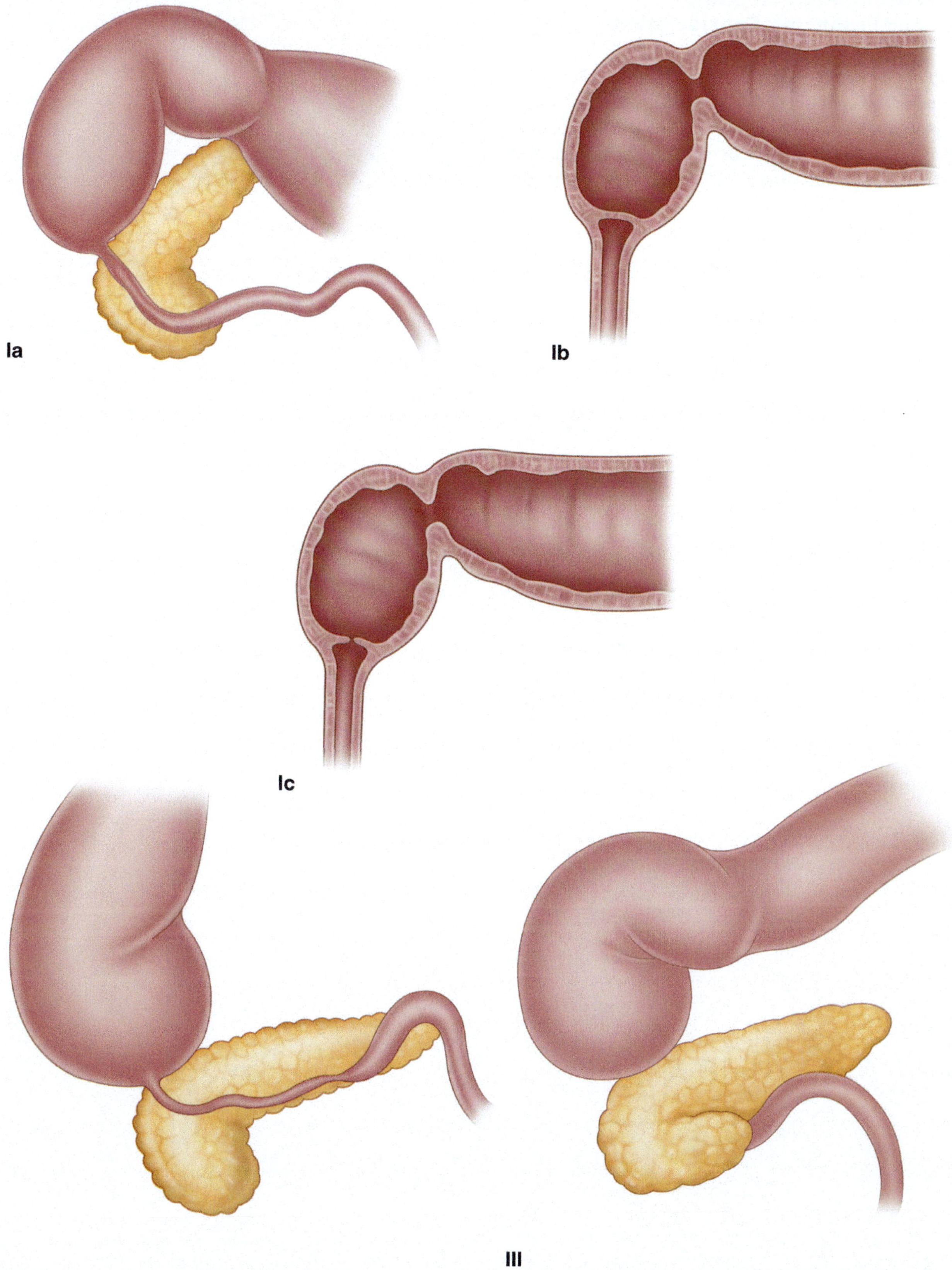

Fig. 20.1 There are three categorised types of duodenal atresia. Type 1 involves either a diaphragm or web that includes submucosa and mucosa. Type 1a is termed the 'windsock' deformity, where the diaphragm has ballooned distally. Type 1b involves a membrane without ballooning, whereas type 1c involves a web between the duodenal segments. Type 2 atresias have a dilated proximal segment, with collapsed distal segment connected by a fibrous cord. Type 3 atresias have no connection between proximal and distal segments. Most atresias occur at the level of D2

20.4 Positioning, Port Siting, and Ergonomic Considerations

The baby is positioned supine with the legs as close to the lower end of the operating table as possible. A 3 mm hasson port is placed at the umbilical fold and two stab incisions are placed at the level of the umbilicus on either flanks (Fig. 20.2). The portless approach is used to introduce the suture with its needle into the abdominal cavity.

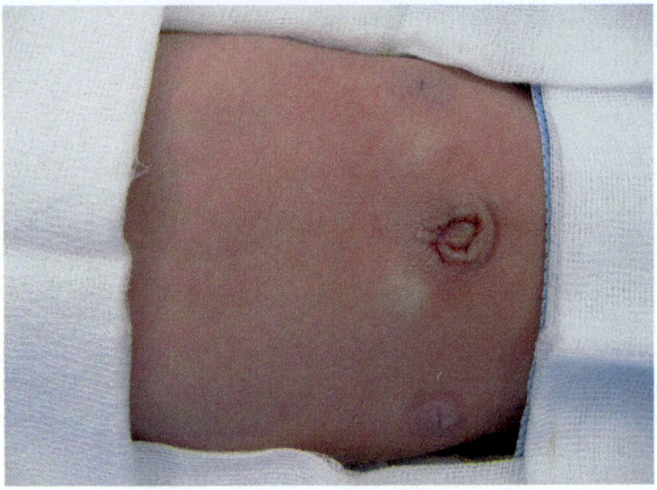

Fig. 20.2 A 3-mm hasson port is placed at the umbilical fold and two stab incisions are placed at the level of the umbilicus on either flanks

20.5 Surgical Technique

The anatomy is first assessed to confirm the diagnosis. We need to ascertain that the proximal duodenum is dilated and that there is no malrotation.

Figures 20.3, 20.4, 20.5, 20.6, 20.7, 20.8, 20.9, 20.10, 20.11, 20.12, 20.13, 20.14, 20.15, and 20.16 illustrate the steps involved in duodenal atresia repair.

Special note on technique: avoid handling the needle point during suturing, which can blunt the needle, making passing the needle into the tissue more traumatic. While performing continuous suturing, the suture should be grasped towards the needle end to avoid weakening it and risk breaking.

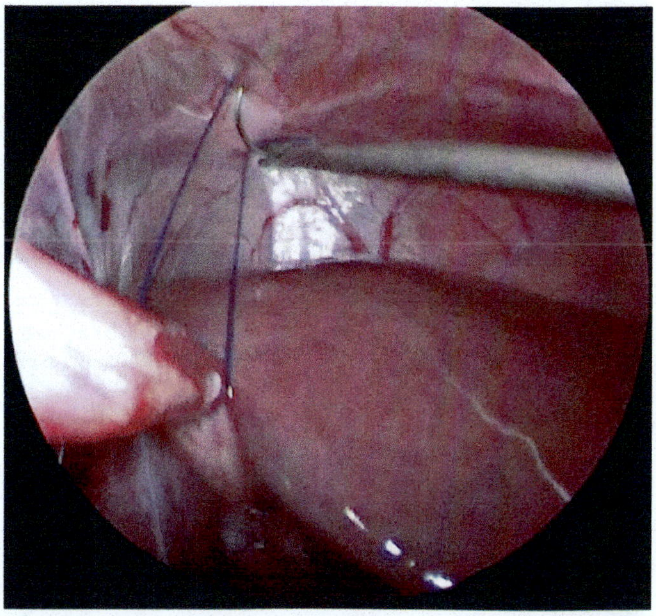

Fig. 20.3 Falciform ligament being retracted via sutures to the anterior abdominal wall. Once duodenal atresia is confirmed, the liver is retracted away from the field of interest by means of a "holding stitch" through the anterior abdominal wall and around the falciform ligament. The suture ends are pulled up and held together with an artery clip and a gauze swab protecting the skin from pressure injury

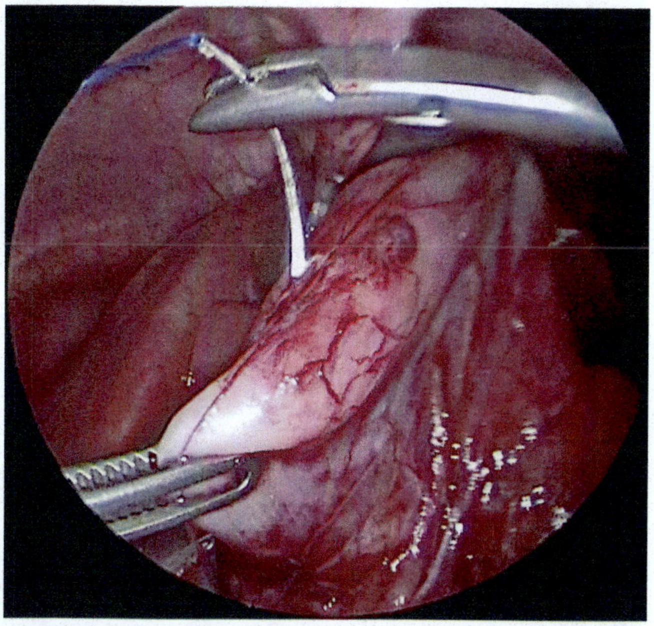

Fig. 20.4 Second suture being placed on the proximal duodenum. The distal part of the proximal dilated duodenum is suspended by means of two sutures suspended form the anterior abdominal wall. This helps to display the anatomy better and to stabilise the duodenum when it comes to suturing later

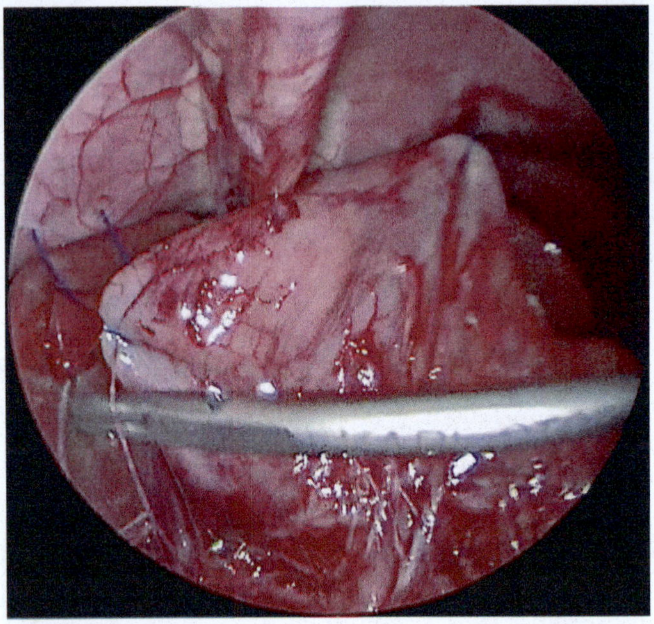

Fig. 20.5 Proximal duodenum suspended from the anterior abdominal wall

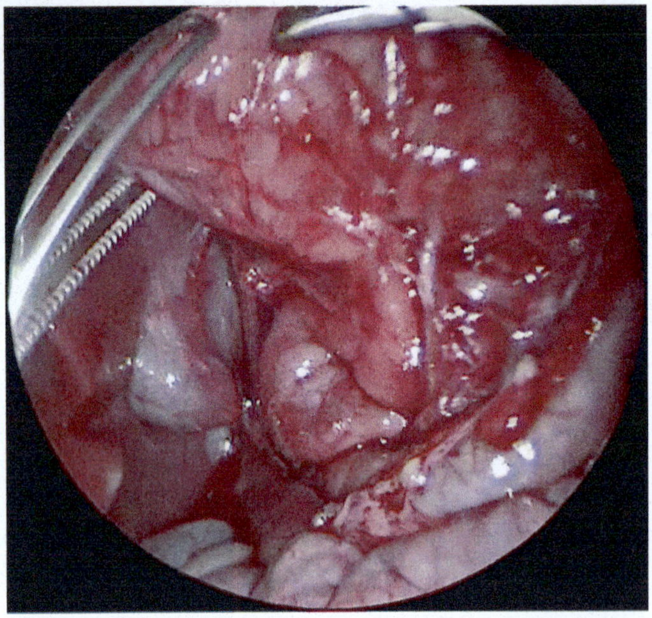

Fig. 20.6 Display of the essential anatomy—proximal duodenum held with soft forceps, distal to it is the pancreas and the distal duodenum

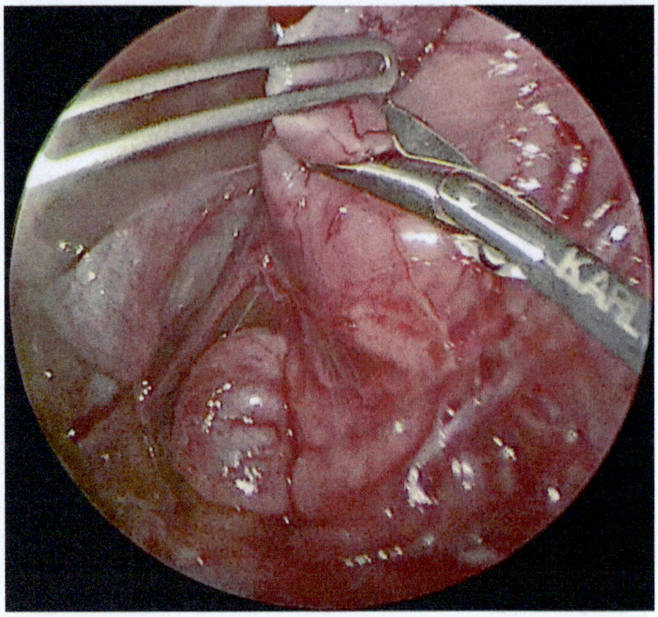

Fig. 20.7 Proximal duodenum enterotomy made transversely for the anastomosis. Scissors or monopolar diathermy can be used to make a transverse incision of the most distal (dependent) part of the proximal dilated duodenum

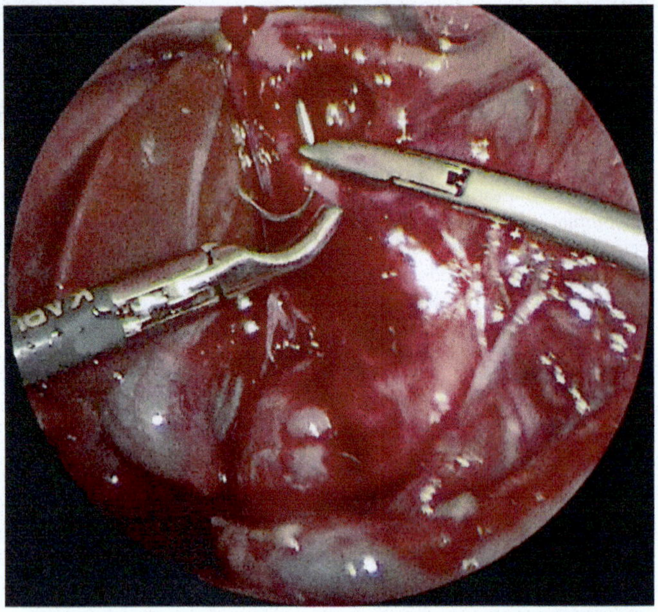

Fig. 20.8 An anchoring suture is placed between the midpoint of the enterotomy and the proximal part of the distal collapsed duodenum without much tension between the tissues

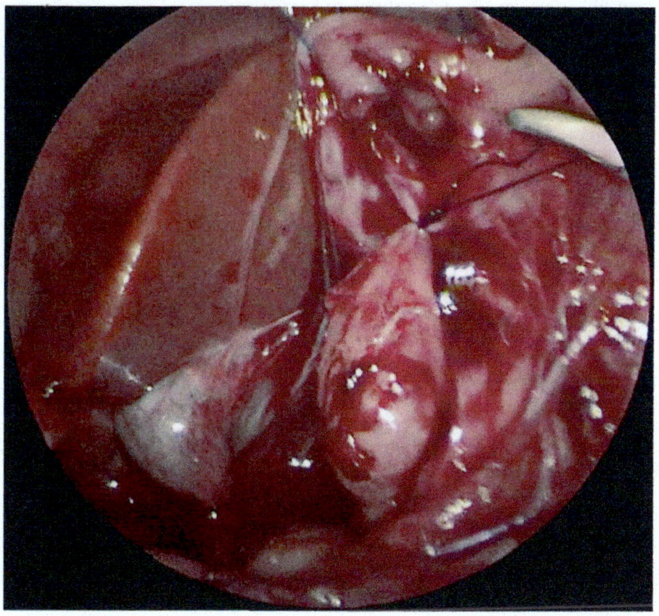

Fig. 20.9 The two ends to be anastomosed are brought together by means of one single serosal suture. This allows for the distal duodenum to be stabilised for an enterotomy and better estimation of the position of the cut. This stitch later becomes the midpoint of the "inner" suture line. It keeps the both edged of the posterior suture line from slipping away and risks anastomotic leaks

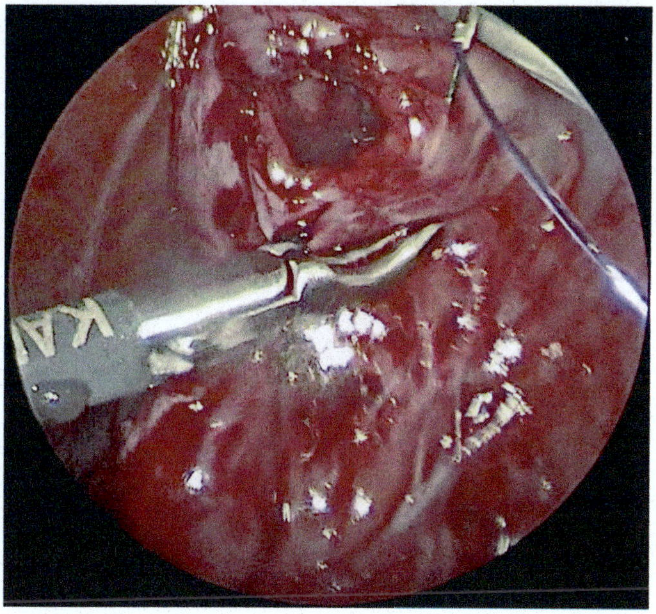

Fig. 20.11 A diamond shaped (Kimura type) anastomosis is prepared. Two further sutures are placed on the two corners of the proximal duodenal enterotomy. These are stitched to the midpoint of the longitudinal incision of the distal duodenum on either side. This image shows a "bite" taken at the proximal corner

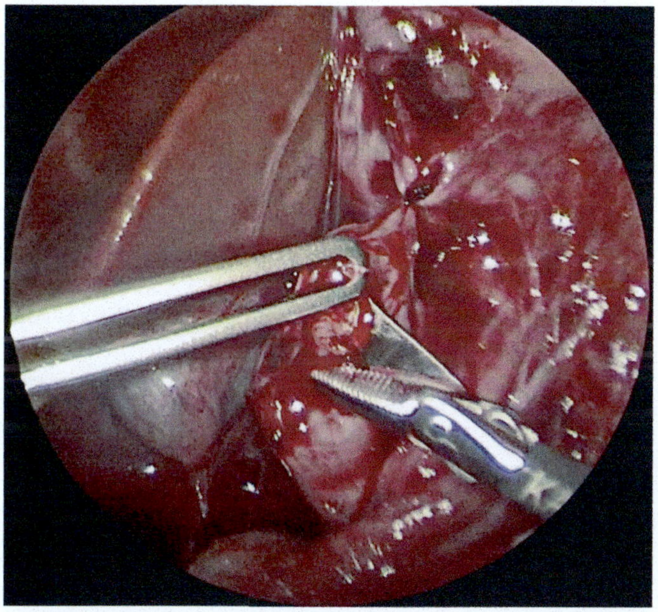

Fig. 20.10 Enterotomy of the distal duodenum, made longitudinally along its length

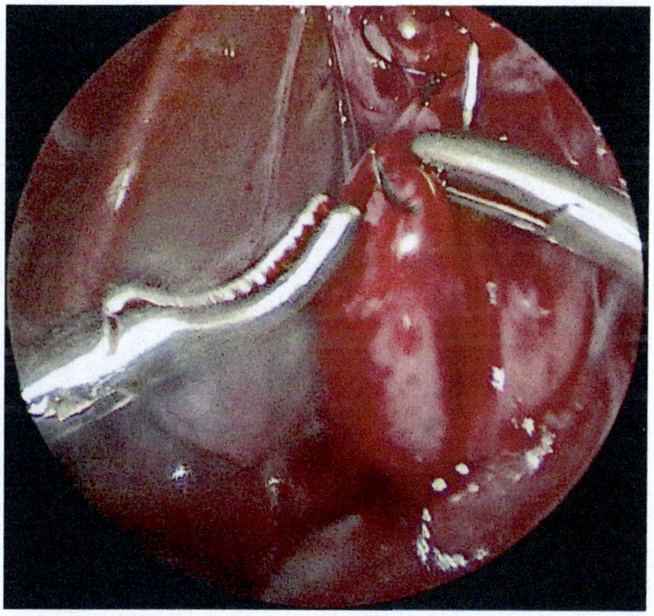

Fig. 20.12 Another "bite" of the same suture at the mid point of the longitudinal incision of the distal duodenum. The same is done on the other side

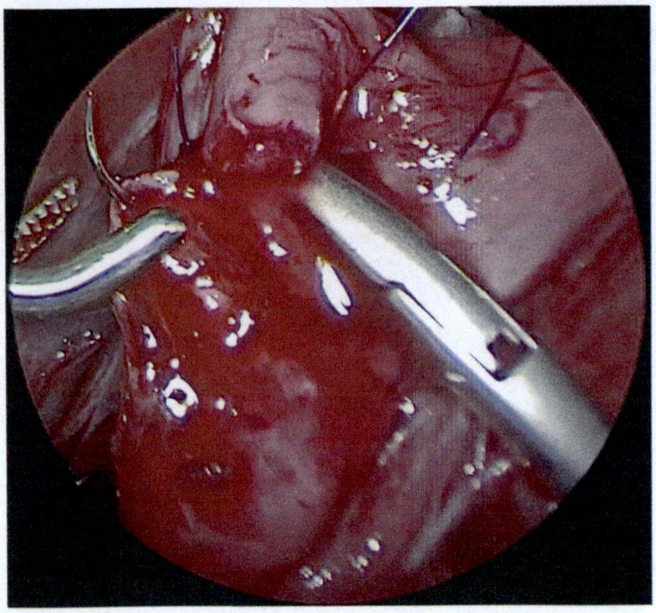

Fig. 20.13 The start of the continuous suture of the posterior wall anastomosis, intraluminally

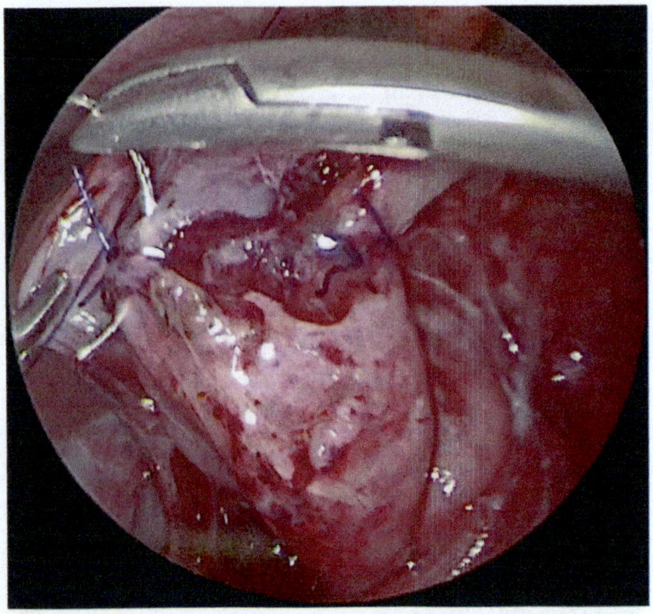

Fig. 20.14 The start of the anterior wall continuous suturing

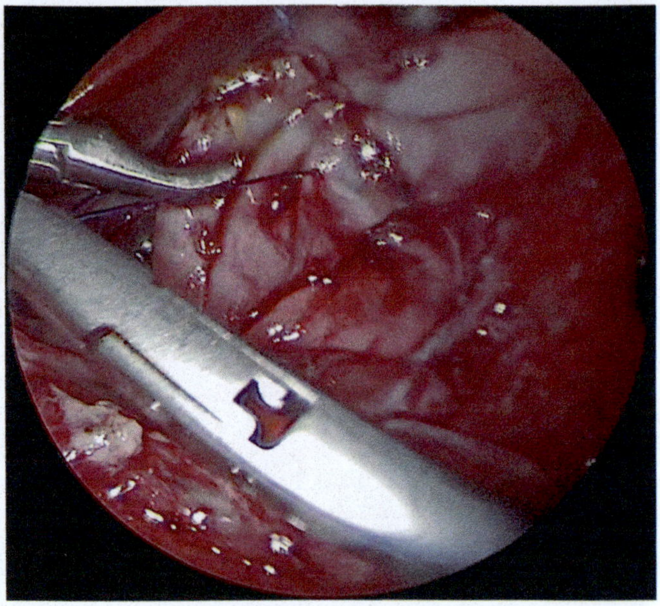

Fig. 20.15 Towards the end of the continuous suturing of the anterior wall of the anastomosis. The holding sutures are removed to complete the procedure. The wound is closed with skin glue. Gentle feeding can commence the next day and gradually increase

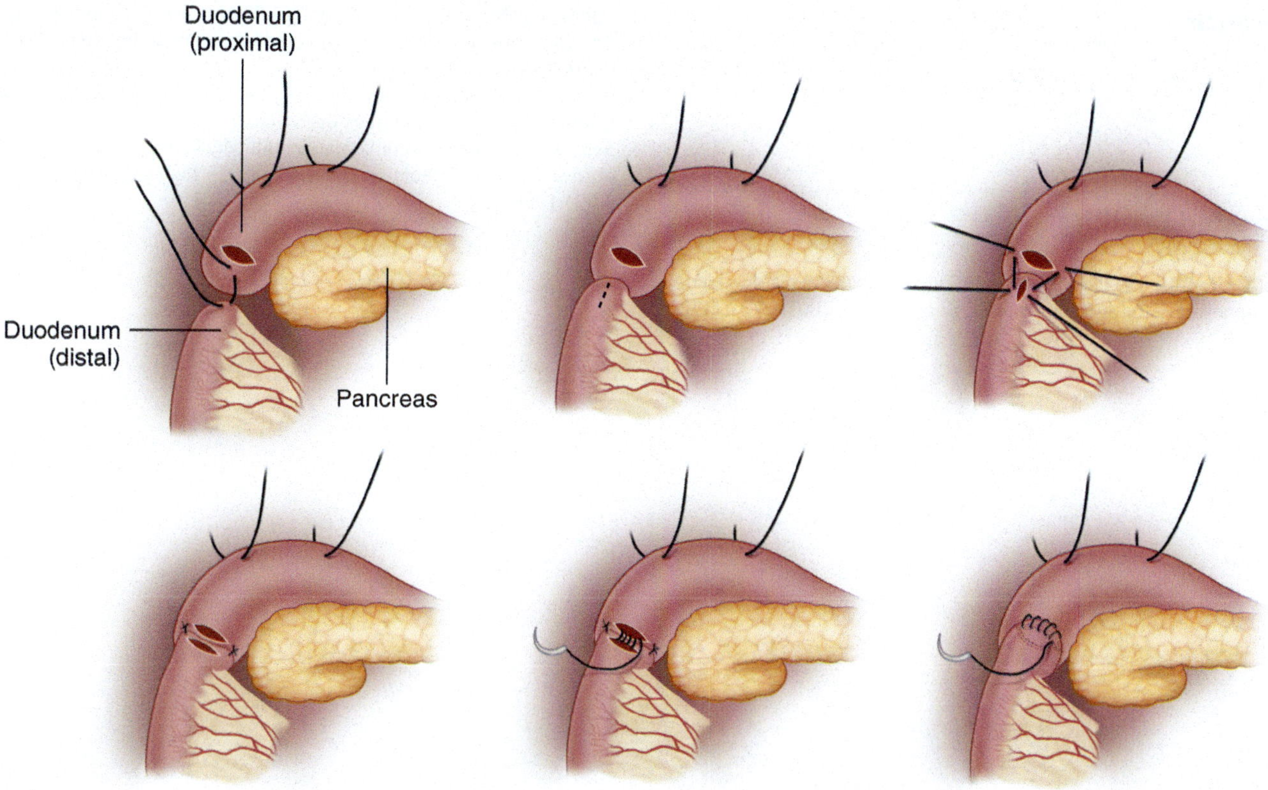

Fig. 20.16 Diagramatic representation of the procedure

Reference

1. Van der Zee DC. Laparoscopic repair of duodenal atresia: revisited. World J Surg. 2011;35:1781–4.

Suggested Reading

Bax NM, Ure BM, Van der Zee DC, van Tuijl I. Laparoscopic duodeno-duodenostomy for duodenal atresia. Surg Endosc. 2001;15(2):217.

Holcomb GW, Murphy JP. Ashcraft's paediatric surgery: duodenal and intestinal atresia and stenosis. 5th ed. Philadelphia: Saunders; 2010. p. 400–4.

McMinn RMH. Last's anatomy: regional and applied. 9th ed. Edinburgh: Churchill Livingstone; 1995. p. 335–6.

Rothenburg SS. Laparoscopic duodenoduodenostomy for duodenal obstruction in infants and children. J Pediatr Surg. 2002; 37(7):1088–9.

Meckel Diverticulum

21

Brian MacCormack and Philip Hammond

Abstract
Laparoscopy facilitates identification of a Meckel diverticulum. It can then be delivered through the umbilical port site and segmental resection with anastomosis performed extracorporeally in a laparoscopic-assisted manner.

Keywords
Meckel diverticulum

21.1 General Information

Laparoscopy facilitates identification of a Meckel diverticulum. It can then be delivered through the umbilical port site and segmental resection with anastomosis performed extracorporeally in a laparoscopic-assisted manner.

21.2 Working Instruments

- 10-mm Hasson port
- Either a 30° or 0° telescope
- 5-mm Atraumatic grasping forceps

B. MacCormack
Department of Paediatric Surgery,
Royal Hospital for Sick Children, Edinburgh, UK

P. Hammond (✉)
Royal Hospital for Sick Children, Edinburgh, UK

M. McHoney et al. (eds.), *Color Atlas of Pediatric Anatomy, Laparoscopy, and Thoracoscopy*,
DOI 10.1007/978-3-662-53085-6_21

21.3 Positioning, Port Siting, and Ergonomic Considerations

A 10-mm umbilical port is used as the primary port, and a pneumoperitoneum is established. Two working instruments are placed in the left iliac fossa and suprapubic region through 5-mm ports in the same technique that is employed for appendicectomy. The operator stands on the patient's left side. A grasper is placed in each hand to "walk" the length of the small bowel from the ileocecal valve (Figs. 21.1, 21.2, 21.3, and 21.4).

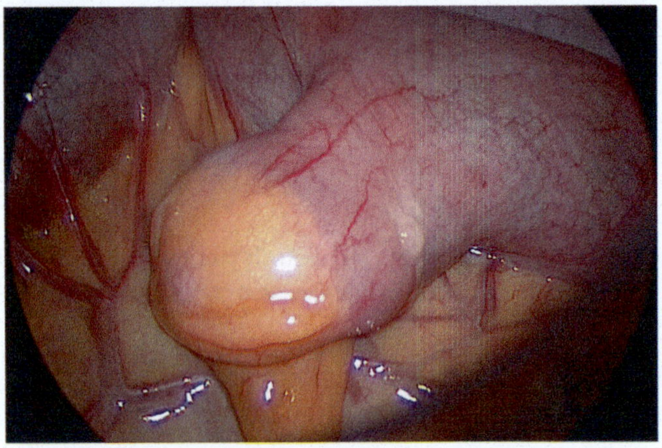

Fig. 21.1 Laparoscopic view of a Meckel diverticulum

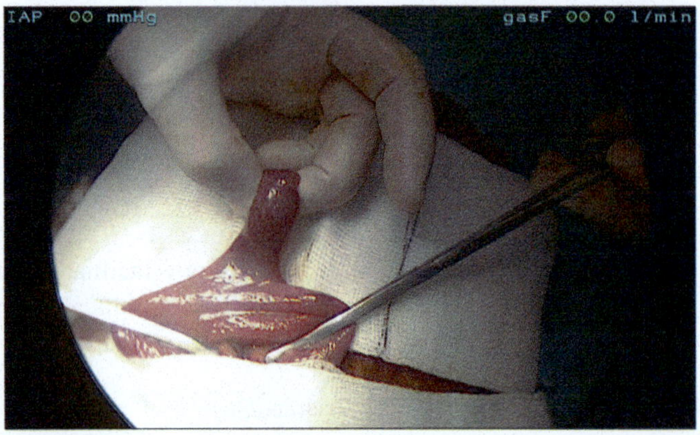

Fig. 21.3 Operative photograph illustrating the use of soft bowel clamps to control the ileum and its contents

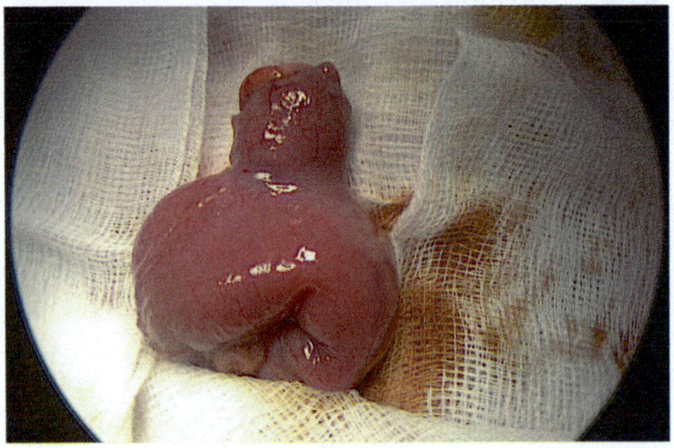

Fig. 21.2 Operative photograph following delivery of the Meckel diverticulum through the umbilical port site

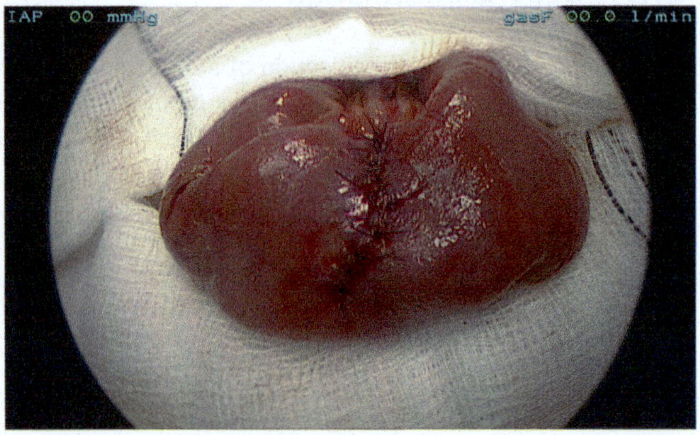

Fig. 21.4 Operative photograph highlighting the completed end-to-end small bowel anastomosis (wedge resection shown here)

21.4 Surgical Technique

The procedure starts with complete visualization of the entire peritoneal cavity followed by identification of the ileocecal segment. The small bowel is examined stepwise from the ileocecal valve proximally using atraumatic graspers. It is important to clearly display both sides of the bowel by crossing the instruments at each stepwise progression.

Once the Meckel diverticulum is identified, the 5-mm telescope is moved to the left accessory port, allowing exteriorization via the umbilical port.

Small bowel resection and end-to-end anastomosis are then completed extracorporeally before returning the bowel to the peritoneal cavity.

Instruments are removed under direct vision. Wounds are closed with an absorbable suture to muscle and subcuticular sutures or skin glue.

21.5 Alternatives

If preferred, the Alexis wound retractor (Applied Medical; Rancho Santa Margarita, CA, USA) can be used at the umbilical port site to reduce the risk of surgical site infection. A Meckel diverticulectomy can be performed by making an elliptical incision around the base of the diverticulum and closing the enterotomy transversely (wedge resection). Use of an Endo GIA stapler (Medtronic-Covidien; Minneapolis, MN, USA) or Endoloop (Ethicon; Cincinnati, OH, USA) suture has been reported for intracorporeal diverticulectomy, although it should be recognized that with this technique there is a risk that ectopic gastric mucosa or a bleeding mucosal ulcer may be left behind. Intracorporeal anastomosis is more challenging and may have a greater risk of peritoneal contamination, whereas the laparoscopic-assisted technique maintains the cosmetic and physiologic advantages of laparoscopy.

21.6 Highlights and Pitfalls

- Care must be taken to visualize both sides of the bowel so as not to miss a Meckel diverticulum.
- The umbilical port may require enlargement to allow delivery of the Meckel diverticulum.
- The use of the Alexis retractor aids visualization and may reduce surgical site infection.

Suggested Reading

Papparella A, Nino F, Noviello C, Marte A, Parmeggiani P, Martino A, et al. Laparoscopic approach to Meckel's diverticulum. World J Gastroenterol. 2014;20:8173–8.

Shalaby RY, Soliman SM, Fawy M, Samaha A. Laparoscopic management of Meckel's diverticulum in children. J Pediatr Surg. 2005;40:562–7.

Laparoscopic Appendectomy

22

Shabnam Parkar and Simon A. Clarke

Abstract

Appendicectomy is the most common laparoscopic procedure performed in children. The laparoscopic approach allows for improved cosmesis, reduced wound infection, and an earlier recovery to activities. Laparoscopy also proves a useful diagnostic tool in investigating other paediatric abdominal pathology that may mimic appendicitis.

Keywords

Appendix • Laparoscopy • Endoloop • SILS

22.1 General Information

Appendicectomy is the most common laparoscopic procedure performed in children. The laparoscopic approach allows for improved cosmesis, reduced wound infection, and an earlier recovery to activities. Laparoscopy also proves a useful diagnostic tool in investigating other paediatric abdominal pathology that may mimic appendicitis.

Laparoscopic appendicectomy can be single-port or multiport. Operative technique varies from simple electrosurgical dissection with endoscopic loop ligation to linear stapling and the use of tissue sealing devices.

S. Parkar (✉)
Department of General Paediatric Surgery, Chelsea Children's Hospital, Chelsea and Westminster Hospital, London, UK

S.A. Clarke
Department of Pediatric Surgery, Chelsea and Westminster NHS Foundation Trust, London, UK

M. McHoney et al. (eds.), *Color Atlas of Pediatric Anatomy, Laparoscopy, and Thoracoscopy*,
DOI 10.1007/978-3-662-53085-6_22

22.2 Positioning and Preparation

Careful preoperative setup is essential to ensure that this common operation is performed efficiently, ergonomically, and economically. The patient should be supine on the table under general anaesthesia and should have appropriate intravenous antibiotics. A nasogastric tube is inserted to decompress the stomach, and a urinary catheter to empty the bladder, enabling an improved view of the pelvis. Patients with a diagnosis of mild appendicitis may be asked to pass urine before the procedure, reducing the need for urethral instrumentation. The monitor should be in line with the surgeon and the right lower abdominal quadrant (Fig. 22.1).

Checking all instruments, ports, and sutures with the nursing team is essential before scrubbing. The patient is draped to allow four-quadrant access to the abdomen. Three ports are required: umbilical (optical) (variable 5–12 mm) and two instrument ports (5 mm). The larger variable optical port is used at the umbilicus to allow the appendix to be removed without wound contact, or if necessary to allow for the insertion of Endo Catch™ retrieval bags (Covidien; Mansfield, MA, USA).

The first instrument port should be inserted in the left iliac fossa (LIF) at the level of the appendix (Fig. 22.2). The second instrument port should be inserted in the suprapubic area to allow good triangulation when operating. The camera can be moved between the LIF port and the umbilicus, depending upon the surgeon's preference.

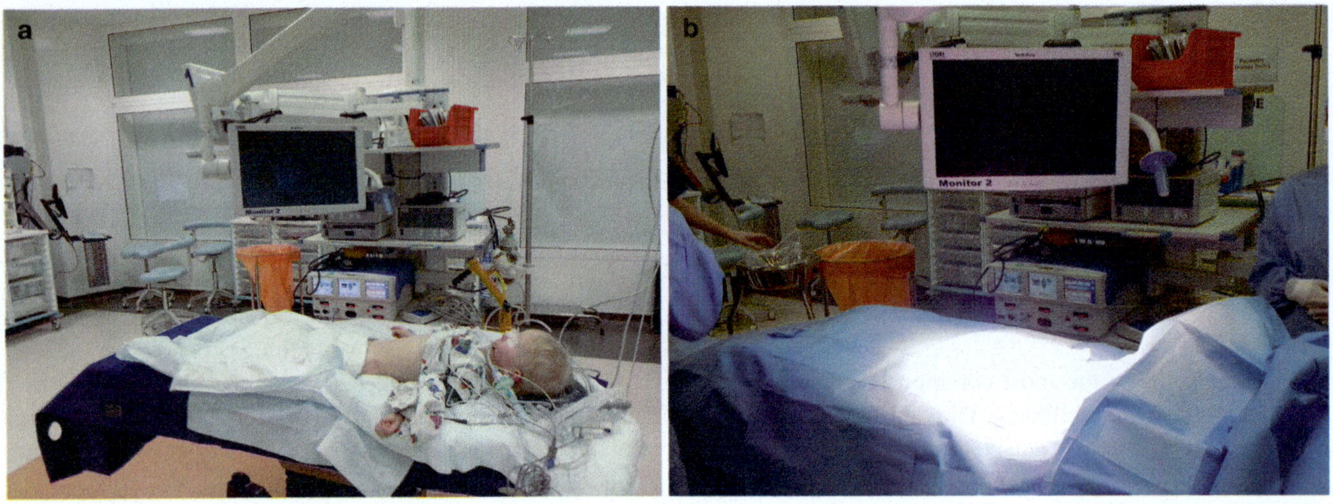

Fig. 22.1 (**a**, **b**) Position of patient and monitor

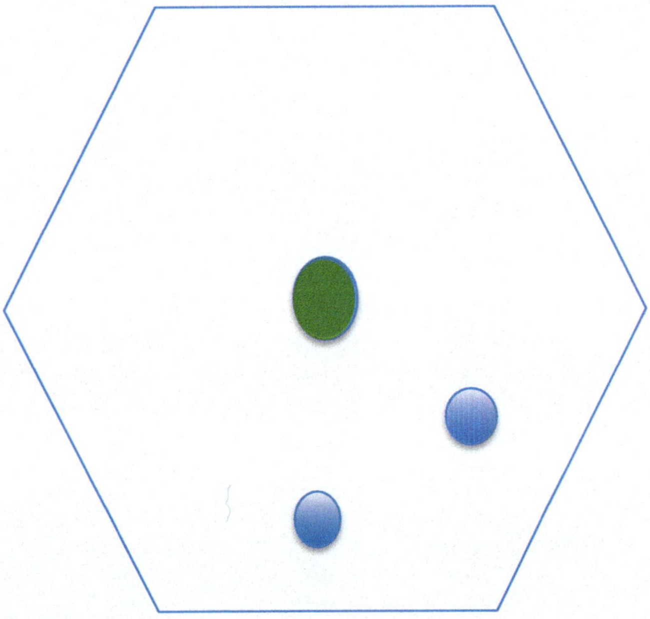

Fig. 22.2 Port positions: Umbilical camera/instrument port (*green*) and instrument ports (*blue*)

22.3 Working Instruments

Other instruments to include in the set (Fig. 22.3):

- 1 × 5-mm 30° camera
- Equipment for suction and irrigation (up to 3 l of normal saline)
- Culture swab (can be placed down a port)
- 5-mm Maryland/Kelly's grasper
- Soft bowel clamp
- Hook diathermy
- Endoloop® ligatures (Ethicon Endo-Surgery; Cincinnati, OH, USA), or surgeon's preferred device for ligating and resecting the appendix

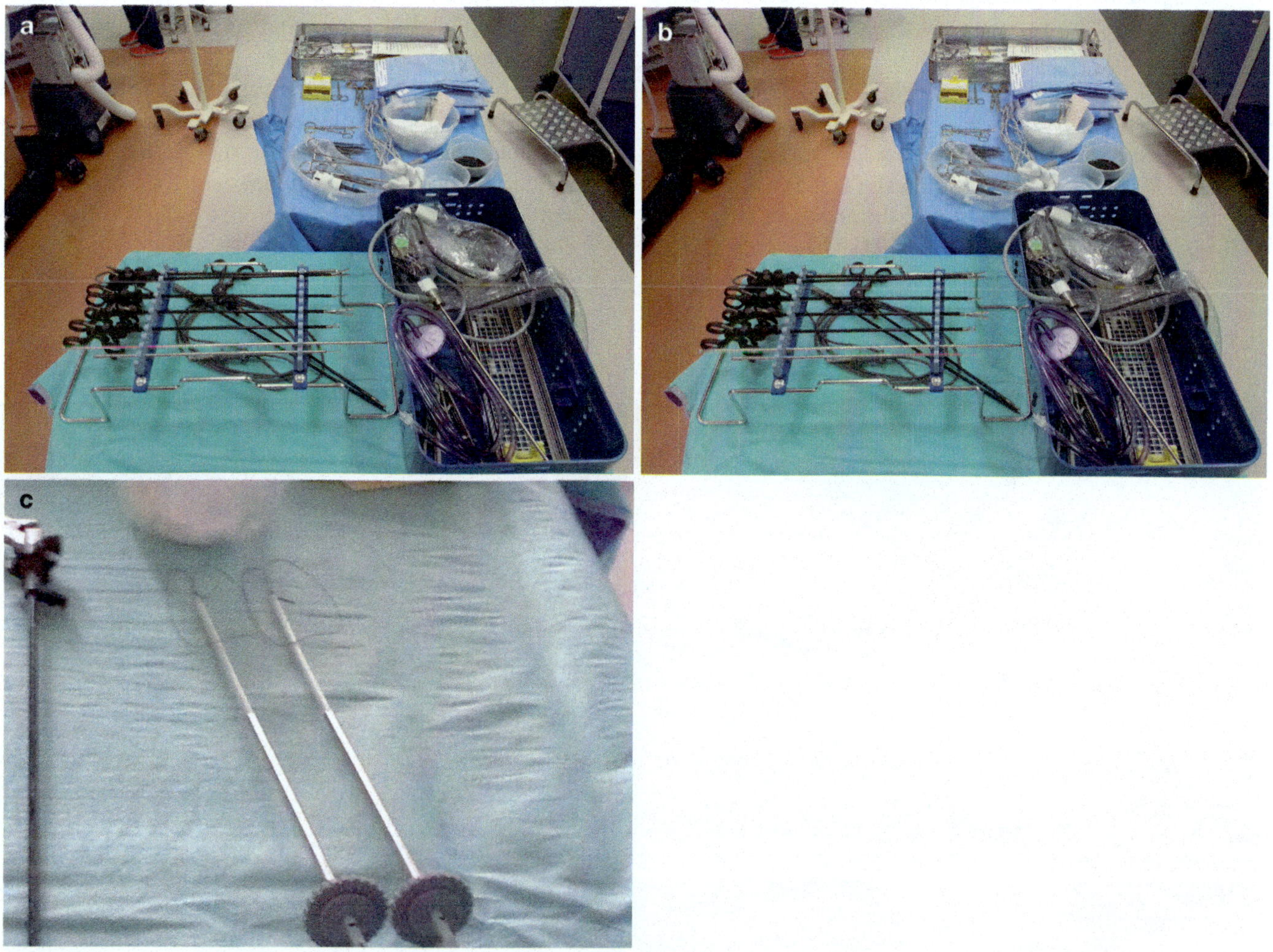

Fig. 22.3 (**a**) Instruments ready for procedure. (**b**) Laparoscopic tools: soft bowel clamps, Maryland grasper, ratcheted grasper. (**c**) Endoloop® ligatures (Ethicon Endo-Surgery; Cincinnati, OH, USA)

22.4 Surgical Technique

1. *Port insertion*. An incision can be made in the superior or inferior umbilical crease and the peritoneum entered by carefully incising each layer between forceps (Fig. 22.4). A pursestring suture is placed in the umbilical fascia. The 5–12 mm optical port is then inserted approximately 2–3 cm under direct vision. The pursestring suture is tightened and tied once before securing to the port. This prevents leak and port dislodgement.

   The anaesthetist should be informed before insufflation and the creation of the pneumoperitoneum. Pressures vary according to the size of the patient but usually range between 8 and 12 mmHg, with a flow rate of 1–3 l per min. Place the patient head down (Trendelenburg position) and right side up. Inspect the abdomen for pathology, position of the appendix, and the presence of pus, and decide upon the placement of the instrument ports.
2. *Placement of instrument ports*. Place the first instrument port in the LIF at the level of the appendix (Fig. 22.5a, b). Insert under direct vision. Insert the third instrument port in the suprapubic position (Fig. 22.5c, d).
3. *Identifying the anatomy*. On first inspection, there may be serous or purulent fluid in the abdomen. Suction should be used to improve visibility. The appendix may be inflamed, perforated, encased in adhesions, or inflamed and adherent to adjacent bowel. There may be an associated abscess.

   The appendix may vary in its anatomical position (retrocaecal, pelvic, inferohepatic, mesocolic, preileal, or postileal). Look for any faecoliths in the abdomen in cases of perforated appendix. If the appendix appears normal, it is important to inspect the abdomen for other possible causes of abnormality such as ovarian pathology or a Meckel's diverticulum, which can be detected only by 'walking the bowel' from the ileocecal junction for approximately 100 cm.
4. *Mobilising the appendix*. A soft bowel clamp and a Maryland or Kelly's instrument are used to gently manipulate and separate the bowel from the adherent

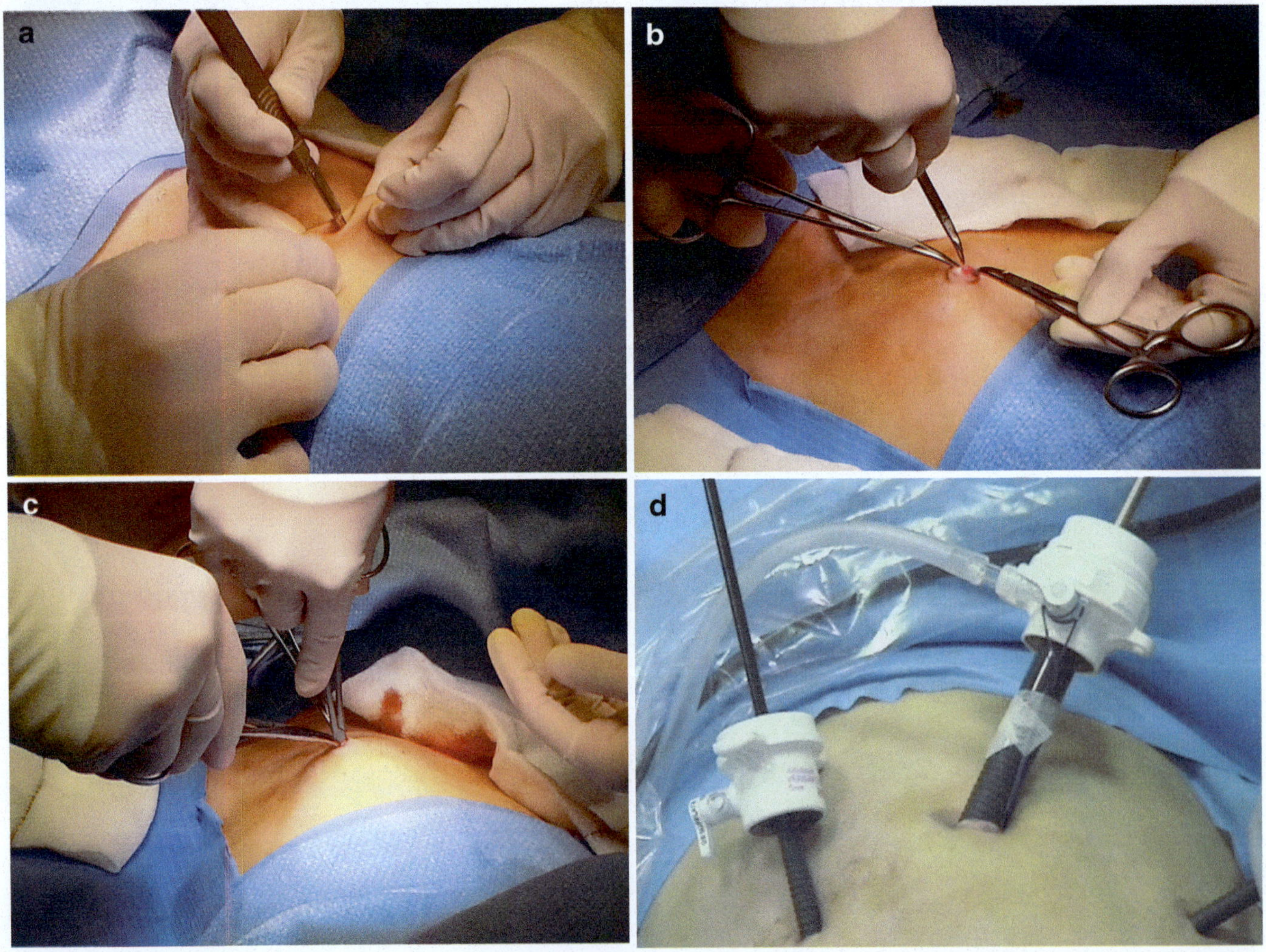

Fig. 22.4 (a–d), Inserting the umbilical port, using a supraumbilical incision. Careful dissection continues down to the peritoneum

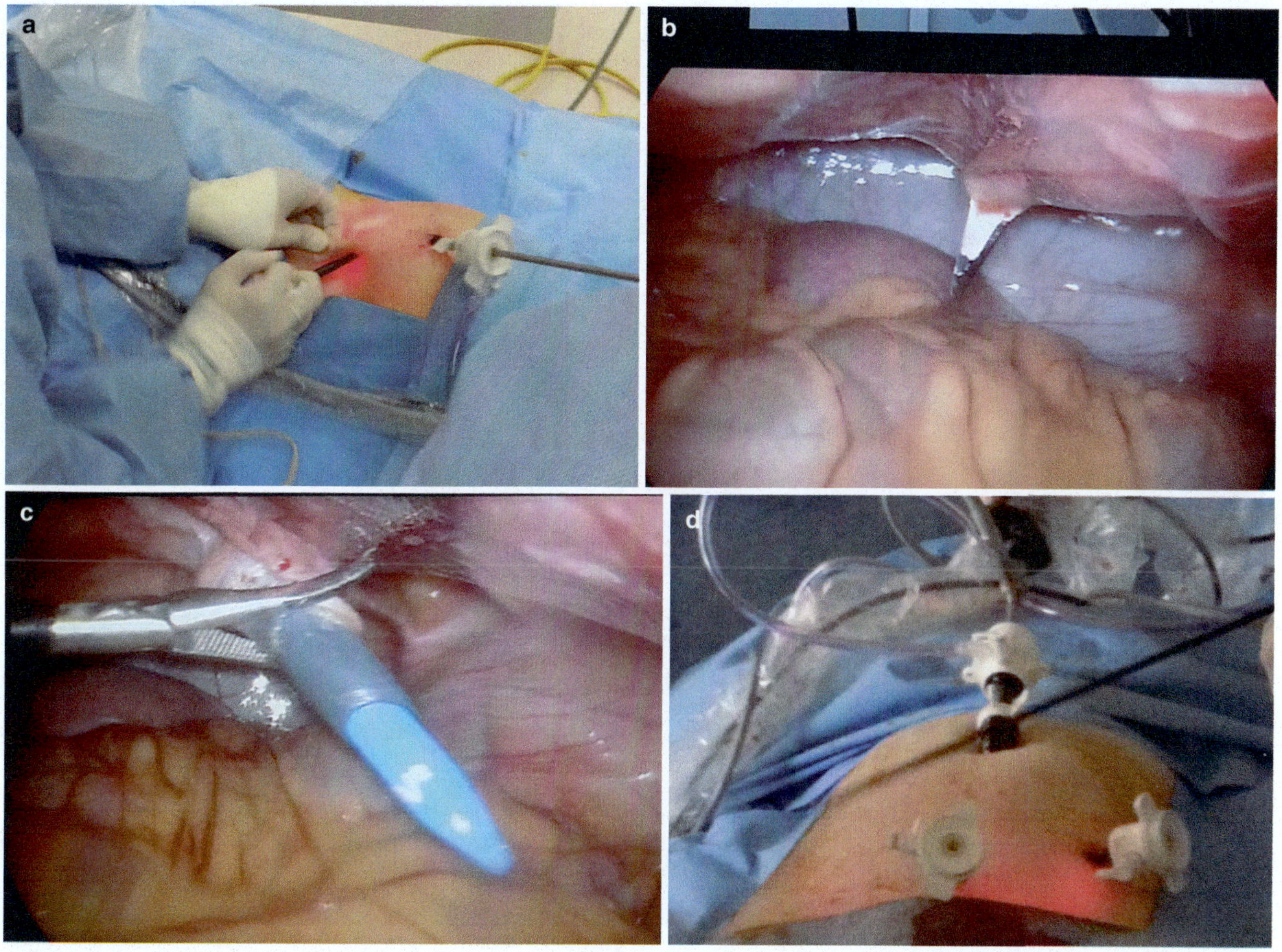

Fig. 22.5 (**a**) Placement of the first instrument port in the left iliac fossa. (**b**) Introducer ports can come with bladed or blunt trocars. (**c**) Use the soft bowel clamp or grasper in the first instrument port to move bowel away when inserting the port. (**d**) Three ports in place

appendix (Fig. 22.6). A suction device can be used as a point of leverage if an abscess is suspected and will enable rapid suctioning if an abscess is incised. Hook diathermy can also be used to carefully release inflammatory adhesions.

Once the appendix is mobilised and the base of the caecum can be seen, the tip of the appendix should be secured in an instrument with a ratchet, and the mesoappendix can then be divided.

5. *Dividing the mesoappendix*. The appendix can now be separated from its mesovascular attachments. This can be done carefully with hook diathermy (Fig. 22.7). Ensure that the appendix is fully retracted, to achieve adequate tension before using the diathermy and to avoid nearby bowel and the abdominal wall. Resect the mesoappendix down to its caecal base. Care should be taken when close to the caecum, which may be friable in severe peritonitis.

   An endoclip on the appendiceal artery prior to hook dissection can minimise bleeding, especially in larger children. A GIA™ stapler (Covidien; Mansfield, MA, USA) can be fired, ligating and resecting mesoappendix and the appendix together. The harmonic scalpel or a plasma kinetic device can also be used to transect the appendicular mesentery. Application of an Nd-YAG laser to the mesoappendix has also been described.
6. *Ligating the appendix: Endoloop® method*. It is advisable to use three Endoloops®: two applied to the base of the appendix and one applied to the appendix itself. The appendix is cut between the proximal and distal sutures. Once the loop is around the appendix base, break off the external tip to expose the extracorporeal tie and pull to tighten the loop internally (Fig. 22.8).
7. *Endoloop® ligation technique*. The first Endoloop® is inserted in the LIF port, along with its cannula. Place the grasper in the other working port and through the loop cre-

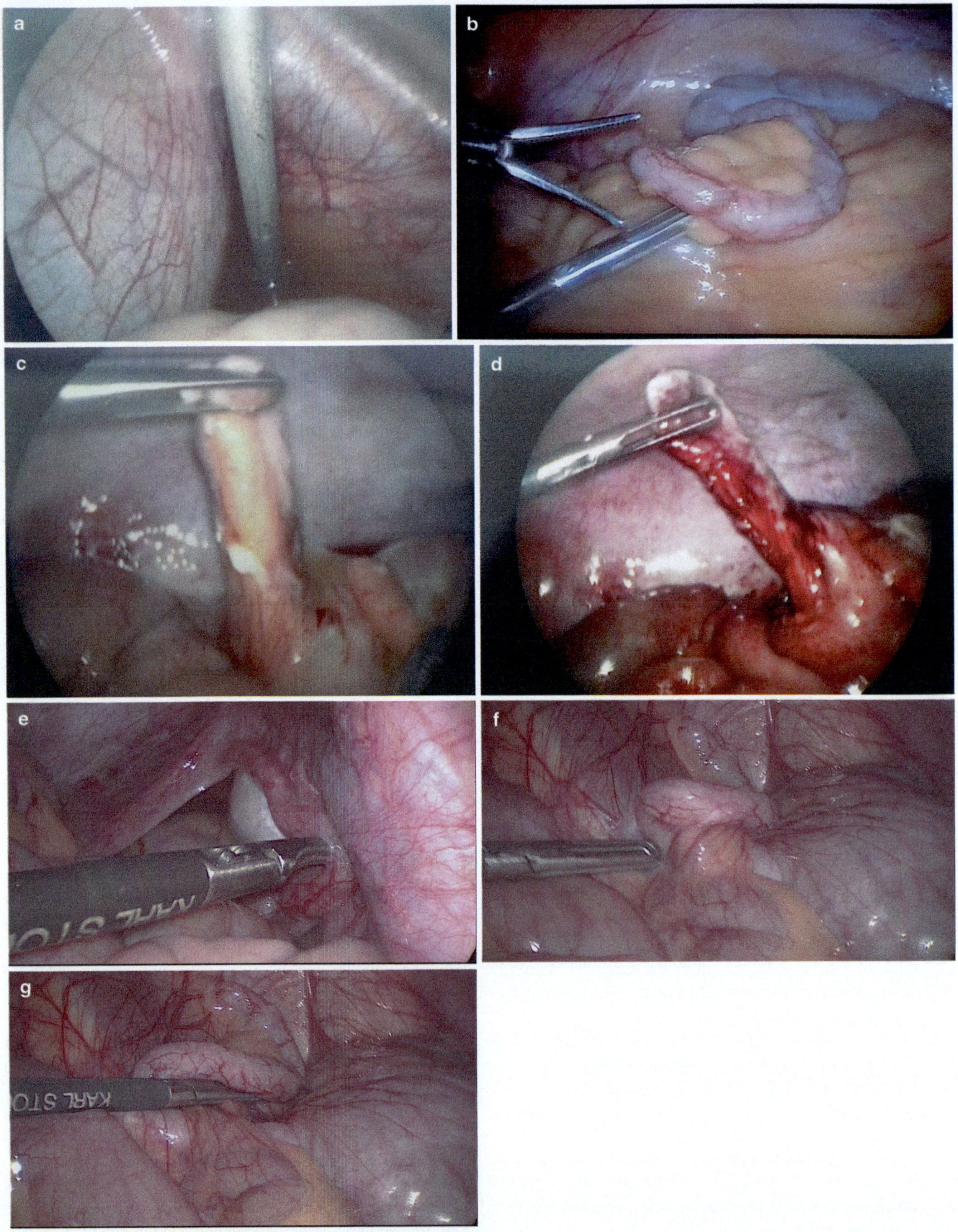

Fig. 22.6 (**a**) Purulent fluid in the pelvis—can use suction. (**b**) Interval appendicectomy. (**c, d**) Inflamed appendix with adhesions released. (**e**) Pelvic appendix: Extracting an adherent inflamed appendix from the pelvis. (**f**) Retrocaecal appendix. (**g**) Mobilising the retrocaecal appendix

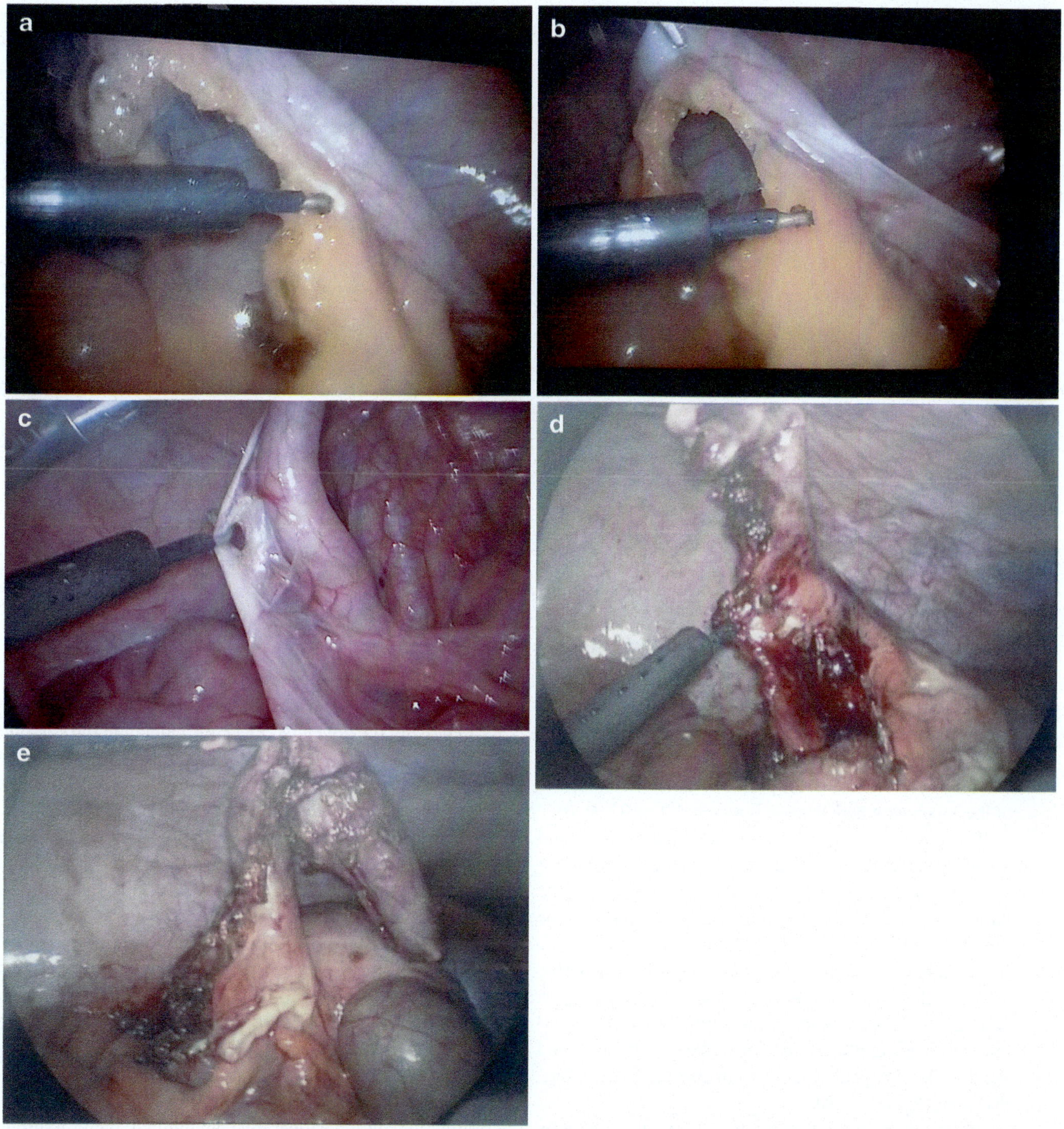

Fig. 22.7 (**a, b**) An example of dividing the mesoappendix in an interval appendicectomy. (**c–e**) The appendix is suspended by a ratcheted grasper, and the mesoappendix is divided by hook diathermy

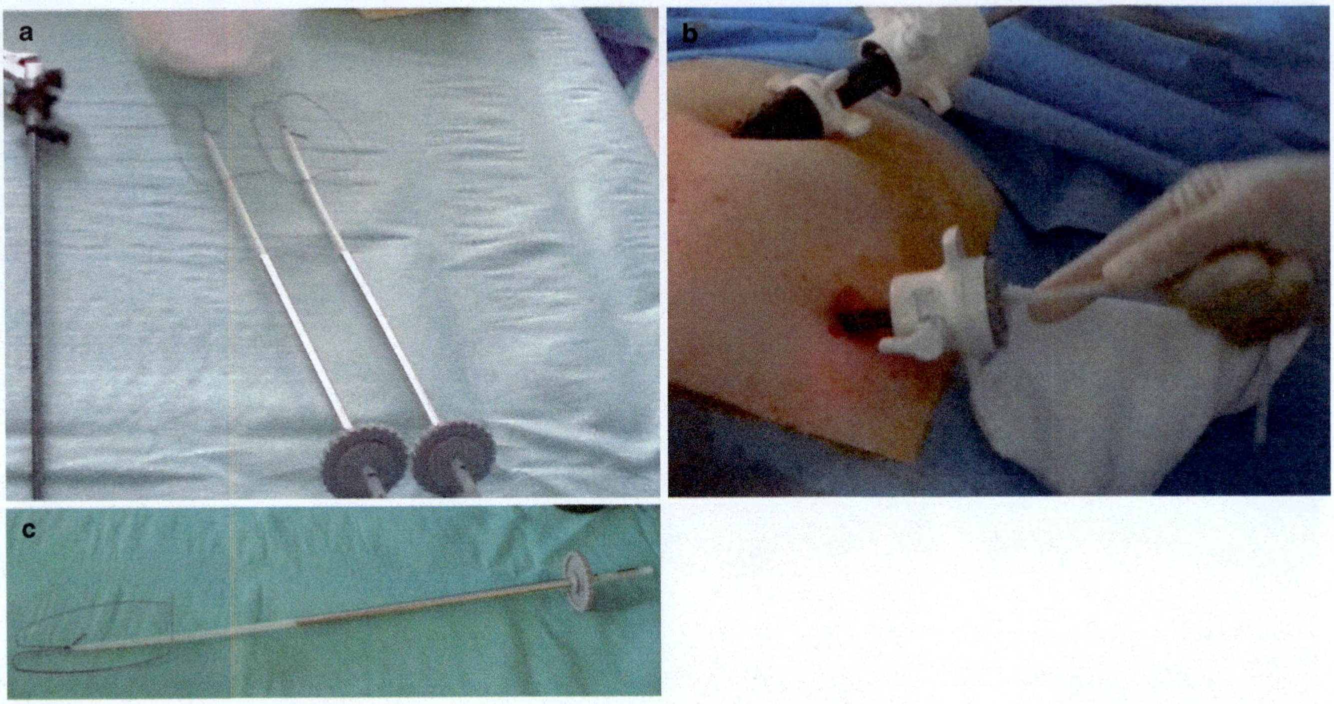

Fig. 22.8 (**a**) Endoloop® ligatures. (**b**) Insertion of Endoloop® through instrument port. (**c**) Endoloop® with introducing cannula, which helps to safely insert the Endoloop® through the port and stabilise it

ated by the Endoloop®, hold the tip of the appendix, and bring it through the loop. Position the loop down to the junction of the appendix and caecum (Fig. 22.9a, b). Reduce the size of the loop by pulling on the extracorporeal tie; carefully keep the base of the plastic sheath fixed on the site of the appendix where ligation is required (Fig. 22.9c, d).

The pushing cannula is then extended forward and the knot is tightened. Ensure that no other visceral tissues are caught in the knot during this process. If the knot does accidentally fix to adjacent bowel, it should be resected and a new Endoloop® inserted. Remove the outer cannula, insert the endoscopic scissors, and cut the suture. This can be done through the same port (Fig. 22.9e, f). Repeat the process twice so that the second endoknot is placed just above the first ligation and the third knot is placed about 1 cm above the two proximal ties (Fig. 22.9g, h).

Insert the endoscopic scissors to resect the appendix in the 1-cm gap between the proximal and distal sutures (Fig. 22.10a–c). Inspect the lumen for the presence of pus, faecolith, or vermicularis. One can use diathermy to seal the mucosa of the exposed base.

8. *Removing the appendix*. Place the camera in the LIF port and a ratcheted grasper in the umbilical port. This arrangement allows the appendix to be removed through the larger port without touching and contaminating the abdominal wound, thus minimising postoperative wound infection. Secure the appendix at the distal end in the ratcheted grasper via the umbilical port and remove it under direct vision (Fig. 22.11). Ensure that no faecoliths are dislodged in the process. Alternatively, the appendix can be placed in an Endo Catch™ bag and removed from the umbilicus without the port, if removal through the port proves difficult. Any dislodged faecoliths should be removed in an Endo Catch™ bag.
9. *After removal of the appendix*. Further inspect the abdomen for faecoliths and remove them, as they can cause infection and postoperative abdominal pain. Aspirate any residual pus with suction, and if possible, send a sample to Microbiology. Alternatively, a culture swab

can often be placed through the lowest abdominal port and into any pelvic pus.

If there is significant free pus and a perforated appendix, then copious irrigation to wash out all affected areas is required. A guide for children up to 10 years of age would be 20 mL/kg of normal saline. Ensure that all is aspirated in order to avoid loculation and potential abscess formation.

Place the patient in reverse Trendelenburg position to allow fluid to run into the pelvis for easier aspiration. Adequate irrigation can take up to 30 min. Laparoscopy allows adequate visualisation of the whole abdomen and appropriate washout and irrigation, minimising postoperative formation of intra-abdominal collections.

10. *The postoperative phase*. Ensure that adequate analgesia is given; nurse-controlled or patient-controlled morphine analgesia may be required. Ensure that the patient is well hydrated. If the appendix was mildly inflamed, sips can be commenced a few hours after the procedure and built up accordingly. If the appendix was perforated, with intra-abdominal pus, the patient may be in ileus for some time and therefore should be managed with a nasogastric tube and intravenous fluids until the ileus resolves.

    Intravenous antibiotics in the postoperative period can range from three doses (in the case of a very mildly inflamed appendix) to at least 5 days for a perforated appendix, depending on the clinical response. In this situation, patients can later complete a 1-week course of oral antibiotics. Cultures of the fluid or pus may assist in directing the antimicrobial course. Follow-up of the histology of the appendix is important.

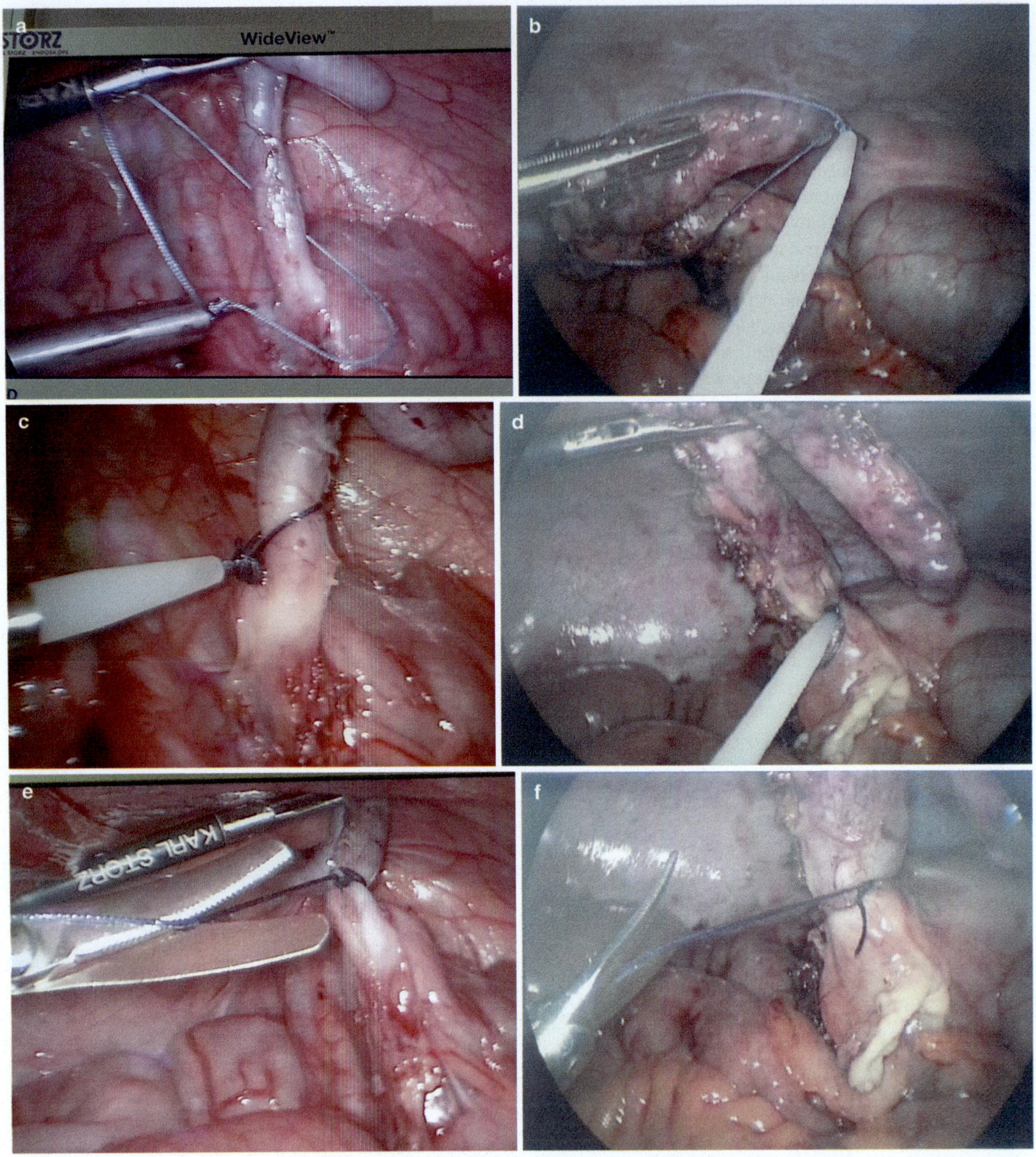

Fig. 22.9 (**a**, **b**) Position the loop at the junction of the appendix and caecum. (**c**, **d**) Reduce the size of the loop by pulling on the extracorporeal tie, carefully keeping the base of the plastic sheath fixed on the site of the appendix where ligation is required. (**e**, **f**) Applying the first Endoloop®. (**g**, **h**) Applying the second Endoloop®

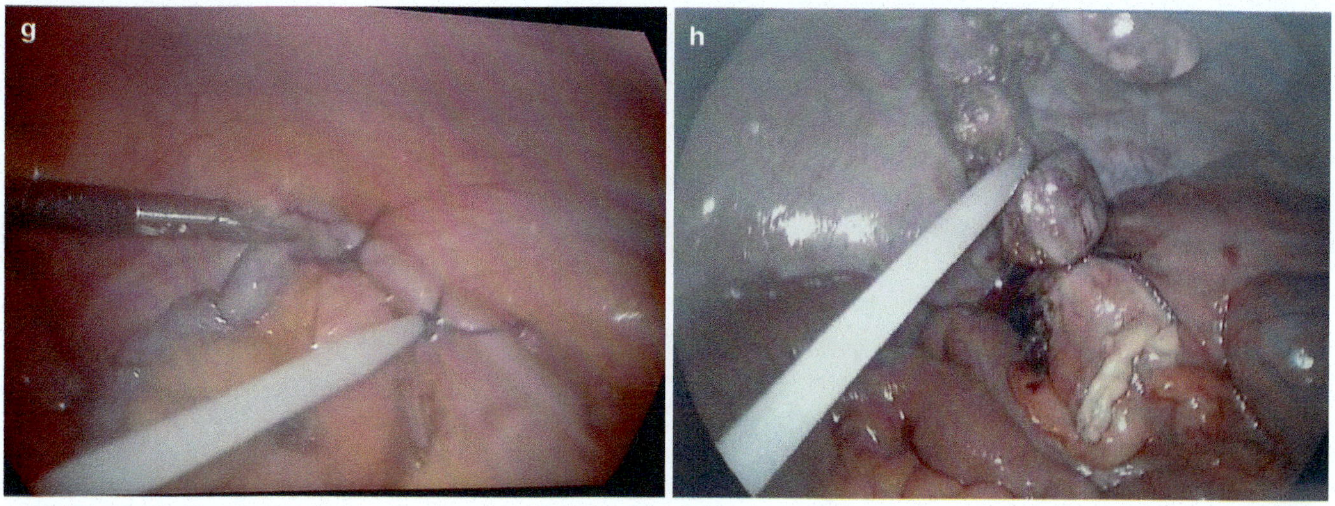

Fig. 22.9 (continued)

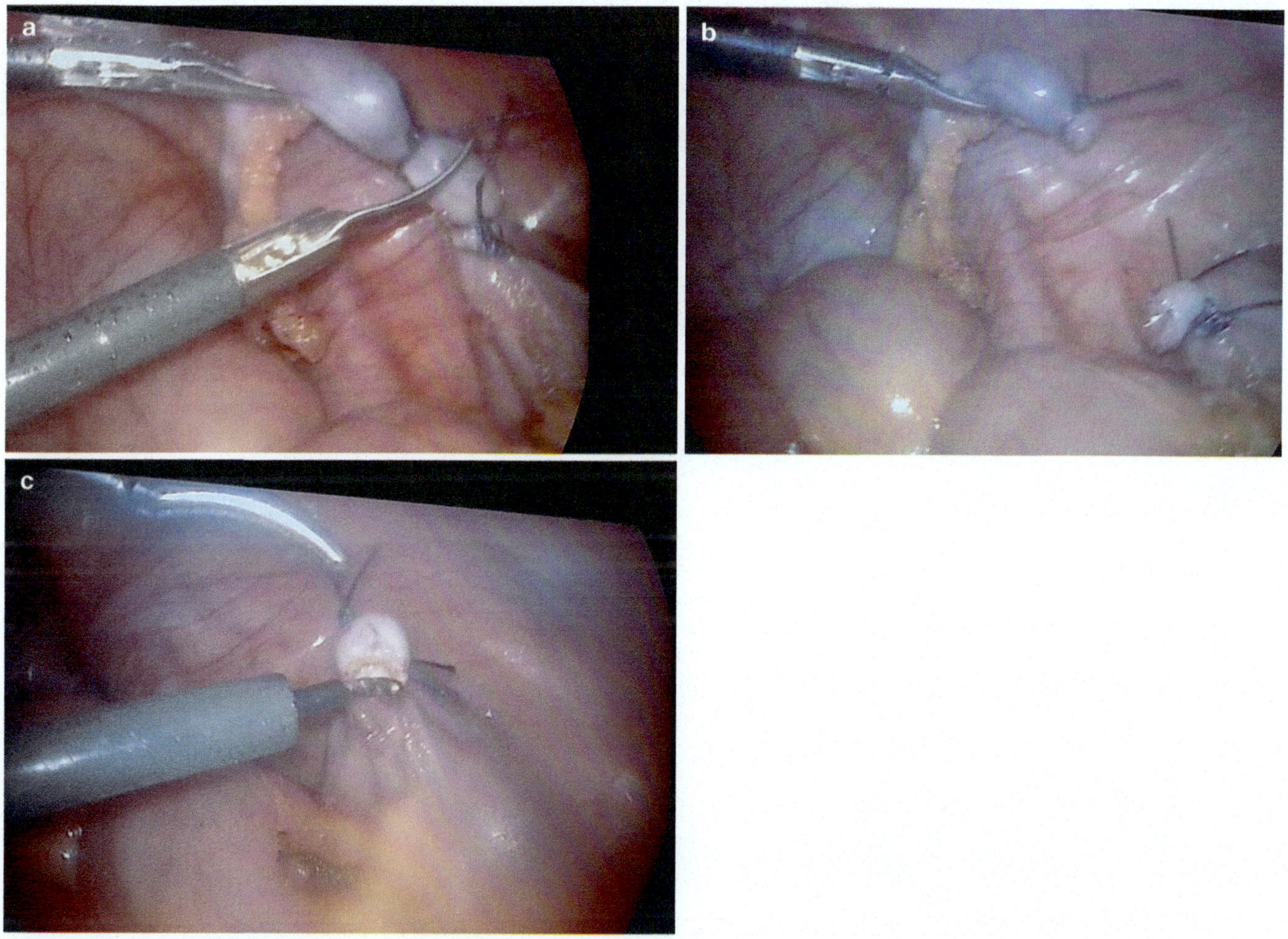

Fig. 22.10 (**a, b**) Resecting the appendix (3 Endoloops in total applied, two to the base and one to the appendix). (**c**) Using diathermy to seal the mucosa of the exposed base

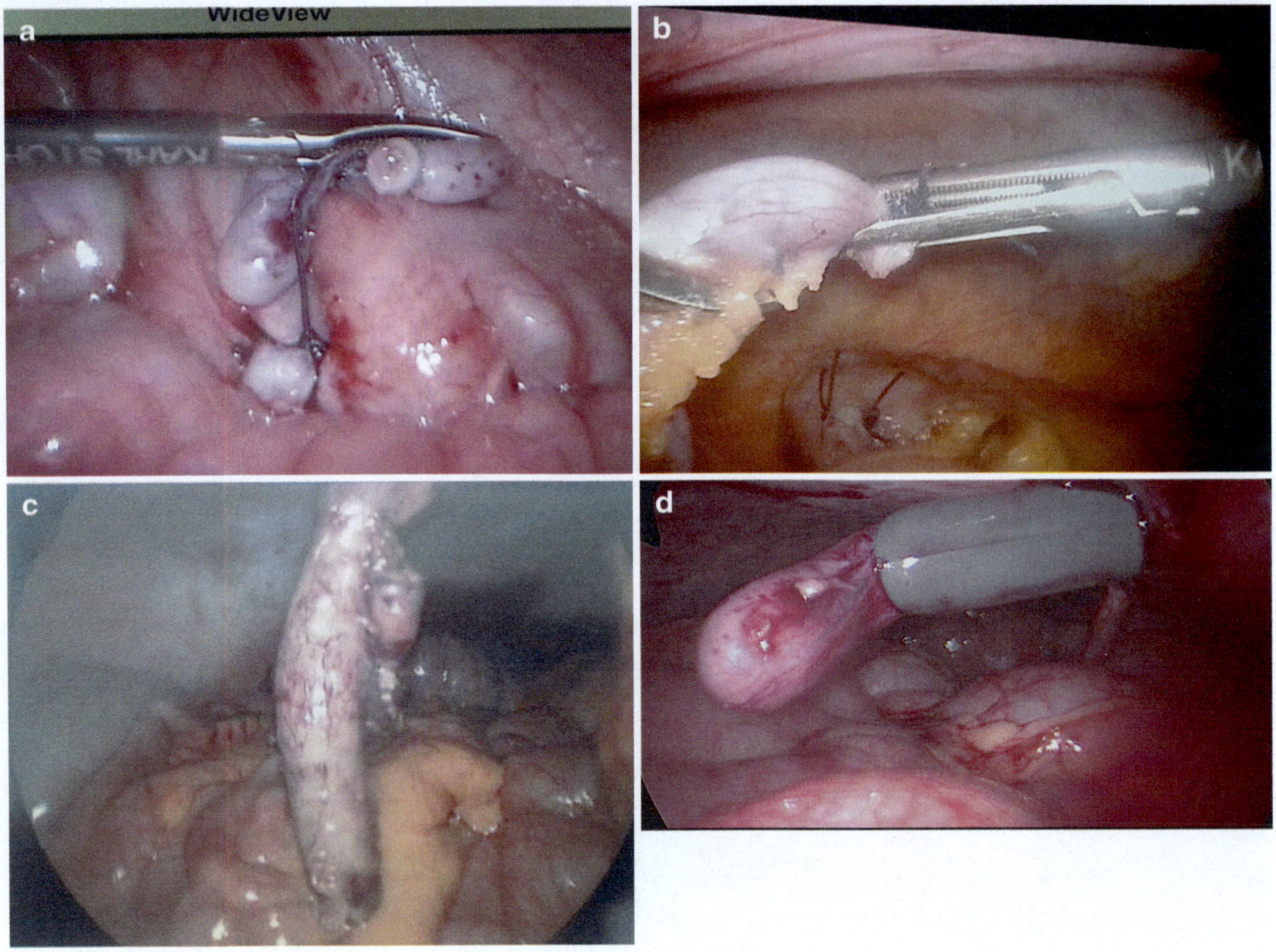

Fig. 22.11 (**a**, **b**) Mobilizing appendix to hold at its base with soft bowel clamp. (**c**, **d**) Removing appendix through umbilical port under direct vision

22.5 Alternative Techniques

Single-incision laparoscopic surgery (SILS) was developed in 1992 to further minimise the invasiveness of laparoscopic surgery (Fig. 22.12). It is increasingly being used as an approach for minimally invasive appendicectomy. Advantages include better cosmesis, less postoperative pain, and an earlier return to activities.

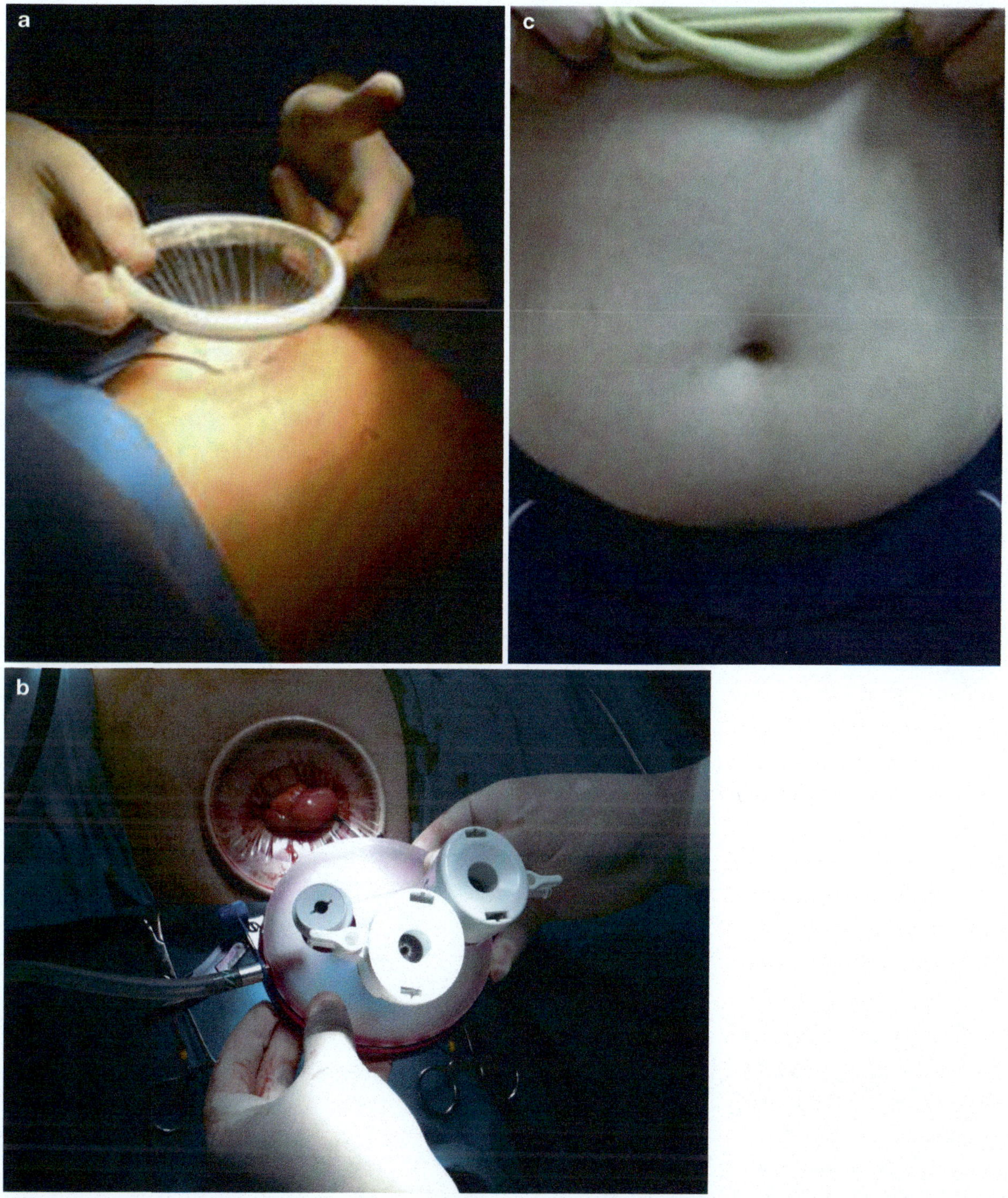

Fig. 22.12 (**a**) Wound protector prior to placement of a single-incision laparoscopic surgery (SILS) port. A 5-mm camera port and two 5-mm instrument ports are used in the same incision. Various devices are commercially available. (**b**) The appendix can be removed via the gelport without contacting the abdominal wall, thereby minimising postoperative wound infection. (**c**) No scar is visible on postoperative follow-up

23 Laparoscopy for Intussusception

Kate Cross

Abstract

Intussusception is a common abdominal emergency in infants and young children, with a peak incidence between 5 and 7 months of age (70% of cases present between 3 and 13 months). Treatment is reduction, usually by pneumatic or hydrostatic enema. However, surgery is required when enema reduction fails or there is radiologic indication of doubt or risk regarding reduction. Laparoscopy may be diagnostic, providing confirmation of reduction or persistent intussusception (where doubt exists), or it may be interventional, allowing a minimally invasive approach to reduction. In the event that laparoscopic reduction is unsuccessful, it also enables a focused and minimal incision for open surgery.

Keywords

Intussusception • Laparoscopy • Diagnosis • Reduction

23.1 General Information

Intussusception is a common abdominal emergency in infants and young children, with a peak incidence between 5 and 7 months of age (70% of cases present between 3 and 13 months). Treatment is reduction, usually by pneumatic or hydrostatic enema. However, surgery is required when enema reduction fails or there is radiologic indication of doubt or risk regarding reduction. Laparoscopy may be diagnostic, providing confirmation of reduction or persistent intussusception (where doubt exists), or it may be interventional, allowing a minimally invasive approach to reduction. In the event that laparoscopic reduction is unsuccessful, it also enables a focused and minimal incision for open surgery.

23.2 Working Instruments

- 5-mm Instruments (3-mm size can be used for a small infants but may be more traumatic during reduction and handling)
- 5-mm 30° Scope
- 5-mm Ports × 2
- 5-mm Johan graspers (atraumatic)

K. Cross
Neonatal and Paediatric Surgery Department, Great Ormond Street Hospital for Children NHS Foundation Trust, London, UK

M. McHoney et al. (eds.), *Color Atlas of Pediatric Anatomy, Laparoscopy, and Thoracoscopy*,
DOI 10.1007/978-3-662-53085-6_23

23.3 Positioning, Port Siting, and Ergonomic Considerations

The patient should be placed with the feet at the end on the table, and the laparoscopic screen should be on the patient's right side with the freedom to be moved from the head to feet end of the table. This movement may be necessary, depending on the initial position and extent of the intussusceptum (which may be as far as the sigmoid colon or rectum) to allow ergonomic positioning for the initial reduction (Fig. 23.1).

A 5-mm umbilical port position is used for the 30° telescope, and the two lateral 5-mm ports are placed in the right upper quadrant (RUQ) and left lower quadrant (LLQ) of the abdomen, opposite and perpendicular to the course of the mesenteric base (Fig. 23.2).

Fig. 23.1 Patient positioning

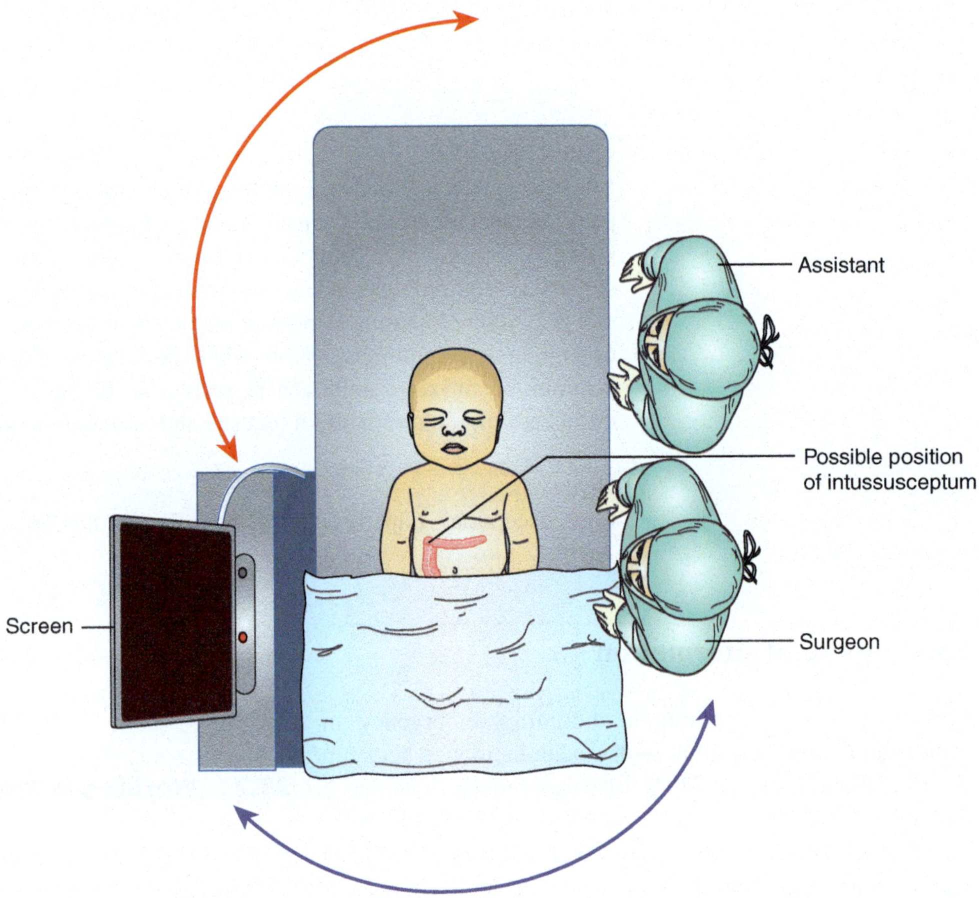

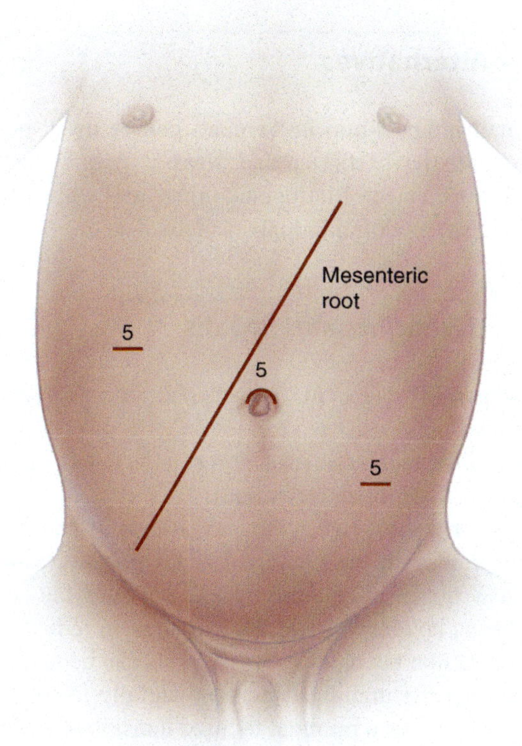

Fig. 23.2 A 5-mm umbilical port position is used for the 30° telescope, and the two lateral 5-mm ports are placed in the RUQ and LLQ opposite the course of the mesenteric base

23.4 Relevant Anatomy

The intussusceptum may be on the patient's left or right, depending on the extent of the passage, and it must be identified as the initial step. When the intussusceptum is quite distal (descending colon), the operator should stand at the right of the patient's legs with the screen at the patient's head to initiate the reduction. As this progresses, the surgeon can move below the patient's feet and to the left-sided position with the screen remaining opposite to complete the reduction at ileocecal valve.

23.5 Surgical Technique

Visual confirmation of the intussusception and identification of the distal point of the intussusceptum can be aided by walking the bowel and placing pressure distally on the collapsed colon. Viability of the intestine can also be confirmed visually. If there is evidence of necrotic or perforated bowel, conversion to an open procedure should be performed at this time.

Reduction should commence with a combination of a distal and proximal approach. The "milking" or pushing action similar to that of the open technique can be performed by using the Johan grasper in the right hand immediately distal to the mass with the jaws completely crossing the bowel and gently squeezing.

Simultaneously a pulling technique on the proximal intussuscipiens with the alternate Johan grasper should be used. The use of 5-mm instruments allows a broader coverage of the bowel diameter and less trauma during this process. Unlike the open procedure, more emphasis may be needed on the pulling action than on the milking technique, and both graspers may be used proximally to allow better traction. The movements should be slow and gradual to avoid serosal tearing of the intussuscepted bowel. These actions can be repeated until complete reduction is performed and seen.

The reduced bowel should then be gently examined to exclude the presence of a lead point such as a Meckel diverticulum that may require a laparoscopic-assisted resection (via the umbilical port).

Port sites can be closed externally under laparoscopic vision, followed by the umbilicus with an absorbable suture and tissue glue applied to the skin.

Postoperatively the child should remain on nothing by mouth until the return of gut function, which is dependent on the duration, extent, and damage to the intestinal mucosa by the intussusception rather than the operative technique. Analgesia should be intravenous until oral fluids are tolerated.

23.6 Alternatives

In a small infant 3-mm instruments can be used; however, they tend to have shorter and sharper jaws, which may increase the chance of iatrogenic injury to the already compromised bowel with minimal cosmetic benefit.

23.7 Highlights and Pitfalls

- Slow gentle movements are necessary to avoid iatrogenic damage.
- The pulling technique usually results in better success (as opposed to the open technique).
- Gentle probing between the intussuscipiens and the intussusception with the blunt end of the Johan grasper may release fibrinous adhesions and pressure between the opposing bowel walls, which may prevent reduction.
- If conversion becomes necessary, this can be achieved by extending the umbilical incision laterally to the right or alternatively by a minimal focused incision placed to provide optimal access, depending on the laparoscopic findings.

Suggested Reading

Bonnard A, Demarche M, Dimitriu C, Podevin G, Varlet F, François M, et al. Indications for laparoscopy in the management of intussusception: a multicenter retrospective study conducted by the French Study Group for Pediatric Laparoscopy (GECI). J Pediatr Surg. 2008;43:1249–53.

Hannon E, Williams R, Allan R, Okoye B. UK intussusception audit: a national survey of practice and audit of reduction rates. Clin Radiol. 2014;69:344–9.

Pierro A, Donnell SC, Paraskevopoulou C, Carty H, Lloyd DA. Indications for laparotomy after hydrostatic reduction for intussusception. J Pediatr Surg. 1993;28:1154–7.

Sklar CM, Chan E, Nasr A. Laparoscopic versus open reduction of intussusception in children: a retrospective review and meta-analysis. J Laparoendosc Adv Surg Tech A. 2014;24:518–23.

Laparoscopic Cholecystectomy

24

Augusto Zani and Niyi Ade-Ajayi

Abstract

Cholelithiasis in the paediatric population is typically secondary to haemolytic disorders, structural biliary tract anomalies, or previous illness/medical interventions of infancy and early childhood. A proportion of cases are idiopathic. More recently, an increasing incidence of stone disease has been reported outside these traditional groups. This may reflect easier access to abdominal sonography and other diagnostic tools such as magnetic resonance cholangio-pancreatography (MRCP) as well as an increase in childhood obesity, particularly in adolescent girls [1]. There is no good evidence for the treatment of asymptomatic gallstones in otherwise well children as complication rates are low [2]. Surgery for a diagnosis of biliary dyskinesia is controversial and incidental cholecystectomy during another procedure is inappropriate in the paediatric population. Conversely, symptomatic patients with background conditions such as sickle cell disease and spherocytosis have a high risk of complications and should, in general, be offered laparoscopic cholecystectomy (LC), which has become the gold standard for intervention where the infrastructure and expertise are available.

Keywords

Laparoscopic • Cholecystectomy • Gall bladder • Bile duct

24.1 General Information

Cholelithiasis in the paediatric population is typically secondary to haemolytic disorders, structural biliary tract anomalies, or previous illness/medical interventions of infancy and early childhood. A proportion of cases are idiopathic. More recently, an increasing incidence of stone disease has been reported outside these traditional groups. This may reflect easier access to abdominal sonography and other diagnostic tools such as magnetic resonance cholangio-pancreatography (MRCP) as well as an increase in childhood obesity, particularly in adolescent girls [1]. There is no good evidence for the treatment of asymptomatic gallstones in otherwise well children as complication rates are low [2]. Surgery for a diagnosis of biliary dyskinesia is controversial and incidental cholecystectomy during another procedure is inappropriate in the paediatric population. Conversely, symptomatic patients with background conditions such as sickle cell disease and spherocytosis have a high risk of complications and should, in general, be offered laparoscopic cholecystectomy (LC), which has become the gold standard for intervention where the infrastructure and expertise are available.

Paediatric LC is predominantly carried out in the elective setting following biochemical and radiological investigations to exclude bile duct stones. Cholecystitis, cholangitis, gall stone pancreatitis, and other acute presentations are

A. Zani
Division of General and Thoracic Surgery, The Hospital for Sick Children, Toronto, ON, Canada

N. Ade-Ajayi (✉)
Department of Paediatric Surgery, King's College Hospital NHS Foundation Trust, London, UK

M. McHoney et al. (eds.), *Color Atlas of Pediatric Anatomy, Laparoscopy, and Thoracoscopy*,
DOI 10.1007/978-3-662-53085-6_24

generally managed conservatively with the aim of delayed surgery. This may be preceded by diagnostic/therapeutic endoscopic retrograde cholangio-pancreaticography (ERCP). LC has been described in the outpatient setting for children and, in expert hands, overall morbidity is low.

24.2 Working Instruments

- 10 mm Hasson port
- 5 mm ports × 3
- 30° 5-mm telescope
- 5 mm bowel graspers × 2
- 5 mm hook diathermy
- 5 mm Maryland forceps
- 5 mm Liga or polymer locking clips
- 5 mm scissors
- 10 mm retrieval bag

24.3 Positioning, Port Siting, and Ergonomic Considerations

The patient is positioned supine and secured to the operating table (Fig. 24.1). There are two widely used setups to perform this operation: the French (Dubois) and the American (Reddick Olsen) positions. In the French position, the patient lies in the lithotomy position, the operating surgeon stands between the patient's legs, the assisting surgeon is on the patient's left side, and the scrub nurse on the right. In the American position, the patient lies supine with arms in abduction, the operating surgeon stands on the patient's left side with the scrub nurse to his left, and the surgical assistant is on the patient's right. From an ergonomic perspective, there is little difference between the two approaches in a modern dedicated minimally invasive surgery suite [3]. The authors favour a modification of the American position: the patient is supine with arms down on either side with one surgical assistant to the left of the operating surgeon for the camera and another (or a robot arm) to the right of the patient for gallbladder retraction.

Port position depends on various factors including patient age and size, liver size, and gallbladder location and, importantly, surgeon preference (Fig. 24.2a, b). A trans- or infraumbilical port is placed using the open technique. In adolescents with a high body mass index, an optical trocar inserted in a suitable peri-umbilical position may facilitate safe access. Once the port is inserted and the camera introduced, the surgeon can judge where to place the working ports. The authors favour three port "subcostal" placement: two ports are positioned in the right subcostal position; the most lateral for gallbladder retraction over the liver to expose the operating field. Medial to that, the surgeon's left-hand operating port and finally an epigastric port, to the left of the falciform ligament for the right operating hand.

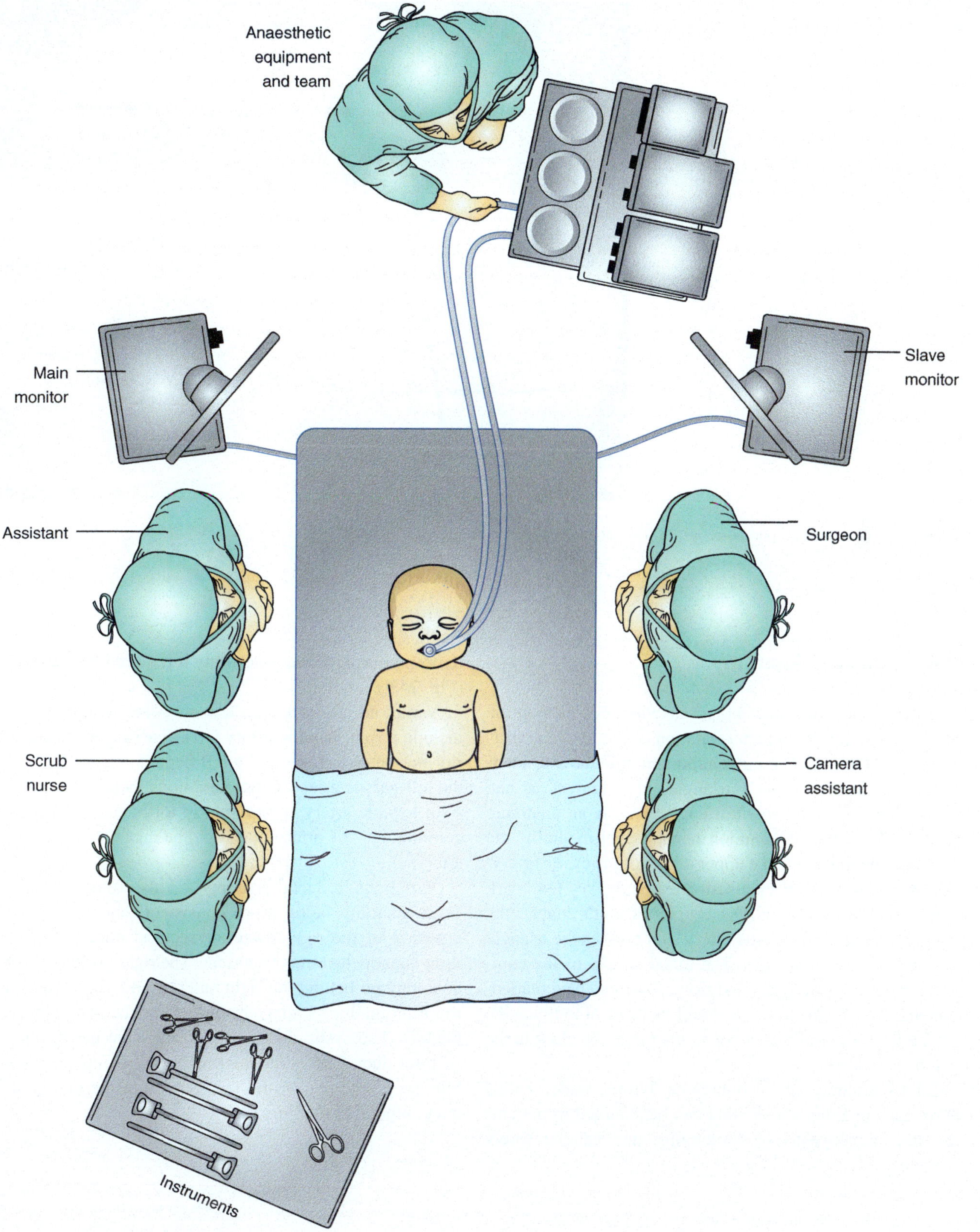

Fig. 24.1 Operating room set-up

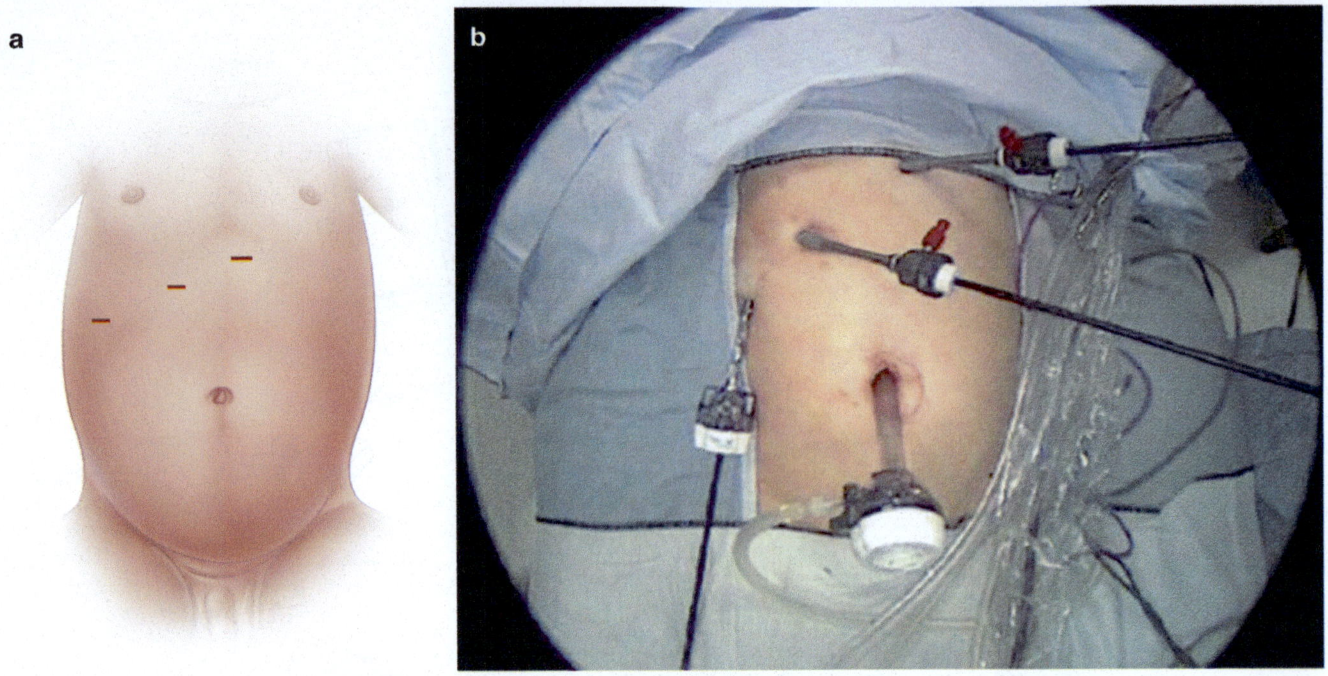

Fig. 24.2 (**a**) Illustration of port positions. (**b**) Digital image of port positions

24.4 Relevant Anatomy

The safety of laparoscopic cholecystectomy depends on the correct identification and interpretation of the relevant anatomy (Fig. 24.3a, b). The most important structures to identify are related to Calot's triangle, also known as the hepatobiliary or cystohepatic triangle. This is an anatomic space bordered by the common hepatic duct medially, the cystic duct laterally, and the upper aspect of the cystic artery/ inferior border of the liver superiorly. It is the key landmark to ensure safety during LC. Dissection of Calot's enables the surgeon to identify the cystic artery as it crosses the triangle from medial to lateral sides. The blood supply of the common bile duct is usually derived from cystic and pancreaticoduodenal artery branches. It is therefore prudent to ligate the cystic artery at its gallbladder end rather than too close to the right hepatic artery.

Calot's triangle also contains a lymph node, called Mascagni's or Lund's node, that can be enlarged in patients who had cholecystitis or cholangitis, and can be safely removed during cholecystectomy if required for a clear view (Fig. 24.4).

Variations in vascular anatomy are common. These include a right hepatic artery (RHA) arising from the superior mesenteric artery and a left hepatic artery (LHA) from the left gastric artery. Other variations include an accessory right hepatic artery, a particularly tortuous right hepatic or common hepatic artery that could be mistaken for the cystic artery.

Variations in ductal anatomy are also frequent with left-sided anatomy being more consistent than the right; low insertion of the right posterior sectoral duct in which the right hepatic duct may be absent. The right posterior sectoral duct may go behind the right anterior and join the left anterior sectoral duct. It may also insert into the neck of the gallbladder. The cystic duct may join the right hepatic or right sectoral duct. The cystic duct may run behind the common bile duct, entering on the left, rendering the common bile duct vulnerable to injury.

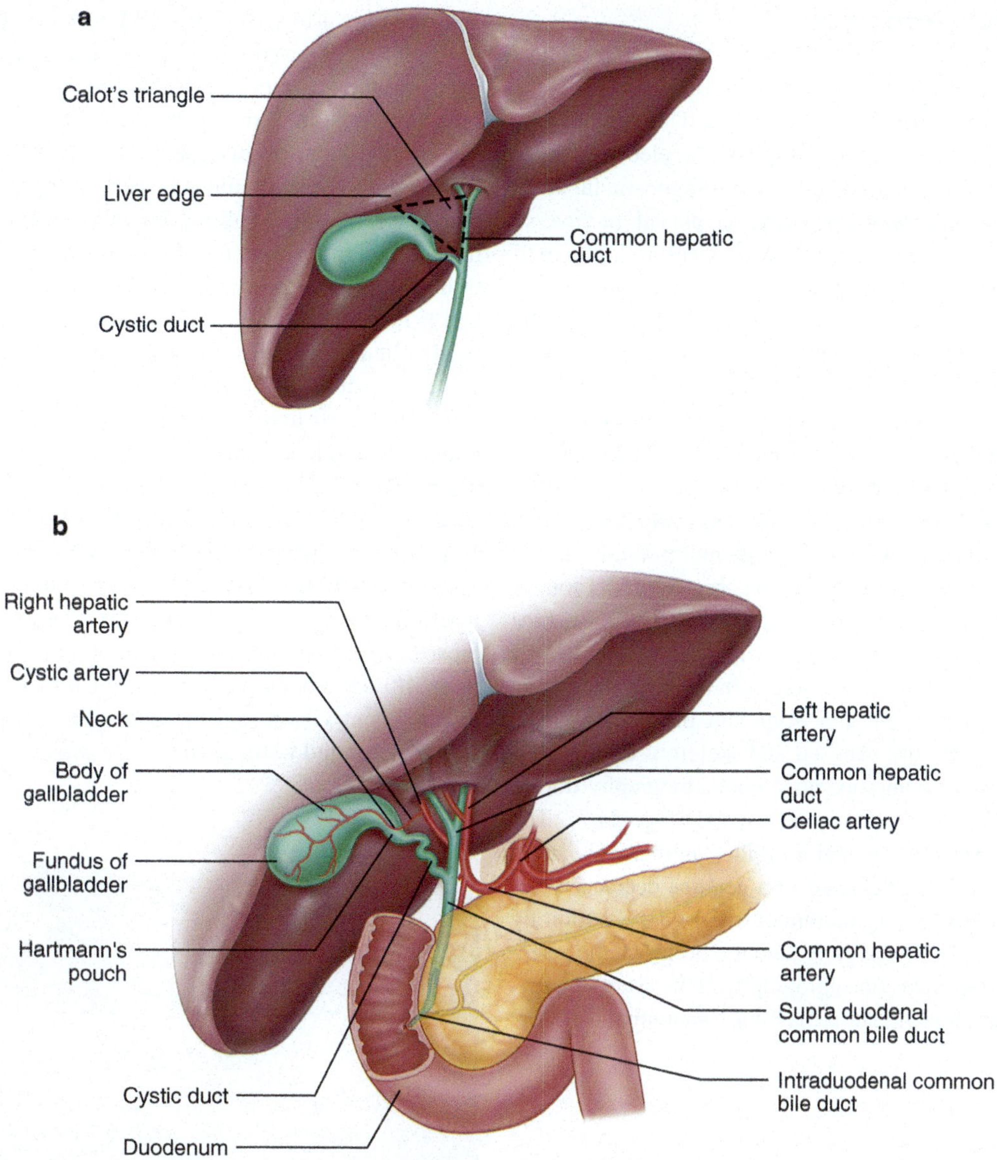

Fig. 24.3 (**a**) Calot's triangle. (**b**) Illustration of essential anatomy of gallbladder, bile ducts and surrounding structures

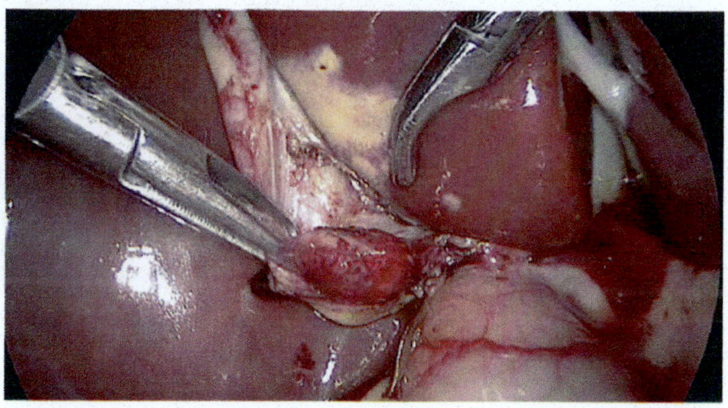

Fig. 24.4 Prominent lymph node of Lund

24.5 Surgical Technique

At timeout, prior to commencement of surgery, the delivery of a bile penetrating antibiotic is confirmed. This step may be omitted in the older child undergoing simple, elective cholecystectomy. Ports are inserted and a brief diagnostic laparoscopy performed. The operating table is placed in reverse Trendelenburg to enable the transverse colon and small bowel to fall away from the operative field. Bowel-grasping forceps are applied to the fundus of the gallbladder via the right lateral port and the gallbladder is retracted cephalad.

Division of peritoneum and adhesions, if present, is carried out anteriorly and posteriorly to expose Hartman's pouch and the related cystic duct (Fig. 24.5). A number of energy sources may be used for this part of the dissection but, carefully deployed, a hook diathermy is effective and inexpensive.

Dissection proceeds along the cystic duct in the direction of the common bile duct (Fig. 24.6a, b). Appreciation of Calot's triangle increases as this part of the dissection proceeds. Calot's triangle is exposed with a combination of hook diathermy and Maryland forceps in the dominant hand and bowel graspers in the non dominant hand. The cystic duct and cystic artery are exposed and skeletonised.

The role of routine intraoperative cholangiography during paediatric cholecystectomy is controversial (Fig. 24.7). The cost benefit in terms of potential morbidity in relatively small ductal systems and the likelihood of a genuine decreased risk of common bile duct injury requires careful consideration by the surgical team [4]. The authors favour a policy of the selective use of intraoperative cholangiography in the event of a particularly difficult dissection or if the biliary anatomy is unclear.

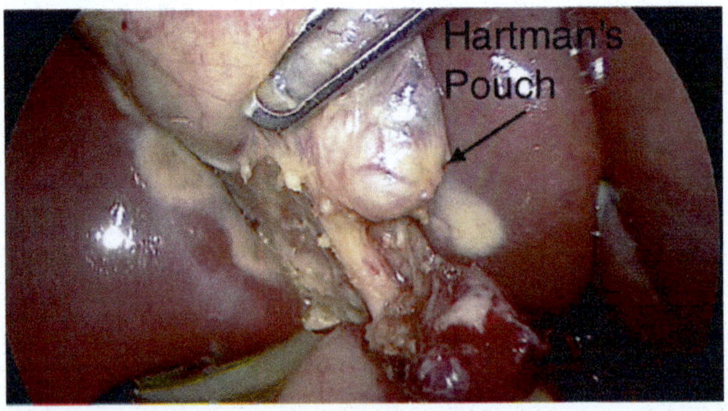

Fig. 24.5 Peritoneal dissection has been carried out to expose Hartmann's pouch

The cystic artery is dealt with by energy device or clipped (two endoclips proximal and one distal) and divided using laparoscopic scissors. The same technique is used for clipping and division of the cystic duct (Fig. 24.8a, b).

The gallbladder is dissected off the liver bed using cautery (Fig. 24.9). Correct tension is important. The spaghetti manoeuvre involves rolling the gall bladder up over the left hand grasper intermittently to optimise tension while dissection with cautery proceeds. At the end of this step, the liver bed is inspected for bleeding and bile leakage. Diathermy is used judiciously for haemostasis. The gallbladder is retrieved using an endo-catch bag. If retrieval through the umbilicus is difficult due to a size discrepancy, the gallbladder can be punctured and bile aspirated within the bag, to avoid spillage in the abdominal cavity. The laparoscope is re-inserted for a final look at the operative field. The use of a drain should be considered at this point but it is rarely required. If needed, a non-suction drain within a sterile system is used. The tube is inserted down a port site and the position adjusted under laparoscopic vision to ensure it is optimal.

A 4/0 absorbable suture is used for deep closure under laparoscopic vision and biologic glue or a 5/0 absorbable suture is applied to the skin.

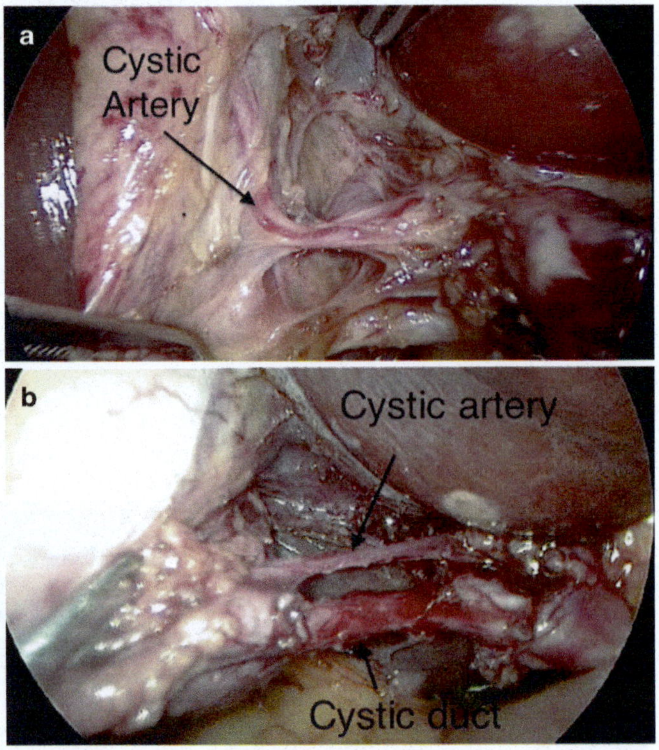

Fig. 24.6 (**a**) Calot's triangle is defined and the cystic duct exposed. (**b**) The cystic duct and artery are skeletonised

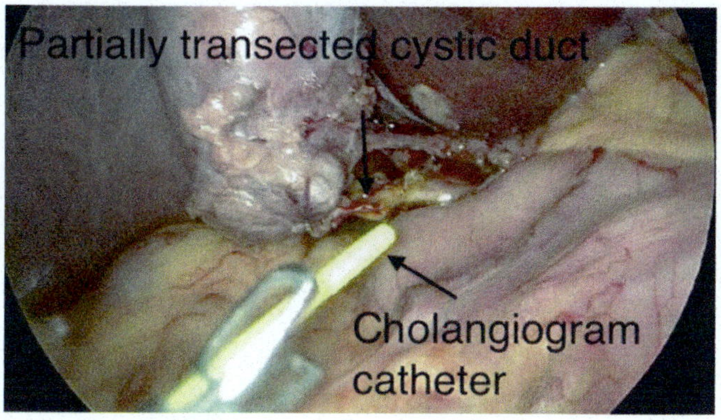

Fig. 24.7 Cholangiography catheter is about to be inserted into the cystic duct

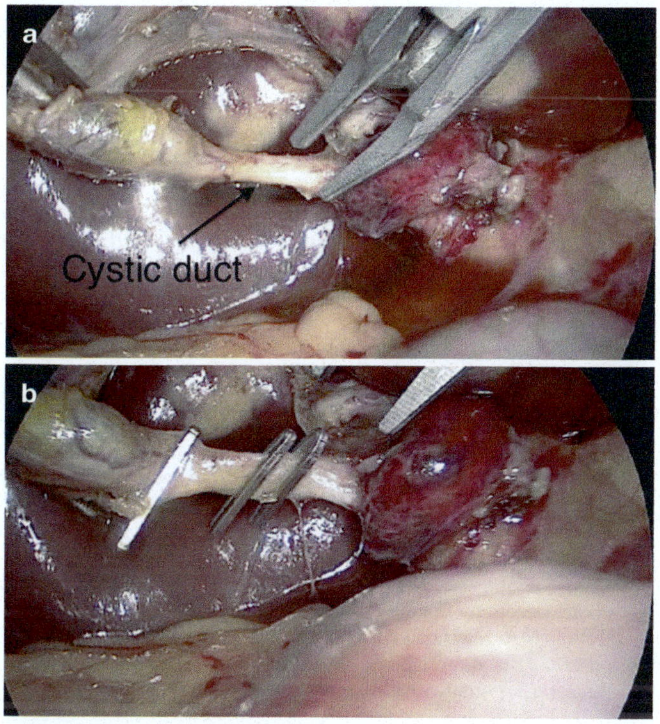

Fig. 24.8 (**a**) The cystic duct about to be clipped. (**b**) The cystic after Liga clips have been applied

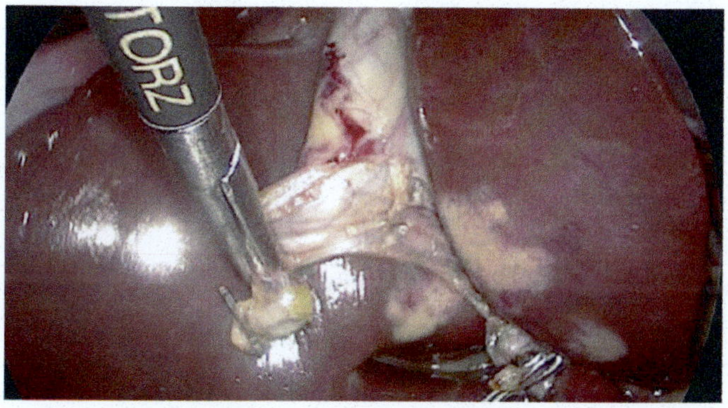

Fig. 24.9 The gallbladder is ready for dissection to remove

24.6 Alternatives

A recent systematic review of the literature has reported that single-port laparoscopic cholecystectomy has become more popular in the paediatric population and is a safe and feasible alternative to multiport laparoscopy [5]. Improved instrument design and disposable equipment such as the Gelpoint® platform reduces the ergonomic difficulties once associated with single port laparoscopy and has increased recent interest in this method for cholecystectomy.

Reduced port and 3-mm cholecystectomy have also been described with slow but steady uptake among paediatric surgeons as emerging technology catches up with the concepts.

24.7 Highlights and Pitfalls

- Careful patient selection and pre-operative evaluation is crucial to determine patients who have bile duct stones. They should undergo diagnostic and therapeutic ERCP prior to LC
- The gallbladder, if contracted at initial view, should raise the possibility of a difficult dissection. A particularly tense gallbladder is usually amenable to aspiration to facilitate handling
- Difficult dissection is anticipated if there is significant inflammation: the response depends on the experience of the individual surgeon and the operating team. Options include early dissection from the fundus towards the body and neck of the gallbladder. Experienced assistance should be sought if difficulties persist. Conversion to open surgery to facilitate safe gall bladder removal should not be considered a failure
- Iatrogenic injury to major biliary or vascular structures is a concern. Clear anatomical definition, intermittent relaxation of fundal retraction to re-establish where key structures are, and the selective use of cholangiography should make this complication a rarity. The technique for cholanagiography needs to be meticulous to avoid air in the system and an erroneous interpretation of these bubbles as stones. A brief pause before clipping the cystic duct and artery may be helpful.
- Commonly, a persistent post-operative bile leak represents a failure of surgical technique or instrumentation as it comes from the stump of the cystic duct clip or tie fallen off. Meticulous technique and deployment of the first Liga clip outside the body help reduce the likelihood of this happening.
- Deeply embedded gall bladder encourages deep dissection and may result in persistent bile leakage from the duct of Luschka, small tributaries of minor intrahepatic radicals that drain directly from the liver into the body of the gallbladder.

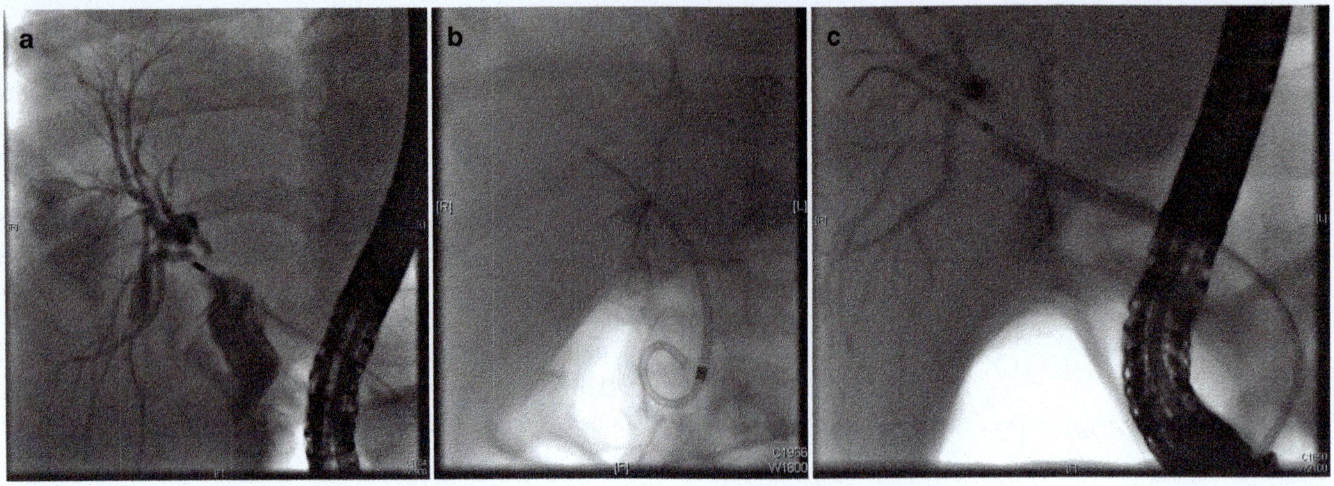

Fig. 24.10 (a) A major ductal injury is confirmed with extravasation of contrast at ERCP. (b) A biliary stent is left in place for management. (c) Intermittent balloon dilatation is undertaken until resultant stricture has resolved

- Threshold for conversion is dependent on individual surgeon and institutional experience.
- Mirizzi syndrome—impactin of one or more stones at Hartman's pouch leading to chronic inflammation and possible erosion at the level of the common hepatic duct or common bile ducts [6]. Erosion may also be in the duodenum, stomach, or colon. If a fistula is present, definitive surgery should only be undertaken by a very experienced surgeon and serious consideration should be given to conversion from LC to an open technique.
- If a significant ductal injury is recognised, referral to an expert hepato-biliary centre is recommended (Fig. 24.10). Interventions to remedy the situation include diagnostic scintigraphy and ERCP with stent insertion and subsequent balloon dilatation. For those not suitable for these techniques, hepatico-jejunostomy may be required.
- Post-cholecystectomy syndrome—a search for unrecognised ductal stones or other anatomical causes is carried out. If found, then ERCP intervention to retrieve stones is the gold standard of care.

References

1. Khoo AK, Cartwright R, Berry S, Davenport M. Cholecystectomy in English children: evidence of an epidemic (1997–2012). J Pediatr Surg. 2014;49(2):284–8.
2. Bogue CO, Murphy AJ, Gerstle JT, et al. Risk factors, complications, and outcomes of gallstones in children: a single-center review. J Pediatr Gastroenterol Nutr. 2010;50(3):303–8.
3. Kramp KH, van Det MJ, Totte ER, Hoff C, Pierie JP. Ergonomic assessment of the French and American position for laparoscopic cholecystectomy in the MIS Suite. Surg Endosc. 2014;28(5):1571–8.
4. Overby DW, Apelgren KN, Richardson W, Fanelli R, Society of American Gastrointestinal and Endoscopic Surgeons. SAGES guidelines for the clinical application of laparoscopic biliary tract surgery. Surg Endosc. 2010;24(10):2368–86.
5. Chrestiana D, Sucandy I. Current state of single-port laparoscopic cholecystectomy in children. Am Surg. 2013;79(9):897–8.
6. Beltrán MA. Mirizzi syndrome: history, current knowledge and proposal of a simplified classification. World J Gastroenterol. 2012;18(34):4639–50.

Laparoscopic Splenectomy

25

Khalid Elmalik and Sean Marven

Abstract

Laparoscopic splenectomy is most commonly performed for those with hereditary red cell membrane disorders such as spherocytosis causing hemolysis. The hemoglobinopathies such as sickle cell and thalassemia causing sequestration or hypersplenism are increasingly common indications. Chronic immune thrombocytopenia (ITP) is a good indication but less effective. Tumors, abscesses, or cysts are rare indications. Nonparasitic splenic cysts should be completely excised by partial splenectomy. Torsion of a wandering spleen is similarly rare.

Keywords

Splenectomy • Laparoscopic • Sickle cell disease • Hereditary spherocytosis • Immune thrombocytopenic purpura • Thalassemia

25.1 General Information

Laparoscopic splenectomy is most commonly performed for those with hereditary red cell membrane disorders such as spherocytosis causing hemolysis. The hemoglobinopathies such as sickle cell and thalassemia causing sequestration or hypersplenism are increasingly common indications. Chronic immune thrombocytopenia (ITP) is a good indication but less effective. Tumors, abscesses, or cysts are rare indications. Nonparasitic splenic cysts should be completely excised by partial splenectomy. Torsion of a wandering spleen is similarly rare.

A preoperative ultrasound is essential to identify the size of the spleen and any accessory splenunculi or gallstones. If concomitant cholecystectomy is indicated, it is usually best performed after the splenectomy. Reliable vessel sealing is the key element of the procedure and can be achieved in various ways. Early splenic artery sealing is recommended to allow the spleen to be emptied of blood. The use of a stapler is less desirable for the hilar vessels because it requires a 12-mm port and may be difficult to deploy in the small patient; therefore it is best used as a backup for troublesome bleeding. Retrieval may be problematic in small patients, and a suprapubic port or pfannenstiel incision to extract the spleen may be necessary.

Risks include recurrent disease from a missed spleneculus or splenosis from an intra-abdominal or port site implantation of tissue. Perforation of the colon or stomach is possible. Diaphragmatic hernia is a recognized complication. Portal vein thrombosis is a potential risk or more recent concern. Pancreatitis seems not to be a significant risk. The risk of overwhelming postsplenectomy sepsis is small but real, and therefore prophylactic daily oral penicillin is recommended.

25.2 Preoperative Preparation

- An ultrasound scan of the abdomen to assess the size of the spleen and to look for any splenunculi or gallstones.
- Immunization against encapsulated organisms, namely, *Streptococcus pneumoniae*, *Neisseria meningitides*, and *Haemophilus influenzae* type B.

K. Elmalik • S. Marven (✉)
Sheffield Children's Hospital, Sheffield, UK

M. McHoney et al. (eds.), *Color Atlas of Pediatric Anatomy, Laparoscopy, and Thoracoscopy*,
DOI 10.1007/978-3-662-53085-6_25

- A "top-up" transfusion may be wise if anemia or a high sickle index is detected.
- Steroid prophylaxis may be required.
- Penicillin prophylaxis is mandatory.
- Thromboprophylaxis should be considered in teenage or obese patients.
- Blood cross-match of one unit of blood is essential.
- Platelet transfusion for ITP.
- ITP may be required.
- Nasogastric tube.

25.3 Working Instruments

- 12- to 15-mm Umbilical port (ideally a "smart port" that is easily replaceable [e.g., SILS port, Medtronic-Covidien, Minneapolis, MN] or such as a balloon or other airseal [SurgiQuest; Applied Medical Resource Co., Milford, CT]).
- 30° 5- and 10-mm Telescopes.
- 5-mm Needle holders.
- 3- or 5-mm Single use scissors.
- 3- or 5-mm Maryland (multiple manufacturers) and fenestrated grasper forceps.
- Laparoscopic stapler 12-mm Ø angulated with 45-mm vascular load cartridge (2- to 2.5-mm staples).
- 3- or 5-mm Hook diathermy
- Ultrasonic shears.
- 5- or 3-mm Electrothermal bipolar vessel sealing device.
- 5- or 3-mm Suction.
- Tissue suspension suture or device (optional).
- 5-mm clip applier, e.g., Hem-o-lok Weck clips (Teleflex; Morrisville, NC, USA).
- Ecosac retrieval bag (Espiner Medical, UK)
- Haemstatic matrix

25.4 Positioning, Port Siting, and Ergonomic Considerations

The patient is placed in a semilateral position with the left flank raised, the table head up 30°, and the right side of the table tilted down 15°. A 12- or 15-mm umbilical port is inserted by open technique with a vertical umbilical incision through the umbilicus. Ideally this would be a port that is easily replaced like a balloon port or a multiple instrument port (e.g., SILS type port) that allows use of a 12-mm stapler if used and a 15-mm retrieval bag. Two or three working instruments are placed as shown in Fig. 25.1. In cases of simultaneous cholecystectomy, the splenectomy ports are positioned first and any additional ports are placed later. The left-hand working port is placed in the midline just below or through the falciform ligament, and the right-hand port is placed in the midclavicular line above the level of the umbilicus. A further port may be placed in the anterior axillary line above the umbilicus for retraction if necessary. The surgeon, assistant, and nurse all stand on the patient's right-hand side, with the table lowered to a comfortable height.

25.5 Operative Procedure (Fig. 25.1)

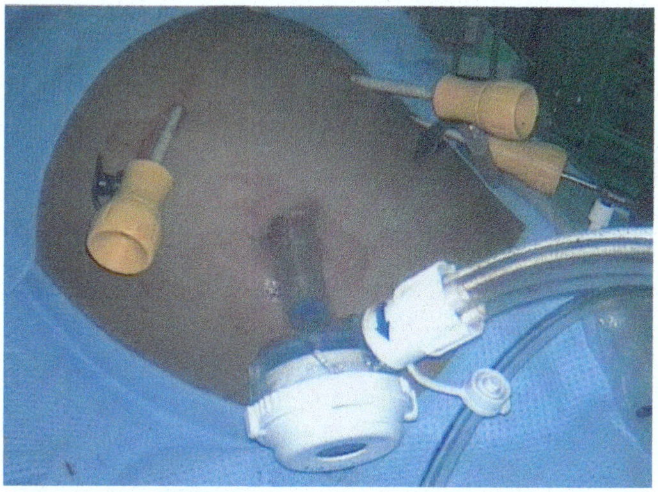

Fig. 25.1 Positioning of patient for splenectomy. In this case, 3-mm ports with a 12-mm Airseal (Surgiquest) port at umbilicus for continuous pneumoperitoneum and smoke evacuation

25.6 Relevant Anatomy (Fig. 25.2)

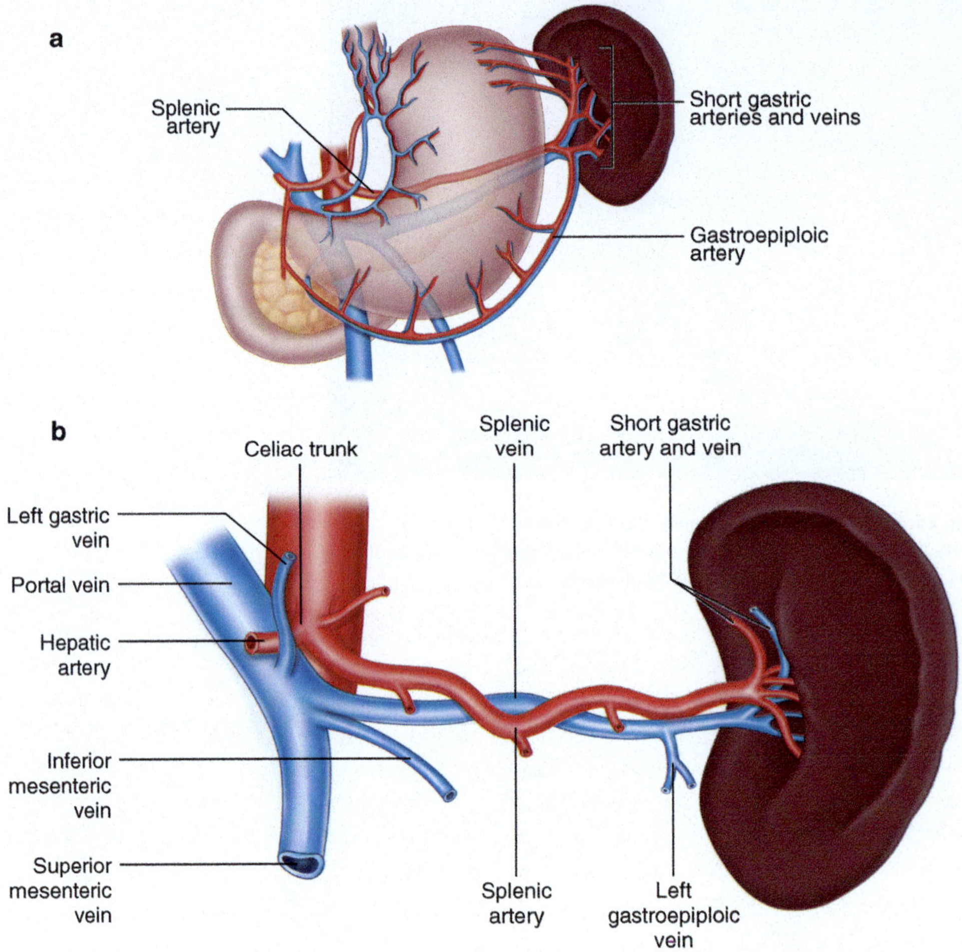

Fig. 25.2 (**a**) Typical blood supply of spleen. (**b**) As many as five segmental vessels may branch from the splenic artery, giving a Medusa-like appearance

25.7 Surgical Technique (Figs. 25.3, 25.4, 25.5, 25.6, 25.7, 25.8, and 25.9)

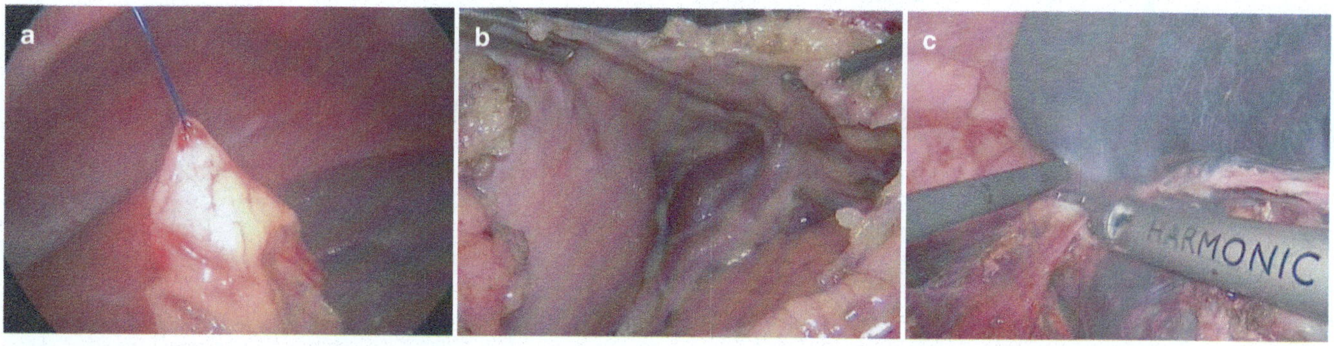

Fig. 25.3 Exposure of the splenic hilum. (**a**) The stomach may be suspended using a percutaneous suture or "T" device. (**b**) The gastrosplenic ligament is opened, the lesser sac entered, and the splenic artery at the upper border of the pancreas is identified. (**c**) The final short gastric vessels are divided using ultrasonic shears or monopolar hook diathermy, which devascularises the upper pole

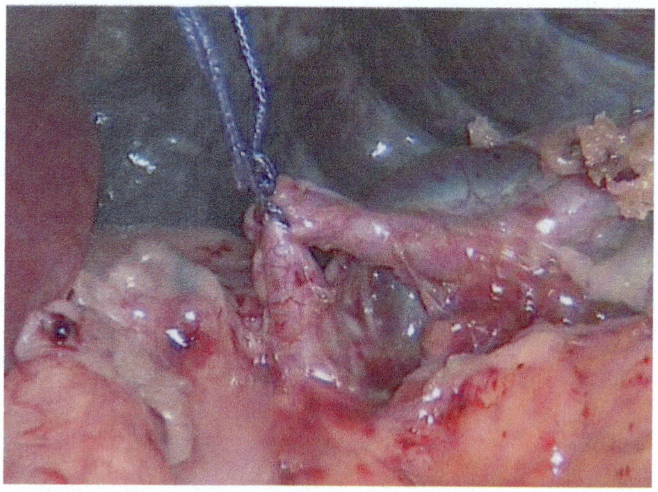

Fig. 25.4 A "two step" approach with early sealing of the main splenic artery is recommended. This allows the spleen to empty and aids retrieval. The splenic artery is sealed using either a ligature of a braided material or 5-mm clips, e.g. Polyester locking clips. The artery can then be divided more distally with an ultrasonic or thermal fusion sealing device. Clipping of the vessels can cause problems if a stapler is subsequently used. If lower pole partial splenectomy is planned, the gastrosplenic vessels should be preserved

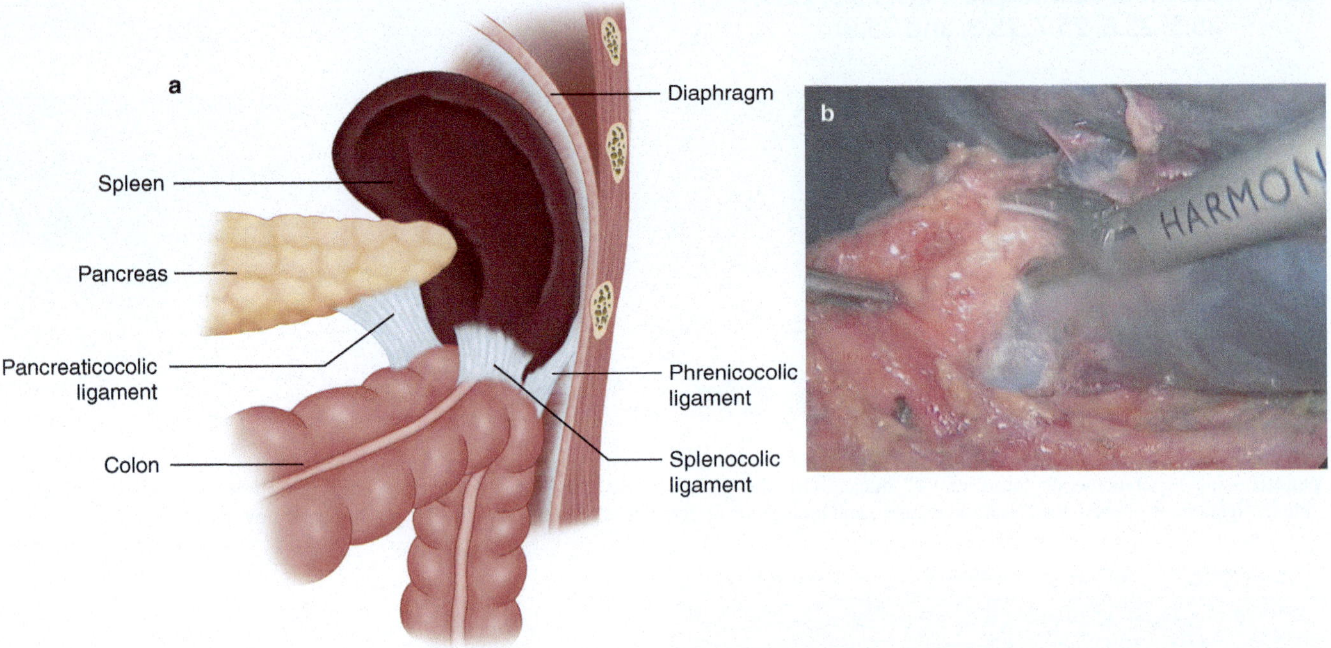

Fig. 25.5 (**a**) Mobilization of the lower pole. The lower pole is freed using the ultrasonic shears to divide the splenocolic, phrenicocolic, and pancreaticocolic ligaments, taking any contained lower pole vessels. (**b**) The pancreaticosplenic ligament is dissected carefully from the hilum to minimize the risk of pancreatitis or pancreatic fistula. This should expose the splenic vein

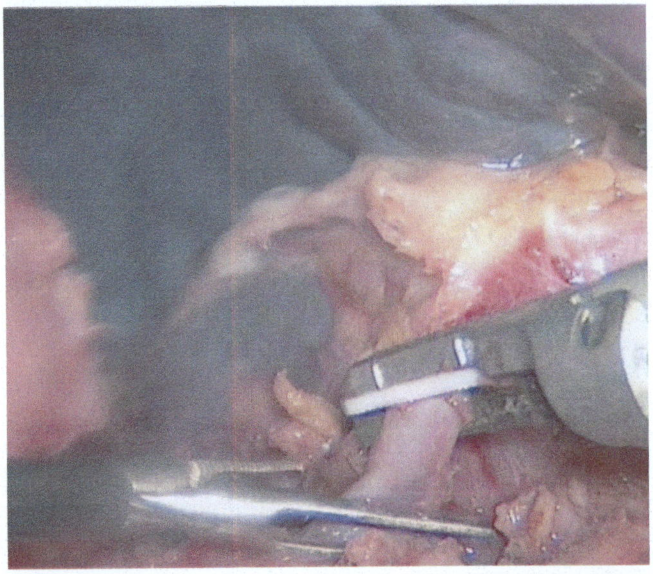

Fig. 25.6 Division of splenic vein. The splenic vein can be sealed and divided again ideally with a two-step approach using either clips or a ligature, then electrothermal fusion, ultrasonic device, or a stapler using a vascular load

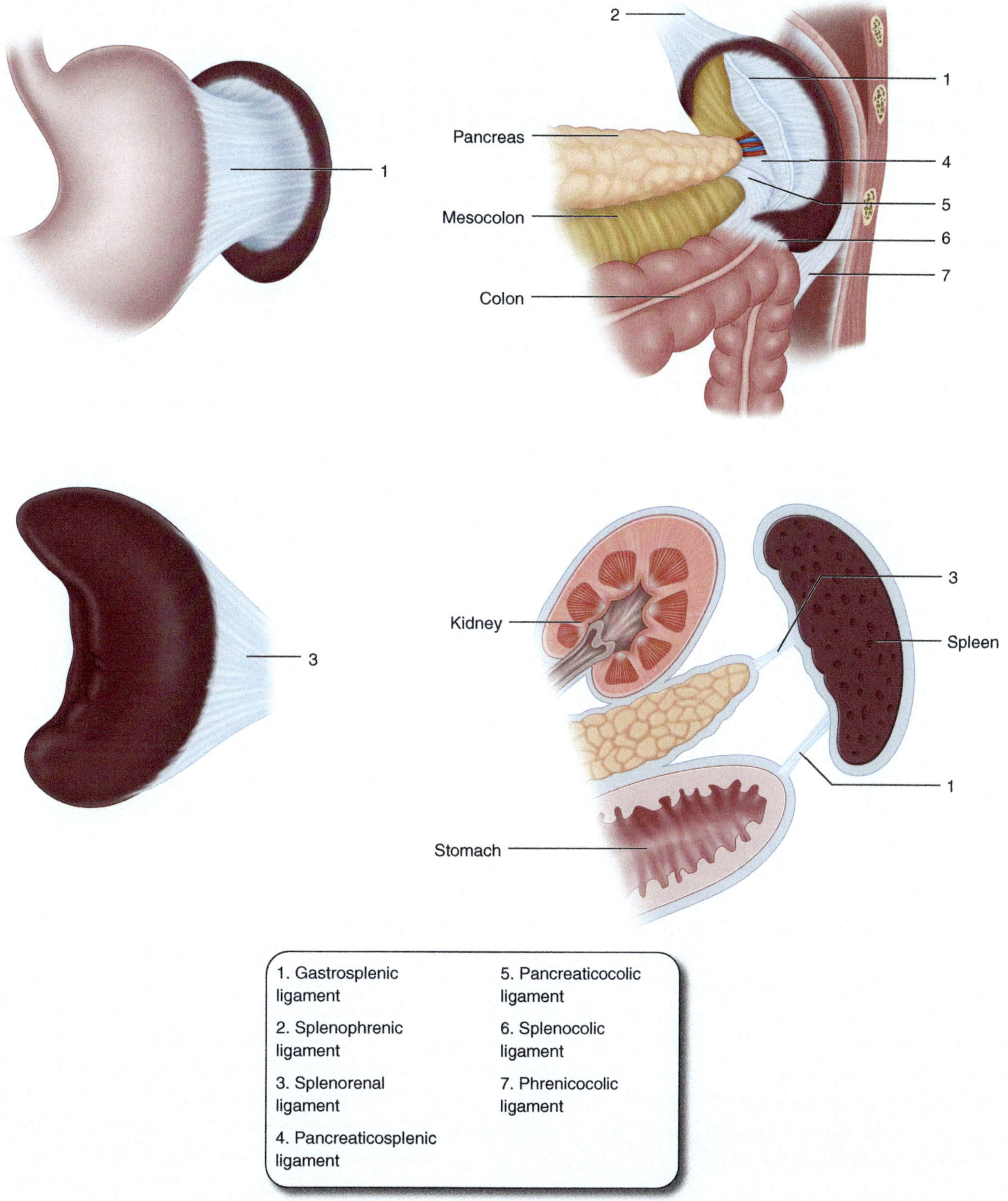

Fig. 25.7 Detachment of the spleen. The spleen is finally freed from the splenorenal and splenophrenic ligaments using ultrasonic shears. The surgeon must be careful at this stage; the tips of the ultrasonic device should not be obscured because this procedure is drawing to an end, and collateral damage may result in traumatic diaphragmatic hernia

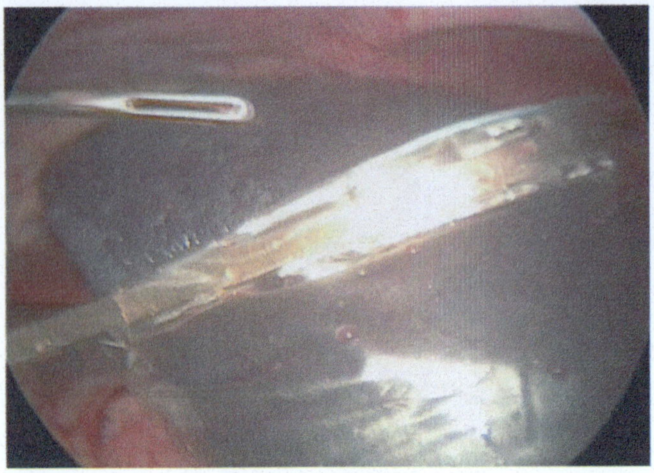

Fig. 25.8 Specimen retrieval and morcellation. The 10-mm 30° scope at the umbilical port is changed to a 5-mm 30° scope at either of the working ports or via the umbilical port alongside the bag for the retrieval. An impermeable ripstop nylon bag is safest to avoid rupture and risk of spillage and splenosis. A "free bag" has advantages over a bag on a stick in a small abdomen, e.g. Ecosac (Espiner Medical). The bag is delivered at the umbilicus, and the spleen is morcellated by using sponge-holding forceps and suction on a guard

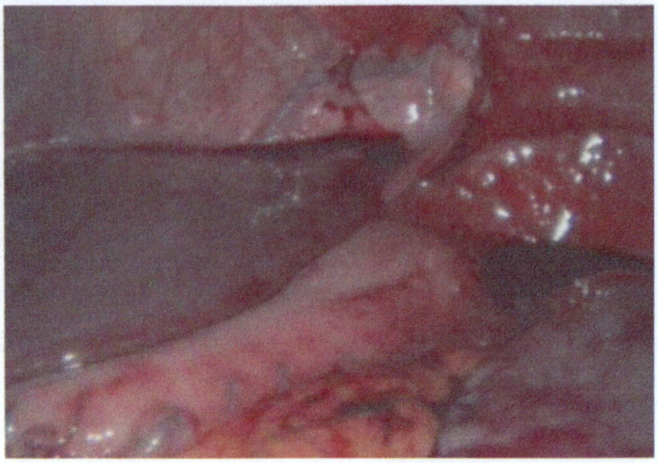

Fig. 25.9 Final inspection of the bed. The bed is revisualized, haemostasis is checked, and a haemostatic matrix can be applied if partial splenectomy is done, e.g. Floseal (Baxter, Newbury, Berkshire, UK). Finally, suction of any residual fluid or blood completes the procedure

25.8 Alternatives

- Single incision splenectomy is a reasonable option for experienced surgeons.
- Three-millimeter instruments are now a realistic choice.
- Preoperative splenic artery embolization is possible.

25.9 Highlights and Pitfalls

- A suprapubic port or pfannensteil incision may be required to deliver the spleen in small patients or those with massive splenomegaly.
- A search for a splenunculus may be fruitful in at least 10 % of cases, but they may be multiple.
- Difficult bleeding may be dealt with by stapling of the splenic hilum.
- A suprapubic port or pfannensteil incision may be required to deliver the spleen in small patients or those with massive splenomegaly.
- Partial splenectomy: complete splenic cyst excision can be achieved by aspiration of the cyst and excision of the majority of the cyst wall using ultrasonic shears. The remaining cyst wall should be removed by either a partial splenectomy using an endoscopic stapler and repeated firings of 45- to 60-mm cartridges and 2.5- to 3-mm staples. Alternatively, stripping of the cyst wall can be achieved by using ultrasonic energy. A hemostatic matrix can be applied to oozing wound edges. Incomplete excision of the cyst wall has a high risk of recurrence.
- A torted wandering spleen is an occasional challenge.

Suggested Reading

Cusick R, Waldhausen JHT. The learning curve associated with pediatric laparoscopic splenectomy. Am J Surg. 2001;181:393–7.

Lansdale N, Marven S, Welch J, Vora A, Sprigg A. Intra-abdominal splenosis following laparoscopic splenectomy causing recurrence in a child with chronic immune thrombocytopenic purpura. J Laparoendosc Adv Surg Tech A. 2007;17:387–90.

Rescorla FJ, West KW, Engum SA, Grosfeld JL. Laparoscopic splenic procedures in children: experience in 231 children. Ann Surg. 2007;246:683–8.

Laparoscopic Surgery for Choledochal Cysts

26

Nguyen Thanh Liem

Abstract

Choledochal cyst is a rare condition in America and Europe. However, it is a common disease in Asian countries, especially Japan, China, and Vietnam. Laparoscopic cystectomy and hepaticojejunostomy for choledochal cyst was first performed by Farello in 1995. Nowadays, laparoscopic surgery is a routine procedure in the management of choledochal cyst in many centers.

Keywords

Choledochal cyst • Laparoscopic • Hepaticoduodenostomy • Hepaticojejunostomy

26.1 General Information

Choledochal cyst is a rare condition in America and Europe. However, it is a common disease in Asian countries, especially Japan, China, and Vietnam. Laparoscopic cystectomy and hepaticojejunostomy for choledochal cyst was first performed by Farello in 1995. Nowadays, laparoscopic surgery is a routine procedure in the management of choledochal cyst in many centers.

26.2 Working Instruments

- 5- and 10-mm Hasson ports
- 30° and 0° telescopes
- 3- and 5-mm needle holders
- 3- and 5-mm scissors
- 3- and 5-mm Maryland forceps

N.T. Liem
Vinmec International Hospital, Hanoi, Vietnam

M. McHoney et al. (eds.), *Color Atlas of Pediatric Anatomy, Laparoscopy, and Thoracoscopy*,
DOI 10.1007/978-3-662-53085-6_26

26.3 Positioning, Port Siting, and Ergonomic Considerations

A nasogastric tube, rectal tube, and urinary Foley catheter are used to decompress the stomach, colon, and bladder, respectively. The patient is placed in a 30° head-up supine position. The surgeon stands at the lower end of the operating table between the patient's legs (Fig. 26.1). A 10-mm trocar is inserted through the umbilicus for the telescope. Three additional 5- or 3-mm trocars are placed for instruments: one at the right flank, one at the left flank, and one in the left hypochondrium (Fig. 26.2). Carbon dioxide pneumoperitoneum is maintained at a pressure of 8–12 mmHg. For small infants, positioning across the table is preferable (Figs. 26.3 and 26.4).

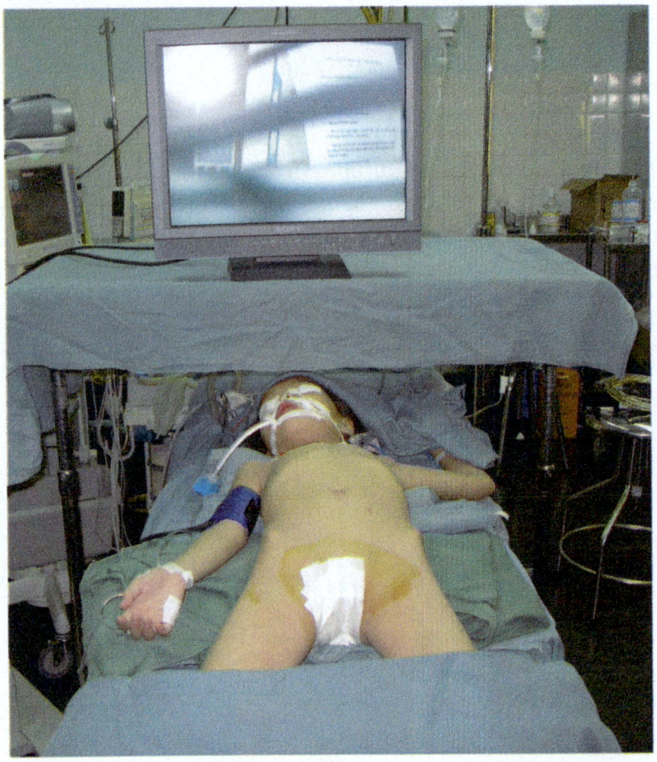

Fig. 26.1 Patient position

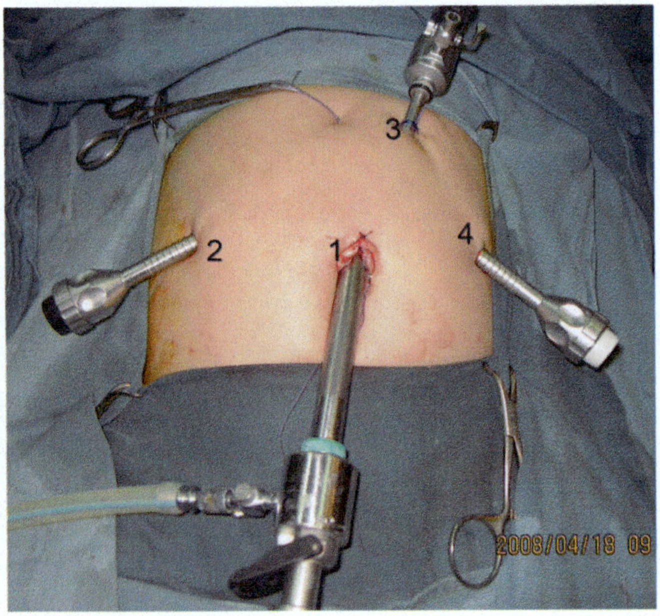

Fig. 26.2 Trocar placement

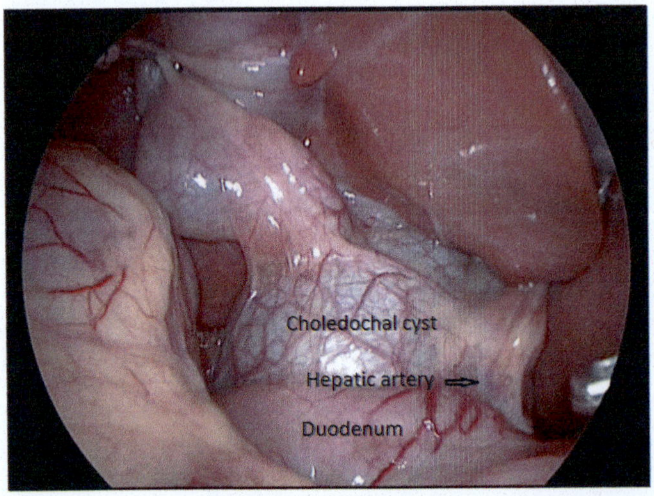

Fig. 26.3 Operative photographs and illustrations of a choledochal cyst

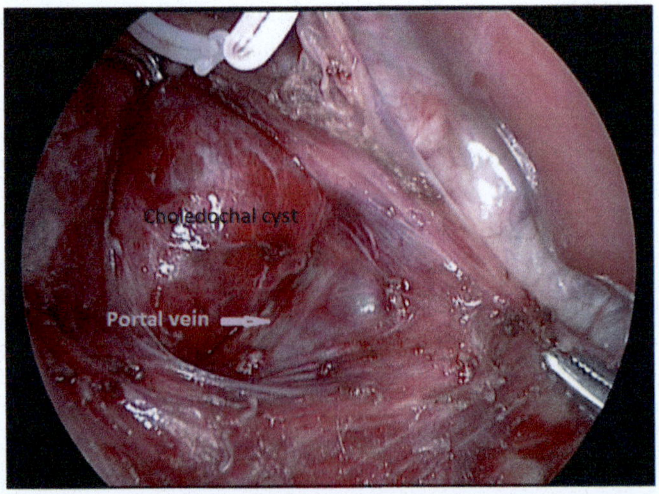

Fig. 26.4 Operative photographs and illustrations of the relationship between a choledochal cyst and the portal vein

26.4 Relevant Anatomy

26.5 Surgical Technique

26.5.1 Cystectomy

The liver is secured to the abdominal wall with a stay suture placed at the round ligament. The cystic artery and duct are identified, clipped, and divided. A second traction suture is placed at the distal cystic duct and the gallbladder fundus to elevate the liver and splay the liver hilum (Fig. 26.5).

The duodenum is retracted downward using a dissector through the fourth trocar site. The midportion of the cyst is dissected circumferentially. The cyst is separated from the hepatic artery and portal vein meticulously until a dissector can be passed through the space between the posterior cystic wall and the portal vein going from left to right (Fig. 26.6). The cyst then is divided at this site.

The lower part of the cyst is detached from the pancreatic tissue down to the common biliopancreatic duct using a 3-mm dissector for cautery and dissection. The distal part of the cyst is removed progressively. Protein plugs or calculi within the cyst and common channel are washed out and removed. The end part of the cyst is opened longitudinally. The inside is inspected to identify the orifice of the common biliopancreatic channel. A small catheter is inserted into the common channel. Irrigation with normal saline via this catheter is performed to eliminate any protein plug until clear fluid comes out and the catheter can be passed down to the duodenum (Fig. 26.7).

A cystoscope may be used to remove protein plugs in the common channel if its diameter permits. The distal choledochal cyst is clipped and divided just above the bifurcation of the choledochus and pancreatic duct (Fig. 26.8).

The upper part of the cyst is dissected further up to the common hepatic duct and then divided therein. The choledochal cyst initially is divided at the level of the cystic duct. After the orifice of the right and left hepatic ducts is identified, a definitive incision is performed, leaving a stump approximately 5 mm from the bifurcation of the hepatic ducts. Irrigation is done with normal saline through a small catheter inserted into the right and then into the left hepatic duct to wash out the protein plugs or calculi until the return fluid from these ducts is clear.

26.5.2 Hepaticojejunostomy

The ligament of Treitz is identified through laparoscopy. A 5/0 silk stay suture is placed 10 cm distal to the ligament of Treitz in newborns, 20 cm in infants, and 30 cm in children. A second 5/0 polydioxanone (PDS) suture is placed 2.0 cm below the first suture to mark the jejunal limb, which will be anastomosed to the hepatic duct. The jejunal segment with two sutures is grasped with an intestinal grasper.

The transumbilical vertical incision is extended 1.0 cm above the umbilicus. The jejunum is exteriorized, and the jejunojejunostomy is carried out extracorporeally. The jejunum then is reintroduced into the abdominal cavity. The extended incision is closed, and the laparoscopic instruments are repositioned.

The Roux limb is passed through a window in the transverse mesocolon to the porta hepatis. The jejunum is opened longitudinally on the antimesenteric border a few millimeters from the end of the Roux. Hepaticojejunostomy is fashioned using two running sutures of 5/0 PDS. (Interrupted sutures are used when the diameter of the common hepatic duct is less than 1.0 cm.) Sutures are placed from left to right with 3-mm instruments.

Mesenteric defects in the transverse mesocolon and the small bowel mesentery are closed with sutures. The gallbladder is detached from its bed and surrounding tissues. Different parts of the cyst and gallbladder are removed through the umbilicus. The operative field is washed with warm saline, and a subhepatic closed suction drain is inserted.

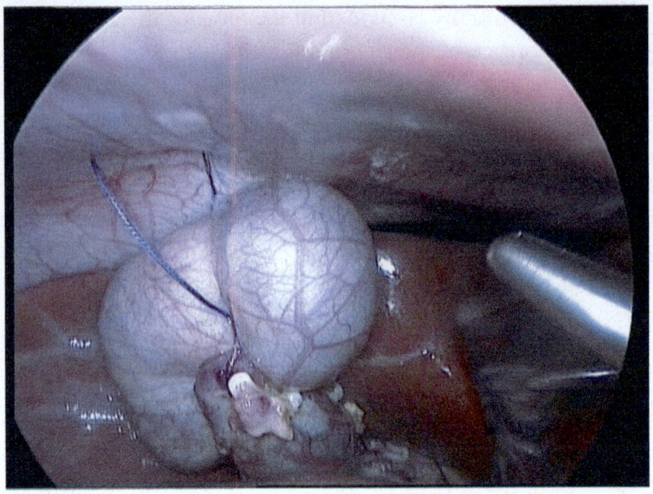

Fig. 26.5 A second traction suture is placed at the gallbladder fundus

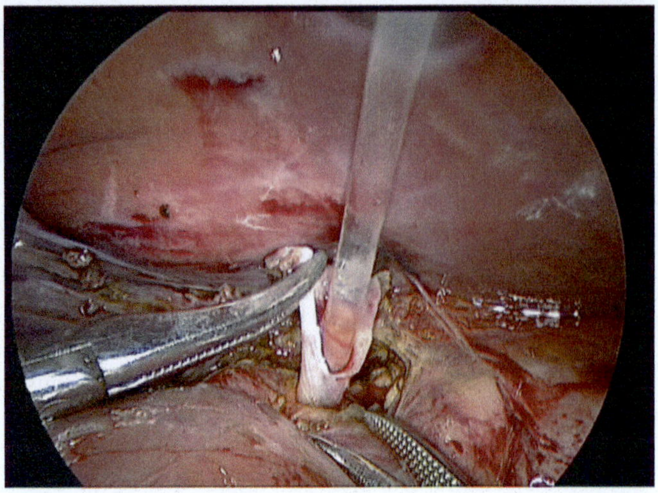

Fig. 26.7 A small catheter is inserted into the common channel

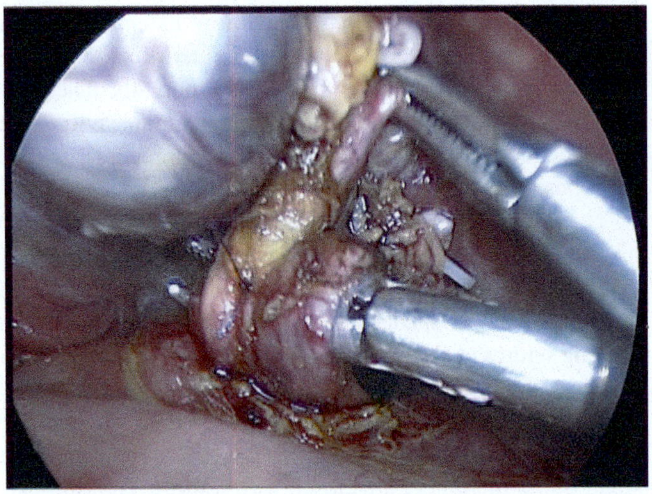

Fig. 26.6 A dissector is passed through the space between the posterior cystic wall and the portal vein

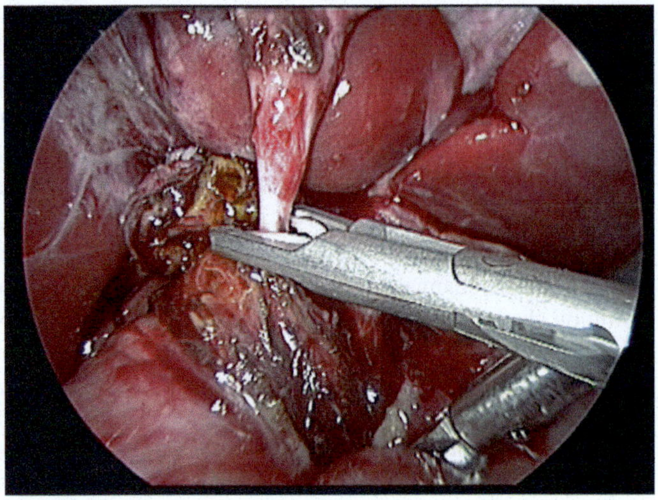

Fig. 26.8 The distal choledochal cyst is clipped just above the bifurcation of the choledochus and pancreatic duct

26.6 Alternatives

26.6.1 Hepaticoduodenostomy

The cyst is excised as described earlier. The duodenum is mobilised as much as possible, and a hepaticoduodenostomy is constructed around 3 cm from the pylorus using two running sutures of 5/0 PDS. Interrupted sutures are used when the diameter of the common hepatic duct is less than 1.0 cm. The rest of the operation is performed the same way using the previous technique.

26.6.2 Single-Port Laparoscopic Cystectomy and Hepaticoduodenostomy

A semi-circumferential infraumbilical incision is made in the skin. A 5-mm trocar is inserted through the middle line under the umbilicus. A 3-mm trocar is inserted in the left side and a 5-mm trocar in the right side.

The liver is secured to the abdominal wall by two stay sutures as mentioned earlier. Another stay suture is placed between the abdominal wall and the anterior wall of the cyst. The cyst is dissected and divided. After the cyst is divided in the middle, one more stay suture is placed to hang the distal part of the cyst to the abdominal wall. The rest of the operation is carried out as in the surgery with three incisions.

26.7 Highlights and Pitfalls

- With a large cyst, the dissection is started at the middle portion, proceeding distally. The distal portion of the choledochal cyst is separated from the portal vein. The cyst is divided above the confluence of the biliopancreatic duct. The cystic stump is managed as described earlier.
- If the cyst is intensely inflamed with extensive pericystic adhesions, it is opened by a transverse incision on the anterior wall. The cystic wall then is dissected from the portal vein carefully while the cyst is viewed from within and without. After the midportion of the cyst is divided, the upper and lower parts of the cyst are removed as described earlier.
- The cyst should be removed at the level of the common biliopancreatic channel orifice at the distal end and approximately 5 mm from the confluence of the right and left hepatic ducts at the proximal end. Postoperative malignancy of residual cyst from the hepatic duct or from the distal part has been reported. Abdominal pain and pancreatitis due to leaving a remnant of the cyst in the pancreatic head was also mentioned.
- Ductoplasty is performed by opening the common hepatic duct and incising the left hepatic duct longitudinally for a variable distance if the diameter of the common hepatic duct is too small.
- The following intraoperative complications may occur:
  - Portal vein injury: This complication may be prevented by keeping the dissection as close as possible to the cystic wall. If severe pericystic inflammation and adhesion are present, the cyst should be opened from the anterior wall; then, the left and posterior wall of the cyst should be separated carefully from the portal vein while being viewed internally and externally.
  - Transection of two hepatic ducts: This situation may happen when the hepatic bifurcation is situated low, far from the liver hilum. This complication may be avoided by identifying the orifice of the right and left hepatic ducts before definitive division of the cyst from the hepatic duct.
  - Pancreatic duct injury: Clear anatomy of the common biliopancreatic channel obtained by magnetic resonance cholangiopancreatography, endoscopic retrograde cholangiopancreatography, or perioperative cholangiography may help the surgeon determine where the distal part of the cyst should be divided.

Suggested Reading

Farello GA, Cerofolini A, Rebonato M, Bergamaschi G, Ferrari C, Chiappetta A. Congenital choledochal cyst: video-guided laparoscopic treatment. Surg Laparosc Endosc. 1995;5:354–8.

Liem NT, Hien PD, Hoan VM. Is the laparoscopic operation as safe as open operation for choledochal cyst in children? J Laparoendosc Adv Surg Tech A. 2011;21(4):367–70.

Liem NT, Pham HD, le Dung A, Son TN, Vu HM. Early and intermediate outcomes of laparoscopic surgery for choledochal cysts with 400 patients. J Laparoendosc Adv Surg Tech A. 2012;22:599–603.

Tanaka M, Shimizu S, Mizumoto K, Yokohata K, Chijiiwa K, Yamaguchi K, et al. Laparoscopically assisted resection of choledochal cyst and Roux-en-Y reconstruction. Surg Endosc. 2001;15:545–52.

Laparoscopic Colectomy

27

Maurizio Pacilli and Hugh W. Grant

Abstract

This chapter describes subtotal and total colectomy for inflammatory bowel disease. Partial colectomy (pull-through) for Hirschsprung's disease is described in Chap. 30. Laparoscopy for colectomy has equivalent results to open colectomy [1]; however, it takes significantly longer to perform. It provides good access and results in better cosmesis and a lower incidence of small bowel obstruction due to adhesions [2].

Keywords

Laparoscopic • Colectomy • Subtotal colectomy

27.1 General Information

This chapter describes subtotal and total colectomy for inflammatory bowel disease. Partial colectomy (pull-through) for Hirschsprung's disease is described in Chap. 30. Laparoscopy for colectomy has equivalent results to open colectomy [1]; however, it takes significantly longer to perform. It provides good access and results in better cosmesis and a lower incidence of small bowel obstruction due to adhesions [2].

M. Pacilli
Department of Paediatric Surgery, Oxford University Hospital, Oxford, UK

H.W. Grant (✉)
Department of Paediatric Surgery, John Radcliffe Hospital, Oxford University Hospitals, Oxford, UK

27.2 Working Instruments

- For children <7 kg:
  - 5-mm 30° scope
  - 3-mm instruments
  - 5-mm ports (in case bigger instruments are necessary) during dissection if using energy devices
  - 12-mm port is necessary for the stapler device (to transect the rectosigmoid junction in subtotal colectomy) and is best placed in the right iliac fossa
- For children >7 kg
  - 5-mm 30° scope
  - 5-mm instruments
  - 12-mm port for the stapling device in the right iliac fossa
  - Energy devices (e.g., bipolar or ultrasonic dissectors) are helpful for dissection and to minimise blood loss

27.3 Positioning, Port Siting, and Ergonomic Considerations

27.3.1 Patient Position

Patients <7 kg should be placed in the supine position on the table with the legs abducted to ensure access to the anus. Children >7 kg should be placed in the modified Lloyd-Davies

M. McHoney et al. (eds.), *Color Atlas of Pediatric Anatomy, Laparoscopy, and Thoracoscopy*,
DOI 10.1007/978-3-662-53085-6_27

position (with access to the anus). Start with the operating table horizontal, but move the table during the operation to maximise visibility: head up for dissection of the transverse colon, head down for rectal dissection, left side up during dissection of descending colon, and right side up for dissection of the caecum and ascending colon.

27.3.2 Port Sites

See Fig. 27.1 for the locations of the port sites. The difference between total colectomy and other laparoscopic procedures is that the operative field keeps changing; the ergonomics may be perfect during some stages of the procedure and dire in other parts, so placing the ports is crucial if the ergonomics are going to be optimal for *most* of the procedure. It is not possible for the ergonomics or port sites to be perfect for the whole procedure. In general, keep the port sites as lateral and peripheral as possible to maximise the working space. The umbilicus is seldom a useful site for a port.

A local anaesthetic (bupivacaine) should be applied to the port sites before the incision and more injected at the end of the procedure, if indicated.

Start by placing the right upper quadrant port site as high and lateral as possible without going above the liver edge. Use a cut-down technique (modified Hasson entry). Place the right lower quadrant port between the anterior superior iliac spine and the umbilicus (slightly lateral to McBurney's point) so that this port site may be extended (medially) to bring out the ileostomy if required. Place the left upper quadrant port as high and lateral as possible, just under the costal margin, and place the left lower quadrant port as low and lateral as possible (just above the anterior superior iliac spine). Place all additional ports under laparoscopic vision to avoid complications.

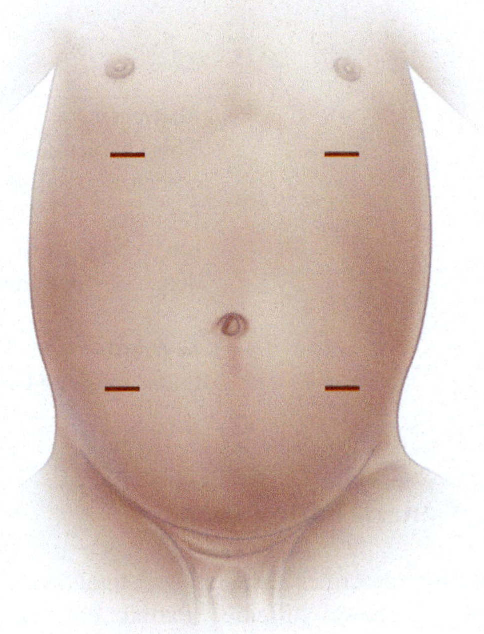

Fig. 27.1 Position for port sites

27.3.3 Insufflation Pressures

For children <7 kg, start with low flow and 7 mmHg CO_2 insufflation pressure. This may be increased as required up to a maximum of 12 mmHg. For older children, e.g. adolescents, start with low flow and 10 mmHg, which may be increased to 15 mmHg. If greater pressures are needed, the surgeon should reassess the situation——i.e., optimise the patient's position, determine the need for more muscle relaxant, and check for any gas leak.

27.3.4 Surgeon's Position

The surgeon's position changes continually. Remember that the surgeon should move to the position that optimises the ergonomics for the part of the dissection he or she is performing.

27.4 Relative Anatomy and Surgical Technique

27.4.1 Sigmoid Colon

It is helpful to start with dissection of the sigmoid colon, as it is the easiest part of the operation and gets the surgeon off to a good start and into the correct tissue planes. For this part of the dissection, the surgeon should stand on the patient's right looking towards the patient's left iliac fossa (Fig. 27.2). The camera is inserted via the right upper quadrant port, and the left upper quadrant and right lower quadrant ports are used as working ports. The assistant places a laparoscopic Babcock forceps through the left lower quadrant port and lifts the sigmoid (Fig. 27.3). Dissection starts by making a window in the middle of the sigmoid mesocolon (quite close to the vascular arcade), which is best done using a hook diathermy. Once the hole has been made in the sigmoid mesocolon, monopolar hook diathermy, bipolar diathermy, or an ultrasonic dissector may be used for the main dissection (Fig. 27.4). The surgeon moves retrogradely up the sigmoid to the proximal sigmoid colon and distally along the sigmoid as far as the pelvic brim, as indicated. Dissection stops at the pelvic brim if a subtotal colectomy is being performed.

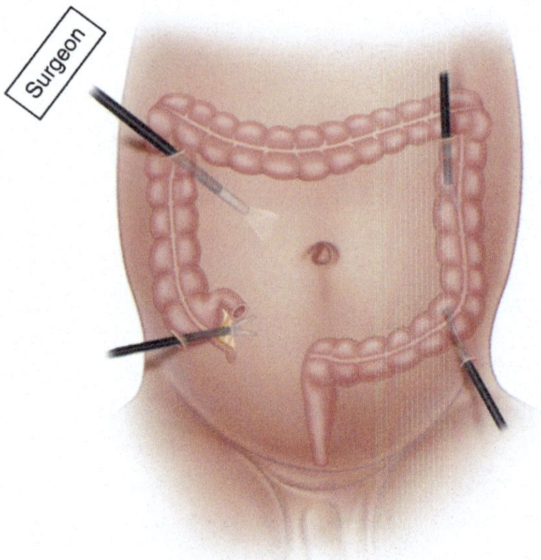

Fig. 27.2 Surgeon and port positions for dissection of the sigmoid colon

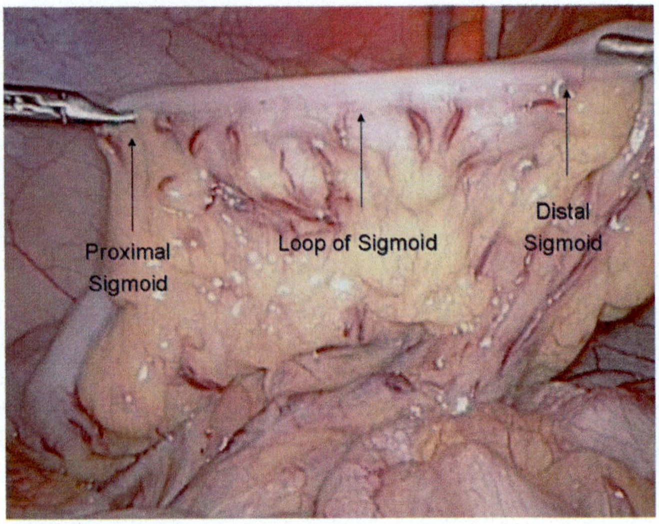

Fig. 27.3 Start of dissection of the sigmoid colon

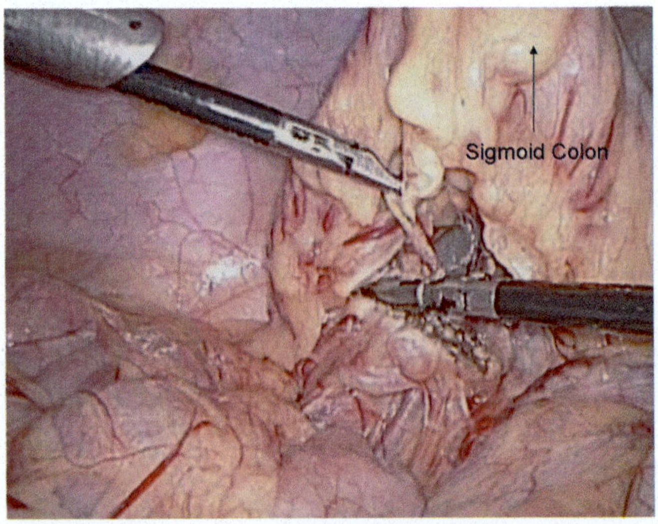

Fig. 27.4 Dissection of the sigmoid colon

27.4.2 Sigmoid Colon and Lower Descending Colon

The surgeon stays in the same position. The peritoneal reflection of the sigmoid and descending colon should then be mobilised retrogradely up the left paracolic gutter. As the dissection proceeds towards the splenic flexure, the surgeon may choose to place the camera through the right lower quadrant incision and use the right upper quadrant and left lower quadrant ports as working ports (Fig. 27.5). The dissection then continues up the descending colon as far as practicable.

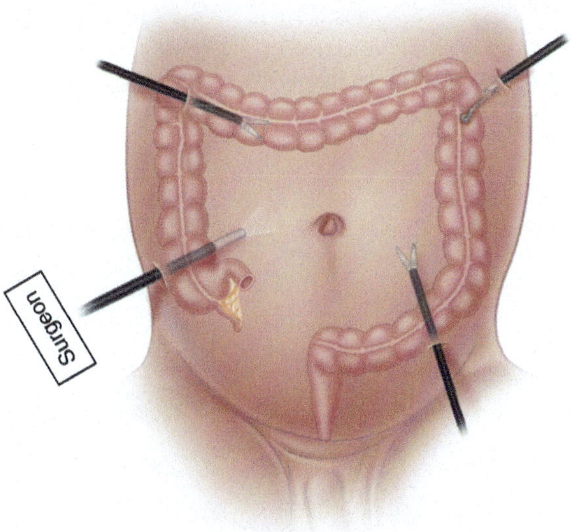

Fig. 27.5 Surgeon and port positions for dissection of the sigmoid colon and lower descending colon

27.4.3 Descending Colon as Far as the Splenic Flexure

The surgeon moves to the right lower quadrant position. It is very important to ensure that the peritoneal reflection of the descending colon is fully divided (Fig. 27.6) to lift the splenic flexure up and away from the tail of the pancreas and the spleen.

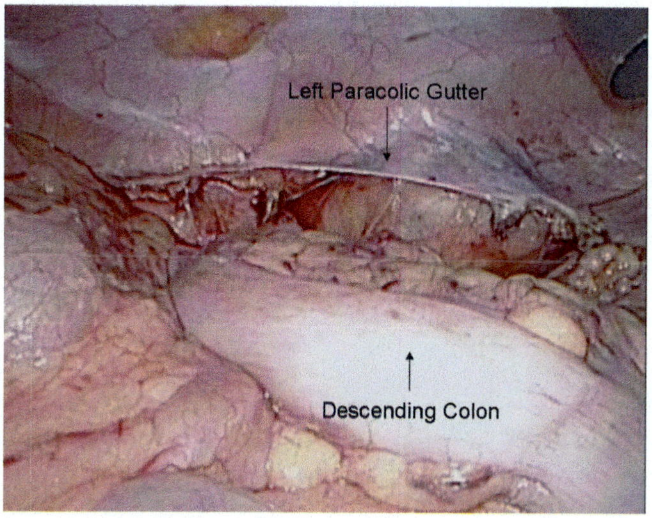

Fig. 27.6 Dissection of the descending colon

27.4.4 Distal Transverse Colon and Splenic Flexure

The surgeon moves the camera to the right lower quadrant port; the working ports should be the right upper quadrant and left lower quadrant ports (Fig. 27.5). The assistant may help by retracting the splenic flexure via the left upper quadrant if necessary (Fig. 27.7). The bloodless plane of Treves is identified; it usually is easier to dissect the greater omentum off the transverse colon. If this is not possible, the greater omentum may be included (but great care must be taken to avoid damage to the stomach). It is best to continue this dissection to the mid-transverse colon (at the level of the falciform ligament).

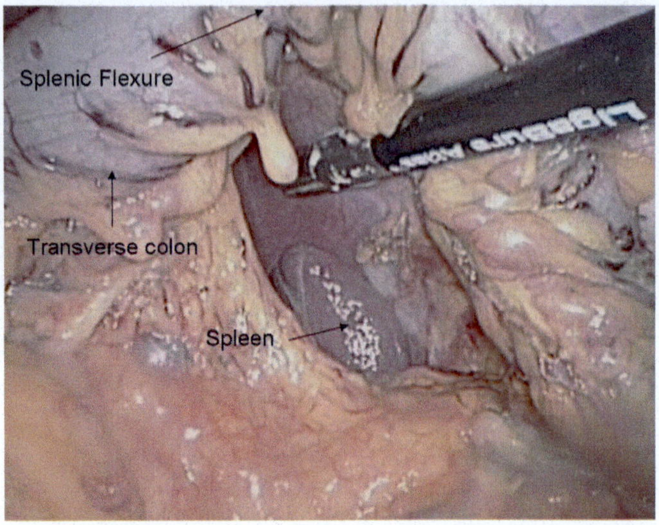

Fig. 27.7 Dissection of the splenic flexure and distal transverse colon

27.4.5 Terminal Ileum, Appendix, and Ascending Colon

The surgeon changes sides and stands on the patient's left. The camera is placed through the left upper quadrant incision, and the working ports are the right upper quadrant and left lower quadrant ports (Fig. 27.8). It is helpful if the assistant lifts the caecum with a Babcock forceps via the right lower quadrant incision. The surgeon begins with the appendicectomy, grabbing the appendix, dividing the appendix mesocolon and lifting it (Fig. 27.9), and making a window in the vascular arcades of the terminal ileum 5 cm proximal to the ileocaecal valve. A bipolar or monopolar hook or an ultrasonic dissector is used to mobilise the caecum and the peritoneal reflection of the right paracolic gutter. The peritoneal reflection is dissected as high as the hepatic flexure and mobilised around the antimesenteric border, allowing the colon to be lifted away from the duodenum (Fig. 27.10). It is very important to visualise the duodenum before dissecting the hepatic flexure; the mesocolon of the right colon should be lifted off the duodenum before the right colic vessels are ligated. The surgeon must stay close to the bowel at all times.

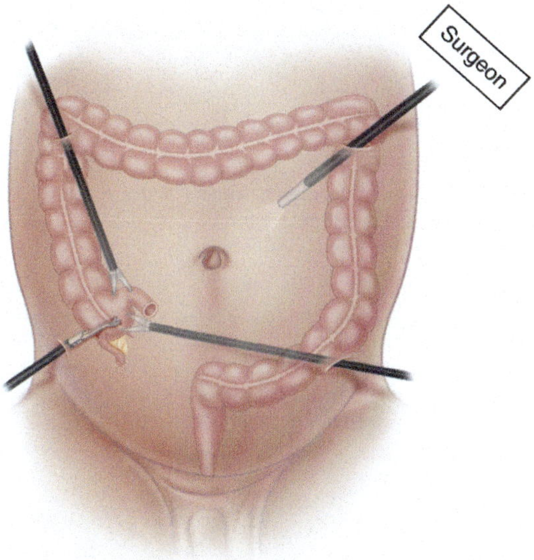

Fig. 27.8 Surgeon and port positions for dissection of the terminal ileum and ascending colon

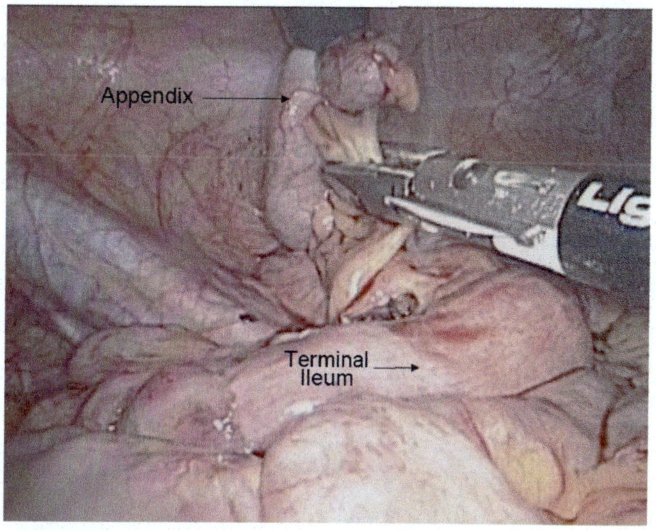

Fig. 27.9 Appendicectomy and mobilisation of the caecum

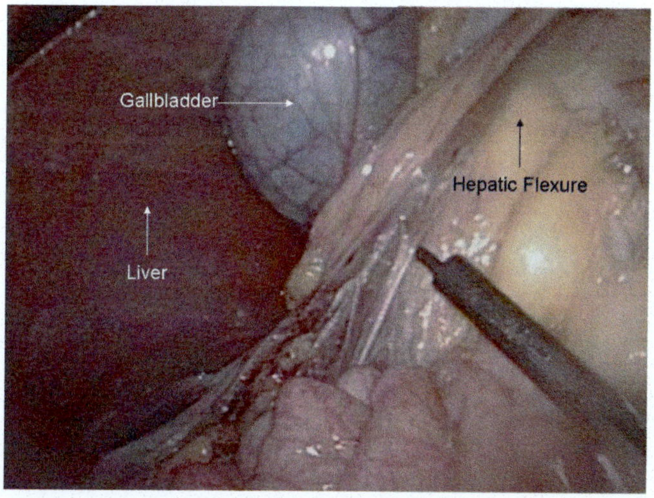

Fig. 27.10 Dissection of the hepatic flexure

27.4.6 Hepatic Flexure and Proximal Transverse Colon

The surgeon stays on the patient's left and places the camera through the left lower quadrant incision (Fig. 27.11). The working ports should be the right lower quadrant and left upper quadrant ports. It is helpful if the assistant lifts the hepatic flexure with a Babcock forceps via the right upper quadrant incision. Dissection of the hepatic flexure and transverse colon continues until it meets the previous dissection in the middle of the transverse colon.

Once the terminal ileum, caecum, ascending colon, transverse colon, descending colon, and sigmoid colon are mobilised, the surgeon may decide whether to stop at the subtotal colectomy stage (e.g., in sick child with ulcerative colitis for whom an ileostomy is planned) or to continue and remove the rectum.

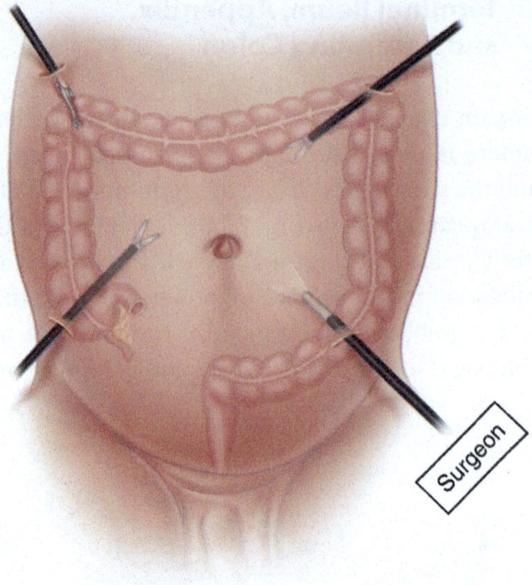

Fig. 27.11 Surgeon and port positions for dissection of the hepatic flexure and proximal transverse colon

27.4.7 Removal of the Rectum

The surgeon stands at the patient's right with the camera in the right upper quadrant port, the working ports are the left upper quadrant and right lower quadrant ports (Fig. 27.2). At this point, it is helpful to position the patient's head down about 20–30°, which keeps the small bowel out of the way. The assistant may help by retracting the bladder and lifting the rectum with a Babcock forceps via the left lower quadrant port. Once the sigmoid is mobilised, the surgeon visualises both ureters and the major vessels entering the pelvis. Dissection should proceed very close to the rectum in the perimuscular plane (Fig. 27.12). Before descending into the pelvis, it is easier for the surgeon to incise the peritoneum over the distal sigmoid mesocolon and then dissect down to the rectum, staying close to the wall of the rectum at all times. This is important as it lifts the rectum anteriorly and allows the rectum to be lifted off adjacent structures and it minimises collateral damage to the hypogastric nerves.

If a primary ileoanal anastomosis is planned, the mucosa (in patients with ulcerative colitis or familial adenomatous polyposis) may be resected from below (according to the Soave procedure), the ileum pulled through the rectal cuff, and the anastomosis fashioned from below.

If a primary restorative proctocolectomy is planned, it is easier for the surgeon to make the J-pouch extracorporeally. The ileum is pulled through the right iliac fossa port (see Sect. 27.4.8). The J-pouch is fashioned with a 12-mm linear stapler, then placed back into the peritoneal cavity, pulled through the short rectal cuff from below using Robert's forceps, and sutured into position from below.

If a subtotal colectomy and (temporary defunctioning) ileostomy are planned, the upper rectum is transected with a 12-mm linear stapler (Fig. 27.13).

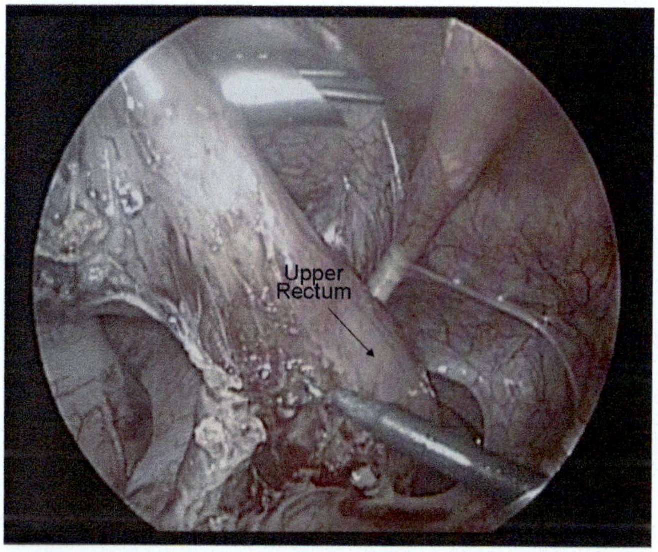

Fig. 27.12 Dissection of the upper rectum

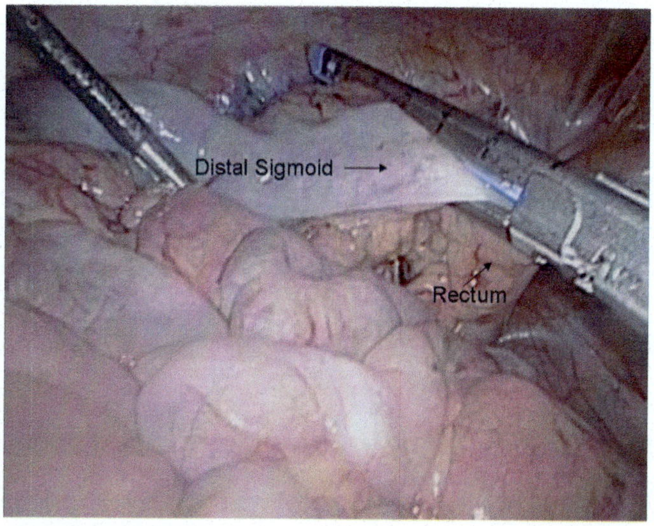

Fig. 27.13 Transection of the upper rectum (in subtotal colectomy) with a linear stapler

27.4.8 Removal of the Specimen Through the Right Lower Quadrant Port Site (Formation of Ileostomy)

In the case of a subtotal colectomy, the surgeon returns to the patient's left side (Fig. 27.8). The assistant grabs the distal end of the sigmoid colon (above the staple line) and pulls the sigmoid through the right lower quadrant port site (Fig. 27.14). This manoeuvre should be done under laparoscopic vision to make sure there is no twisting; it is very easy to twist the ileostomy, which leads to postoperative problems. Once the terminal ileum has been pulled through, it may be transected (extracorporeally) and sutured to make a Brooke's ileostomy.

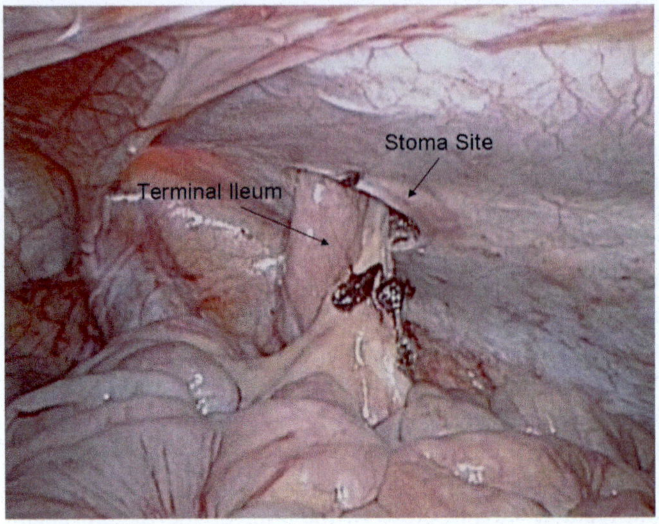

Fig. 27.14 Removal of the colon through the right iliac fossa port site

27.5 Highlights and Pitfalls

- Remember that the operative field changes as the surgeon moves around the colon.
- Maximise port sites for triangulation for as much of the dissection as possible.
- Good triangulation is not possible for the entire procedure.
- Use the patient's position to help with the dissection: e.g., head up when dealing with the transverse colon or right side up and head down for the ascending colon.
- Remember to be flexible and take a break.

References

1. Proctor ML, Langer JC, Gerstle JT, Kim PC. Is laparoscopic subtotal colectomy better than open subtotal colectomy in children? J Pediatr Surg. 2002;37:706–8.
2. Angenete E, Jacobsson A, Gellerstedt M, Haglind E. Effect of laparoscopy on the risk of small bowel obstruction. Arch Surg. 2012;147:359–64.

28 Laparoscopic Duhamels Pullthrough for Hirschprung Disease

Merrill McHoney

Abstract

Laparoscopic Duhamel pull-through is a surgical option in typical Hirschsprung disease (HD). Primary laparoscopic Duhamel pull-through is possible in the newborn who responds to washouts. Effective and ongoing washouts are imperative in management preoperatively and in preparation for surgery. Limited bowel prep (twice daily washouts and low-residue clear fluid diet for 24 h) is sufficient bowel preparation in most cases. The presence of a stoma does not preclude a laparoscopic approach but does make it more technically difficult and the advantages are fewer (e.g., previous laparotomy scar or adhesions). If sited advantageously, the stoma can be brought down for the pull-through. Otherwise a separate stoma closure is also required. Intraoperative frozen section is required to determine the correct level of ganglionic bowel; the appearance of a visible transition zone is not always seen, nor is accurate indication of the level of ganglionic bowel. Antibiotics are given at induction.

Keywords

Hirschprung's disease • Laparoscopy • Duhamel pull-through • Aganglionosis

28.1 General Information

Laparoscopic Duhamel pull-through is a surgical option in typical Hirschsprung disease (HD). Primary laparoscopic Duhamel pull-through is possible in the newborn who responds to washouts. Effective and ongoing washouts are imperative in management preoperatively and in preparation for surgery. Limited bowel prep (twice daily washouts and low-residue clear fluid diet for 24 h) is sufficient bowel preparation in most cases. The presence of a stoma does not preclude a laparoscopic approach but does make it more technically difficult and the advantages are fewer (e.g., previous laparotomy scar or adhesions). If sited advantageously, the stoma can be brought down for the pull-through. Otherwise a separate stoma closure is also required. Intraoperative frozen section is required to determine the correct level of ganglionic bowel; the appearance of a visible transition zone is not always seen, nor is accurate indication of the level of ganglionic bowel. Antibiotics are given at induction.

M. McHoney, MD, PhD
Department of Paediatric Surgery,
Royal Hospital for Sick Children,
9 Sciennes Road, Edinburgh EH9 1LF, UK

M. McHoney et al. (eds.), *Color Atlas of Pediatric Anatomy, Laparoscopy, and Thoracoscopy*,
DOI 10.1007/978-3-662-53085-6_28

28.2 Working Instruments

- Ports of 5-, 10-, or 12-mm (depending on the size of the endoscopic stapler; a Step-port (Medtronic; Minneapolis, MN, USA) can be used to incrementally increase the size as required).
- 3-mm or 5-mm Port × 2
- 30° telescope
- Hook diathermy
- Needle holders
- Dissection scissors
- Johan (or other fenestrated grasping) and Maryland forceps
- Suction and irrigation are helpful
- Endoscopic articulating stapler (for abdominal anastomosis)
- Endoscopic linear stapler (for rectal anastomosis) 45 mm
- Harmonic ultrasound dissector or LigaSure bipolar (Medtronic-Covidien, Minneapolis, MN, USA) (alternative energy sources)

28.3 Positioning, Port Siting, and Ergonomic Considerations

The infant is placed in a supine position and should be put at the foot end of the operating table. The legs should be prepped and draped to allow them to be raised and positioned for the transanal dissection. Urethral catheterization is required.

A 5-mm supra- or subumbilical incision is used for a primary port. Pneumoperitoneum is established to 8 mmHg pressure; this can be increased to 10 mmHg if required and tolerated. A small degree of head-down tilt increases the exposure of the pelvis and rectum.

Two further working instruments are placed on the right and left sides of the abdomen to allow triangulation. The right-sided port may need to be either a 5- or 10-mm size port, depending on the diameter of the reticulating stapler used.

28.4 Relevant Anatomy (Figs. 28.1 and 28.2)

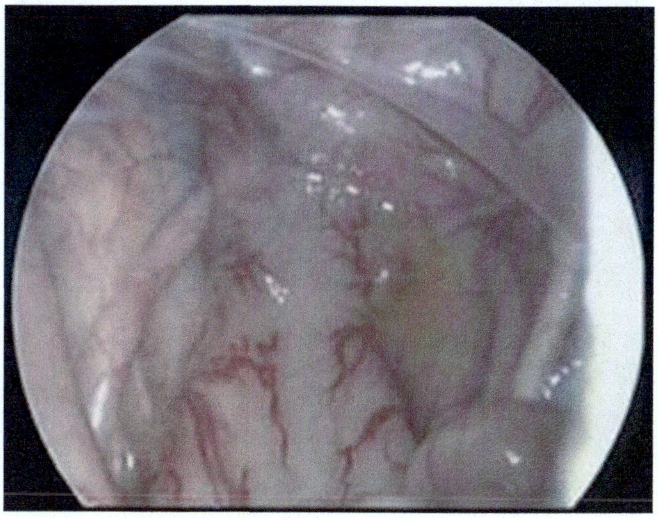

Fig. 28.1 Anatomy of the pelvis. The anatomy of the male pelvis showing the rectum and bladder. The ureters, approaching the base of the bladder, are seen

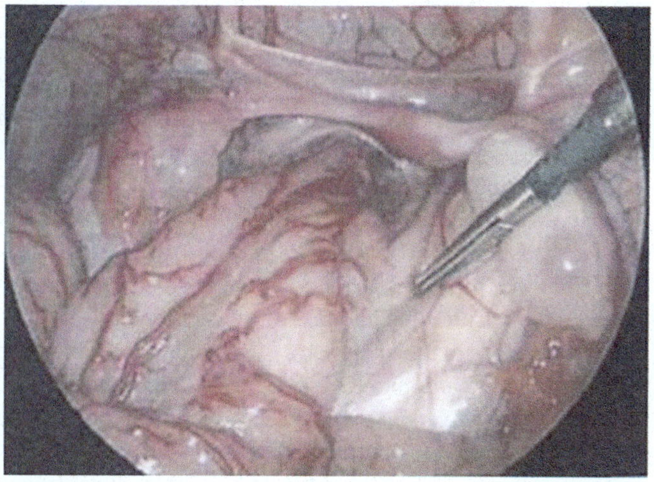

Fig. 28.2 The initial view in a female pelvis. The aganglionic contracted rectum is obvious in this patient. The conical shaped transition zone is also seen leading up to the sigmoid colon

28.5 Surgical Technique

28.5.1 Seromuscular Biopsies

Unless the level of aganglionosis is previously determined, biopsies are taken (Fig. 28.3). Normally, the biopsies should be taken from the upper rectum, the apex of the sigmoid colon, and at the junction with the descending colon. The sites of additional biopsies can be guided by imaging in combination with intraoperative interpretation (accepting that the classic transition zone may not be present owing to elimination from the ongoing washouts). The taenia coli provide an excellent site from which to take the seromuscular biopsies. The taenia close to the biopsy site is grasped with a Maryland forceps (taking care not to crush the tissue) and scissors are used to incise into the submucosal plane. The scissors are then used to excise an approximately 1-cm long strip of seromuscular tissue, which is subjected to immediate frozen section.

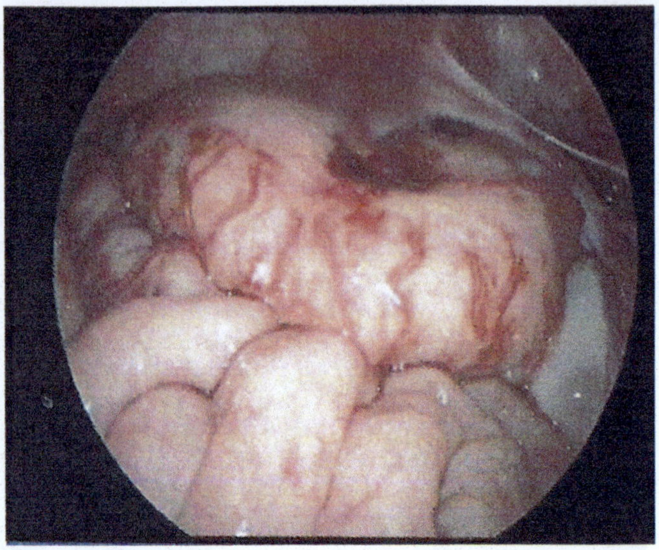

Fig. 28.3 Seromuscular biopsy

28.5.2 Closing the Biopsy Site

The sites of the biopsies are closed either by an interrupted or continuous intracorporeal suture using a 4-0 vicryl or other absorbable suture. The ends of the suture are left slightly long to enable identification of the biopsy sites when the bowel is pulled through. Watertight closure should be ensured in case the upper biopsy sites are left in situ after the pull-through. The rest of the operation requires the results of frozen section to confirm the level of ganglionic bowel before proceeding and is therefore delayed until these are available.

28.5.3 Retrorectal Dissection

The retrorectal space is developed by blunt dissection of the space between the rectum and sacrum by anteriorly displacing the bowel to identify the space below the peritoneal reflection. The combination of spreading with dissecting forceps and occasional hook diathermy can be used. This dissection is maintained close to the posterior bowel wall, isolating and dividing the feeding vessels but preserving the inferior rectal vessels and carried down to the pelvic floor. Anteriorly the peritoneal reflection is divided from the rectum to allow mobilization. The dissection is minimized to prevent nerve injury.

28.5.4 Mobilization of the Rectosigmoid Colon

A small window is created in the mesentery at the junction of the rectum and sigmoid colon (Fig. 28.4). This can be initiated with hook diathermy in an appropriate bloodless window close to the bowel. After this window is created, the rectosigmoid is mobilized by division of the vessels with monopolar, bipolar (e.g., LigaSure) or the harmonic scalpel. The dissection carries on to the level of ganglionic bowel, while keeping appropriate vascular anatomy to maintain perfusion to this level.

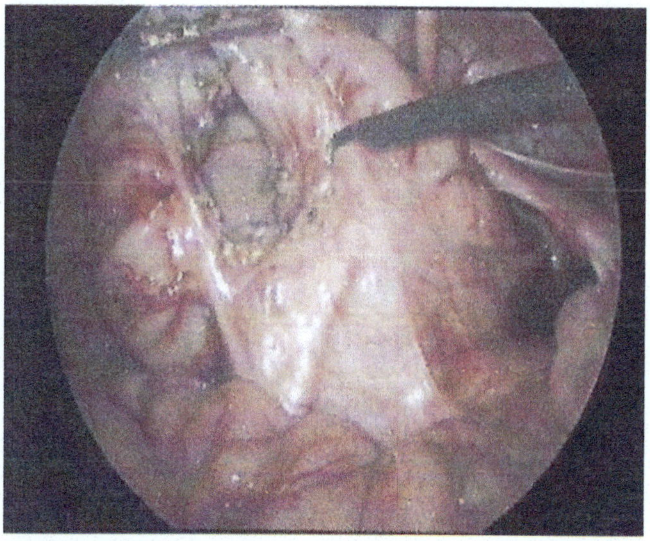

Fig. 28.4 Mobilization of the rectosigmoid colon

28.5.5 Division of the Colon at the Rectosigmoid Junction

The rectum is elevated out of the lower pelvis, and the upper rectum is then divided at the peritoneal reflection using the reticulating stapler (Fig. 28.5). The stapler is placed perpendicular to the bowel and incorporates the entire width if possible. More than one firing of the stapler can be used if needed. The staple line is inspected to ensure that the entire length is secure and water-tight.

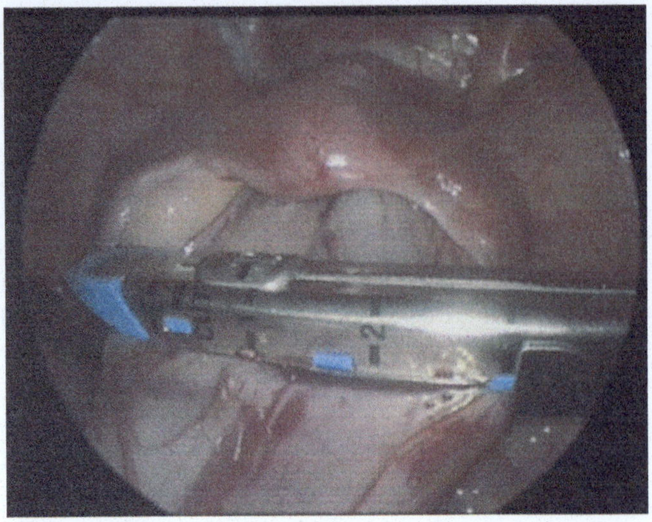

Fig. 28.5 Division of the colon at the rectosigmoid junction

28.5.6 Posterior Incision in the Anal Canal

Attention is then turned to the anal canal from the perineum. An incision is made in the posterior rectal wall 5 mm above the dentate line.

28.5.7 Retrieval of the Rectum and Colon Through the Anal Canal and Excision of the Aganglionic Bowel and Coloanal Anastomosis

Ensuring that the mesentery is posterior and the bowel is not twisted, a grasper is passed into the pelvis from the rectum to grasp and pull down the rectum. Excision of aganglionic bowel just proximal to the level of the ganglionic biopsy is performed. Bleeding from the bowel confirms good vascularity.

After pulling through the colon, the anterior wall is incised and anastomosed to the posterior wall of the native rectum to the corners using interrupted absorbable sutures. Finally the posterior wall of the pull-through colon is incised to complete excision of the aganglionic bowel and anastomosed to the posterior wall of the rectum.

28.5.8 Stapling of the Common Wall of the Rectum and Colon

A linear bowel stapler is then inserted into the anal canal with one limb on the posterior wall of the native rectum and the other in the anterior of the pulled-through colon. After ensuring appropriate deployment from below, the laparoscope is reinserted to inspect and visualize the rectal stump and colon with the staple between them (Fig. 28.6). The tip of the stapler can be confirmed to be at the apex of the rectal stump, as seen, to eliminate any rectal spur formation. The stapler is then fired to excise the common wall. Occasionally more than one firing of the stapler is required to excise the entire common wall and ensure that no spur or blind pouch is present.

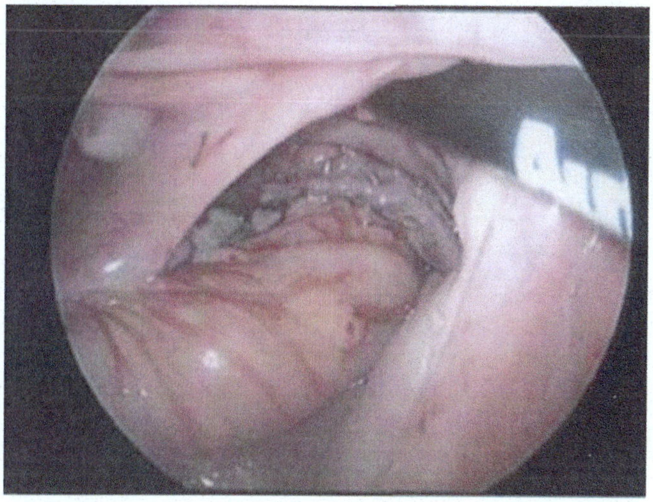

Fig. 28.6 Stapling of the common wall of the rectum and colon

28.5.9 Inspection of the Common Rectal Wall and Pull-Through

A final inspection of the pull-though and anastomosis is performed to ensure that the bowel is well vascularized and there are no potential leak points (Fig. 28.7). The perfusion and orientation of the colon are confirmed before closing.

Postoperatively the infant is kept on nothing by mouth, and feeding is started when bowel motions have occurred and the patient is stable. Antibiotics can be given for 24–48 h postoperatively.

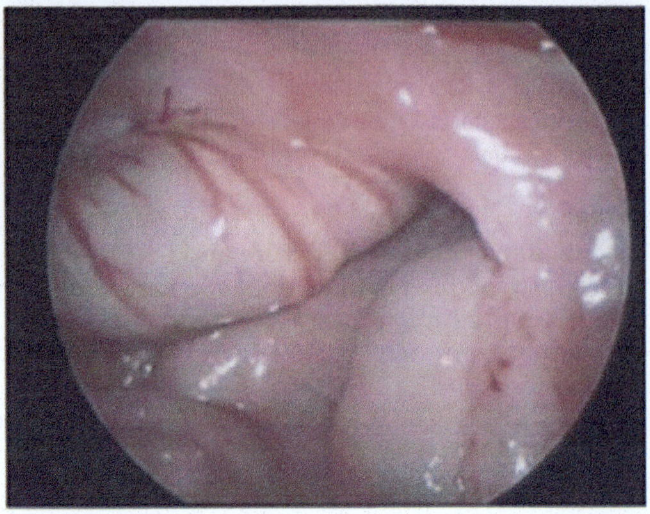

Fig. 28.7 View of the pull-through bowel at the end of the operation

28.6 Alternatives

- A 5-mm primary port can be placed in the right upper quadrant in very small children to achieve increased working space. The right-sided port can then be of appropriate location and diameter for the articulating stapler.
- An intracorporeal sutured closure of the rectal pouch is an alternative to the stapled closure but has a slightly higher incidence of leakage. However, this is an alternative for those with advanced intracorporeal suturing skills if a suitable articulating stapler is not available.
- The mobilization of the colon and the division of the vessels can be performed with hook diathermy to the mesentery if no other energy source is available. However, this takes longer. If a long mobilization is required, a harmonic scalpel or bipolar dissection is advantageous.

28.7 Highlights and Pitfalls

- Identification of the ureters (and the vas deferens in boys) is crucial in avoiding any risk of damaging them.
- 5-mm articulating linear staplers are now developed and available for smaller infants, allowing intra-abdominal stapler use even in smaller children.
- An excessively long pouch should be avoided. However, in total colonic Hirschsprung disease a longer side-to-side colorectal anastomosis is the aim to improve the absorptive capacity of the rectum.
- It is important to ensure that the entire rectal stump is stapled and included in the anastomosis of the common wall with the pulled-through colon. This avoids leaving a spur and a shelf above the pouch, which can be a space for retained stool and formation of a fecaloma.

Suggested Reading

Lamas-Pinheiro R, Henriques-Coelho T, Carvalho JL, Correia-Pinto J. Duhamel pull-through assisted by transrectal port: a hybrid natural orifice transluminal endoscopic surgery approach. J Pediatr Surg. 2012;47:1962–5.

Teitelbaum D, Coran A. Hirschsprung's disease. In: Spitz L, Coran A, editors. Operative pediatric surgery. 11th ed. Boca Raton/London/New York: CRC Press/Taylor & Francis Group; 2013.

29 Laparoscopic-Assisted Soave Pullthrough for Hirschsprung Disease

Amanda J. McCabe

Abstract

Surgical correction of Hirschsprung disease can be achieved with a primary, one-stage procedure. The laparoscopy-assisted approach allows for levelling colonic biopsies with proximal colonic and mesorectal dissection. It prevents excessive retraction of the sphincter mechanism, and direct visualisation ensures that the pulled-through portion of bowel has not been twisted.

Keywords

Laparoscopy • Soave pullthrough • Hirschsprung disease • Serosubmucosal biopsies

29.1 General Information

Surgical correction of Hirschsprung disease can be achieved with a primary, one-stage procedure. The laparoscopy-assisted approach allows for levelling colonic biopsies with proximal colonic and mesorectal dissection. It prevents excessive retraction of the sphincter mechanism, and direct visualisation ensures that the pulled-through portion of bowel has not been twisted.

29.2 Working Instruments

- 5-mm Hassan port
- 30° short telescope
- 3.5-mm port (×2)
- 3-mm Johan forceps
- 3-mm Maryland forceps
- 3-mm Metzenbaum scissors
- Hook diathermy

A.J. McCabe
Department of Paediatric Surgery, Royal Hospital for Sick Children, Edinburgh, UK

M. McHoney et al. (eds.), *Color Atlas of Pediatric Anatomy, Laparoscopy, and Thoracoscopy*,
DOI 10.1007/978-3-662-53085-6_29

29.3 Preparations, Positioning, Port Siting, and Ergonomic Considerations

All patients undergo rectal irrigation preoperatively. General anaesthesia is induced, a bladder catheter is inserted, and parenteral antibiotics are given.

The patient is positioned transversely across the operating table, close to the bottom edge, to allow for a full range of movement of laparoscopic instruments on the patient's right side. The head rotates towards the left side, and all equipment related to anaesthesia should flow back towards the top end of the table (Fig. 29.1).

A diathermy plate is placed high on the infant's back, and total body skin preparation is applied from nipples to knees, extending onto the perineum and lower back. The lower limbs are wrapped in sterile wool placed in sterile sticky plastic sheets, and the feet are permitted to dangle over the table edge (Fig. 29.2).

The surgeon and assistant stand above the infant's head, facing the monitor, which is placed beyond the infant's feet. The scrub nurse stands behind the surgeon and assistant (Fig. 29.3).

Port positions: A single rubber-shod 5-mm port is inserted in the right upper quadrant by direct open dissection through the rectus muscle. Mini lap dissectors are particularly helpful in this confined space, and sutures are placed in each abdominal layer as it is identified and incised (Fig. 29.4). The dissection must be kept perpendicular to the skin. Two further 3.5-mm ports are inserted on the right and left sides under laparoscopic vision (Fig. 29.5). A pneumoperitoneum is achieved with a flow rate of 1l/min and 8 mmHg pressure.

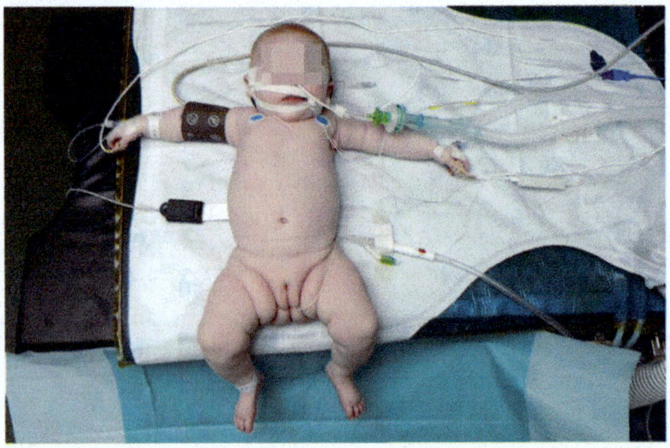

Fig. 29.1 Patient positioning

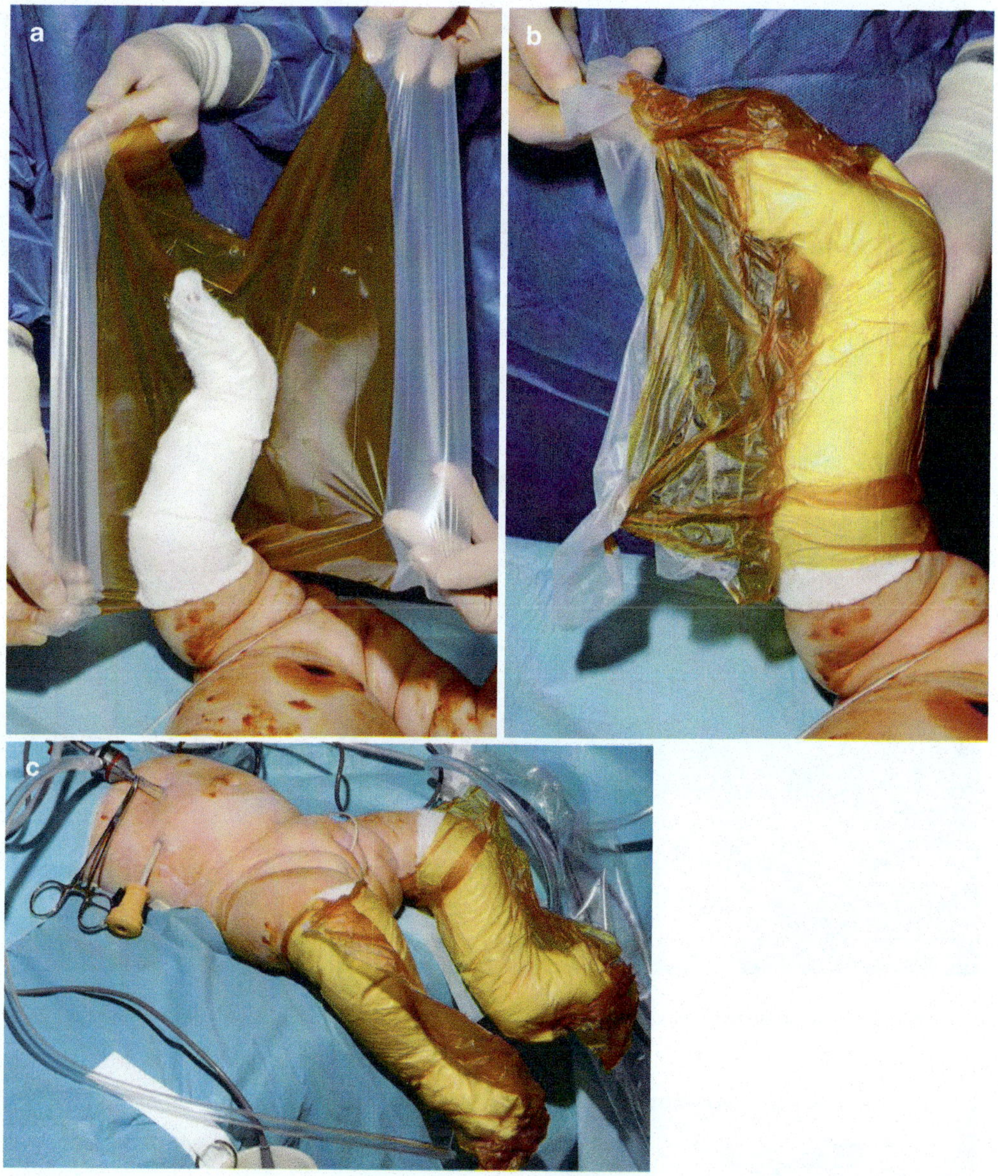

Fig. 29.2 (a–c) Preparation of the patient

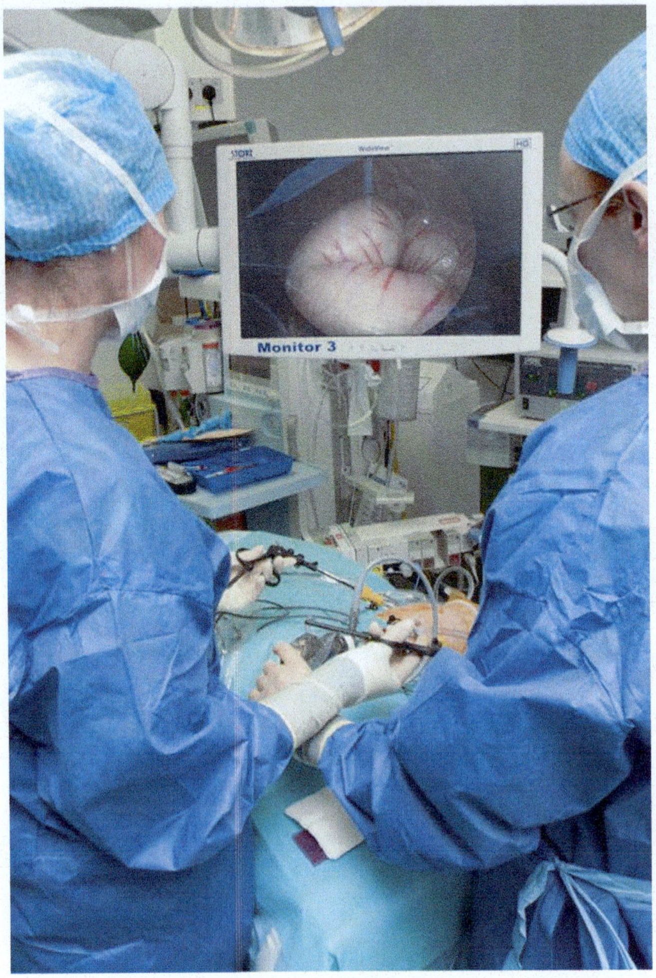

Fig. 29.3 Position of the surgeon and assistants

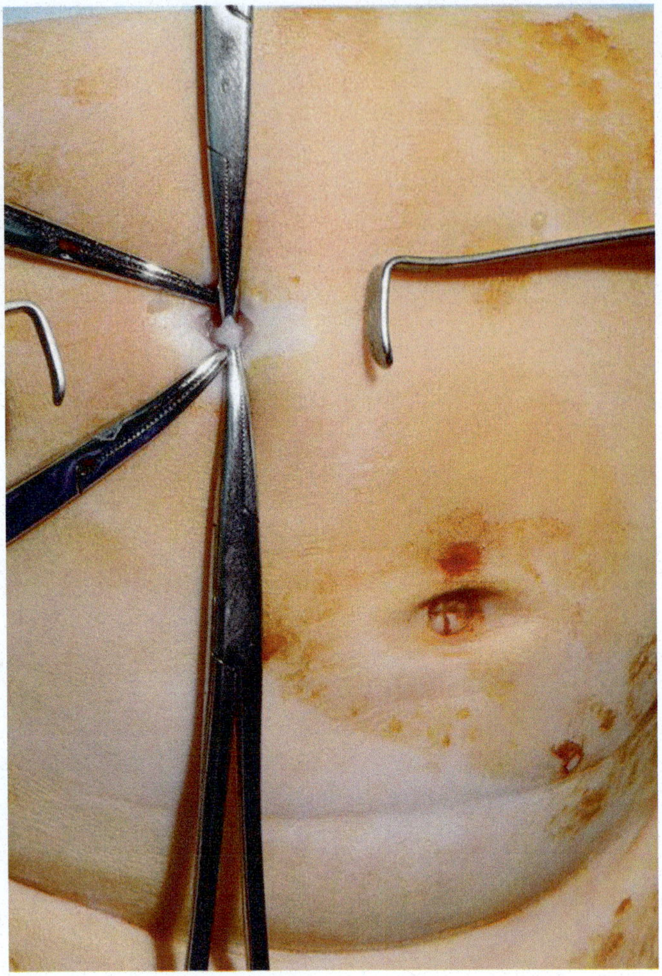

Fig. 29.4 Placement of a 5 mm port in the right upper quadrant

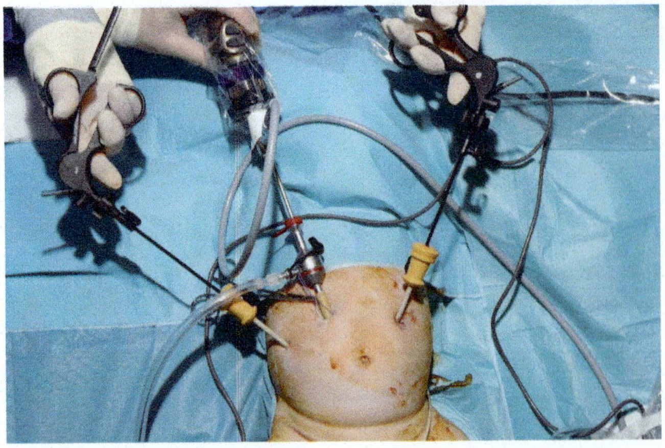

Fig. 29.5 Placement of two 3.5 mm ports on the right and left sides

29.4 Relevant Anatomy

A clear view of the pelvis is needed to view the distal colon and rectum. Identify the bladder (with catheter), peritoneal reflection, anterior wall ligaments, and the uterus and ovaries in female patients. Proximal dilated bowel helps to localise the possible transition zone and sites where biopsy specimens should be taken (Fig. 29.6).

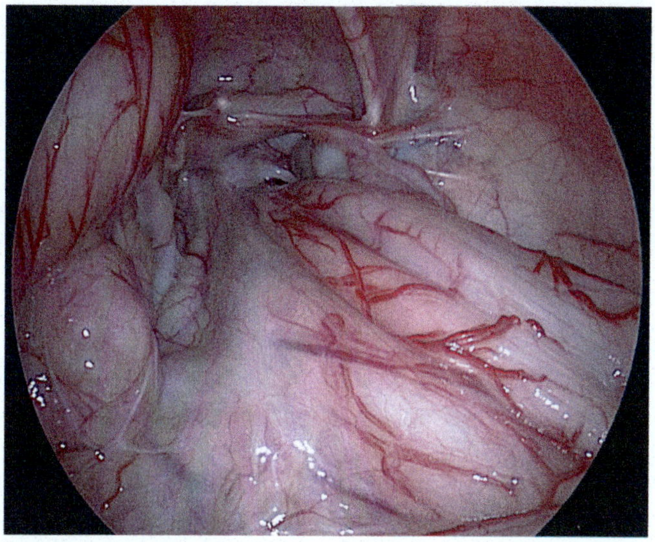

Fig. 29.6 Proximal dilated bowel helps to indicate sites for biopsy

29.5 Surgical Technique

1. *Levelling biopsies.* Three serosubmucosal biopsies are taken: above, within, and below the possible transition zone. The wall of the colon is lifted with fine-tipped forceps (Maryland) in the left hand (Fig. 29.7a), and scissors are used to make a single bite (Fig. 29.7b) in this elevated portion. A V-shaped tongue of colonic wall is seen, and the tip of the V is lifted, with precision, by the fine-tipped forceps in the left hand. Develop the biopsy on each side of the V using small snips and spreading, undermining movements of the scissors. When enough colonic wall has been dissected, cut across the base and send for frozen section (Fig. 29.7c)
2. *Colonic mobilisation* is begun where ganglionated bowel is confirmed in the rectosigmoid region. Sigmoid colon is lifted towards the anterior abdominal wall with a grasper in the left hand (Fig. 29.8). The mesentery is inspected for a suitable mesenteric "window," and the hook diathermy, in the right hand, is used to develop a small hole in the mesentery, through which the left-handed forceps can then be placed closed and lifted upwards to hold the colon more easily (Fig. 29.9). From the window, the mesenteric vessels are individually diathermied with the hook, staying close to the colonic wall, working towards the pelvis (Fig. 29.10). When the peritoneal reflection is reached, a circumferential incision is made and the mesorectum is dissected as far down into the pelvis as possible (Fig. 29.11). It is worth checking that the vascularised proximal bowel to be pulled through, reaches comfortably into the pelvis (Fig. 29.12). During dissection, take time to ensure that both ureters are seen, and not inadvertently pulled up into the dissection plane.
3. *Repositioning.* The legs are elevated over the head and secured with clips onto the drapes (Fig. 29.13).
4. *Submucosal dissection.* The anus is everted with eight circumferential 4-0 silk sutures (Fig. 29.14). Identify the dentate line and mark/score, (with diathermy tip), a circumference on the mucosa 5 mm more proximal (Fig. 29.15). Beginning at the 12 o'clock position, the needle-tipped diathermy is used to incise the mucosa, and interrupted silk sutures are placed from side to side in the mucosal layer and tied. This seals the mucosal "tube" as the full circumference of mucosa is developed (Fig. 29.16). The plane between the submucosa and the white circular smooth muscle is developed inwards, with point diathermy controlling bleeding points (Fig. 29.17). Firm but gentle traction is placed along the 'tube' length. A point will be reached when the dissection begins to separate cleanly with minimal effort, requiring only cotton pledgets to tease the layers apart.
5. *Breakthrough and prolapse.* The mucosal tube is developed and the rolled circular muscle cuff is identified (Fig. 29.18). The cuff is grasped posteriorly with two Allis clamps. It is incised circumferentially, the bowel is prolapsed, and the biopsy sites are noted as they come through (Fig. 29.19). The cuff is split posteriorly in the midline with scissors.
6. *Anastomosis.* The bowel is cut full thickness and an anastomosis is fashioned directly between the neorectum and the circumference just proximal to the dentate line, using absorbable interrupted sutures. Place the 3, 6, 9, and 12 o'clock positions first, leaving the sutures long as retractors clipped onto the drapes (Figs. 29.20 and 29.21). Within each quadrant, three further absorbable sutures are placed (Fig. 29.22). The everting stay sutures are then removed, allowing the tissue to gently retract inwards (Fig. 29.23).
7. *Recheck.* When the anastomosis is secured, the pulled-through portion of bowel is viewed laparoscopically to ensure that it is not twisted and it is lying without tension (Fig. 29.24).

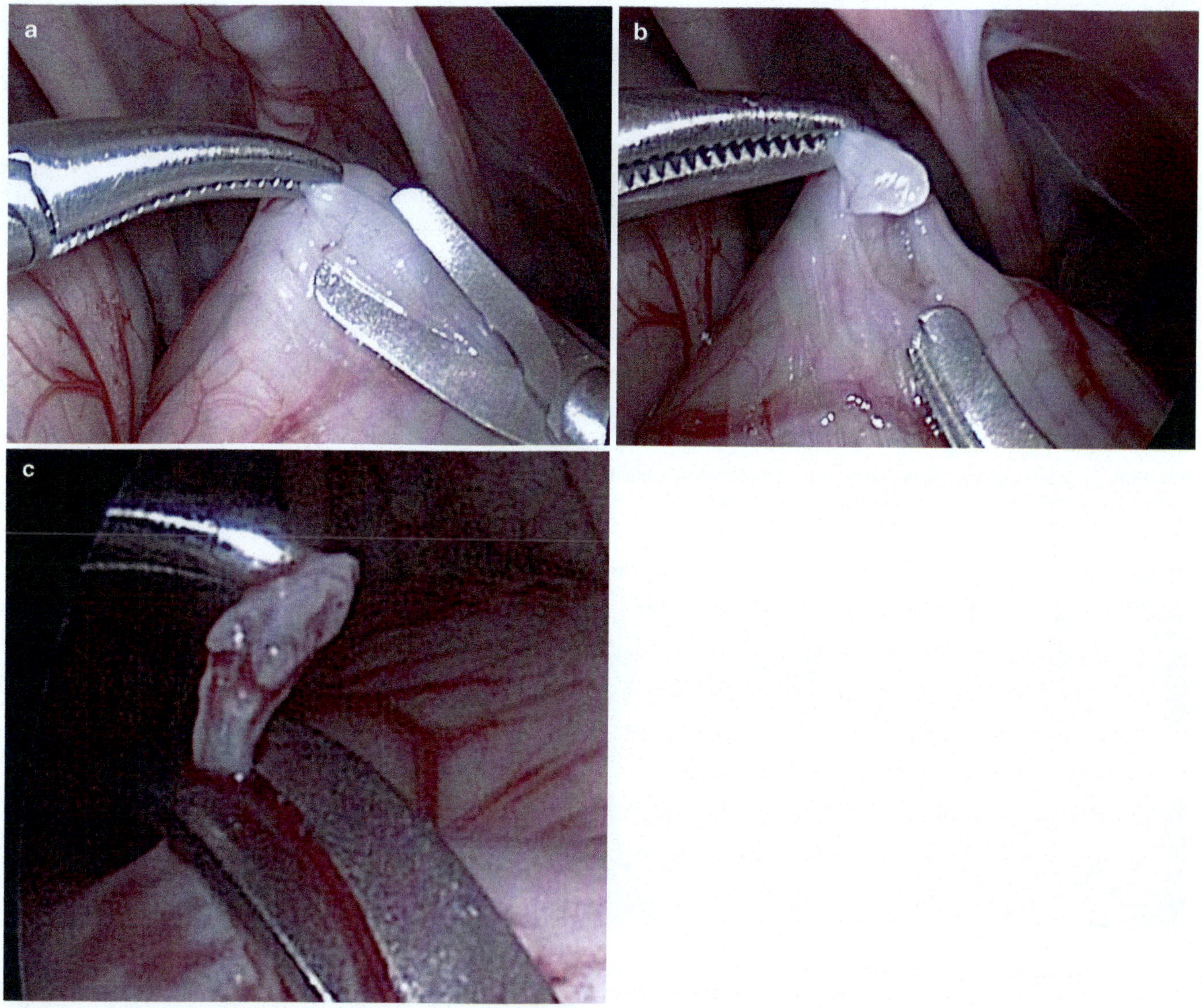

Fig. 29.7 To take levelling biopsies, the wall of the colon is lifted with fine-tipped forceps (Maryland) in the left hand (**a**), and scissors are used to make a single bite (**b**) in this elevated portion. The tip of a V-shaped tongue of colonic wall is lifted by the fine-tipped forceps in the left hand, and the biopsy is developed on each side of the V using small snips and spreading, undermining movements of the scissors. When enough colonic wall has been dissected, cut across the base and send for frozen section (**c**)

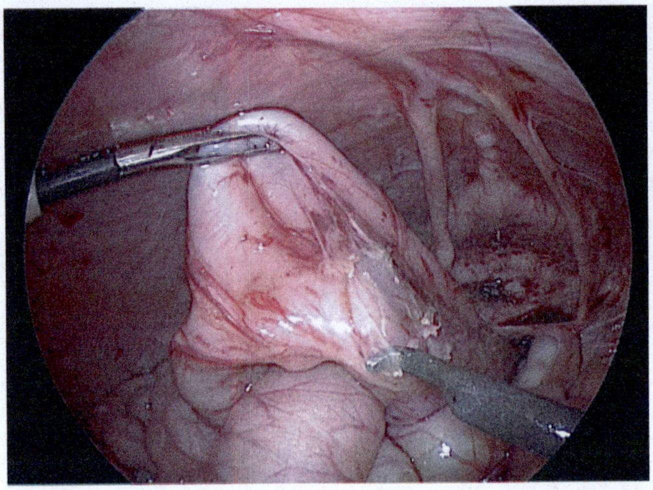

Fig. 29.8 To achieve colonic mobilisation, the sigmoid colon is lifted toward the anterior abdominal wall, using a grasper in the left hand

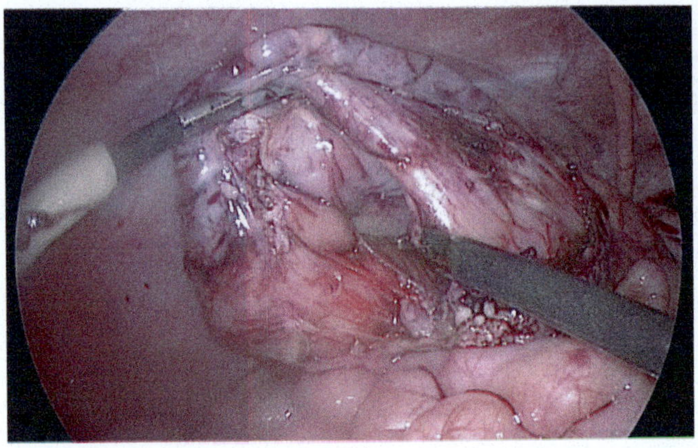

Fig. 29.9 A suitable mesenteric window is identified, and the hook diathermy in the right hand is used to develop a small hole in the mesentery, through which the left-handed forceps can be placed (closed) and lifted upwards to hold the colon

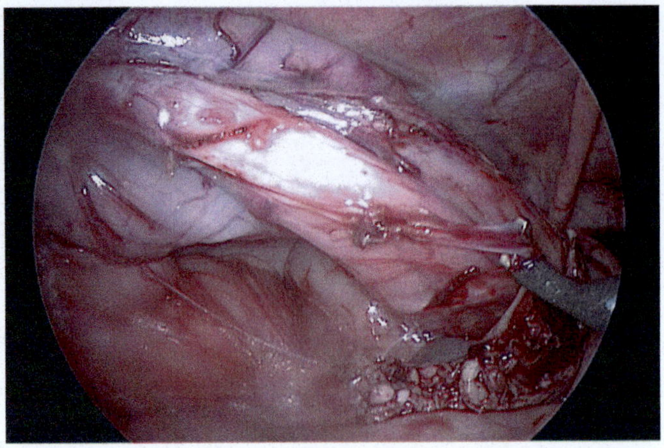

Fig. 29.10 From the window, the mesenteric vessels are individually diathermied with the hook, staying close to the colonic wall and working towards the pelvis

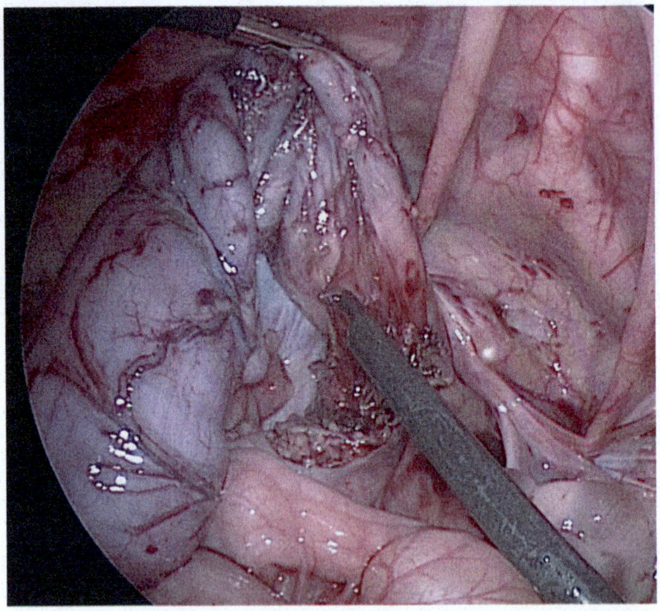

Fig. 29.11 When the peritoneal reflection is reached, a circumferential incision is made and the mesorectum is dissected as far down into the pelvis as possible

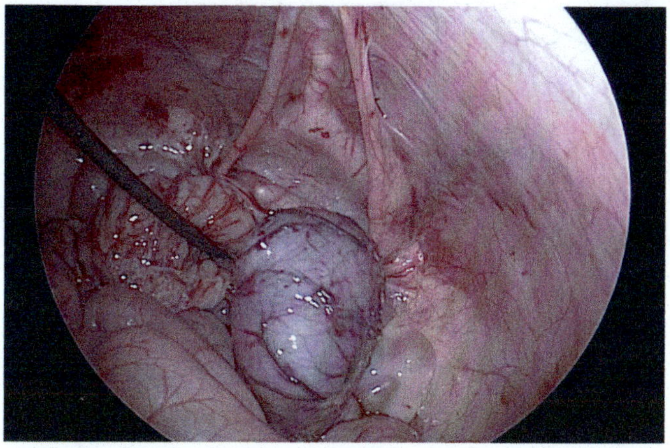

Fig. 29.12 The surgeon should check that the vascularised proximal bowel to be pulled through reaches comfortably into the pelvis and should make sure that both ureters are seen

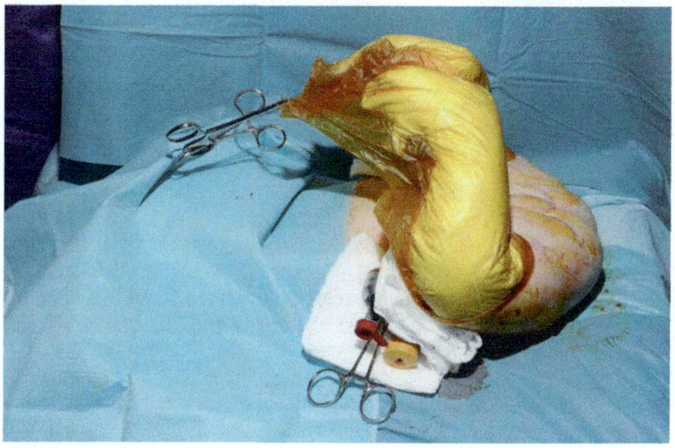

Fig. 29.13 The patient is repositioned with the legs elevated over the head

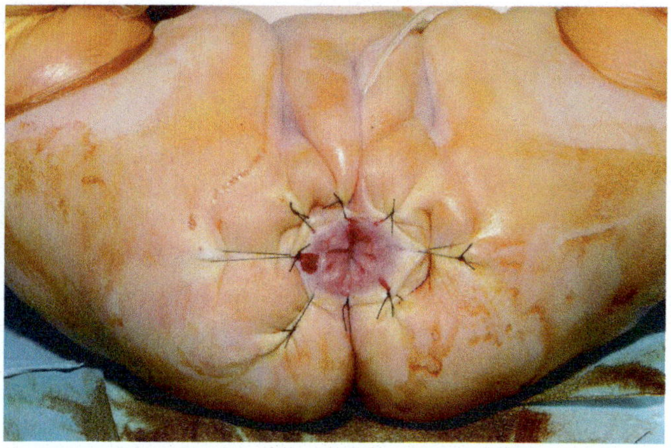

Fig. 29.14 Submucosal dissection begins with eversion of the anus using eight circumferential 4-0 silk sutures

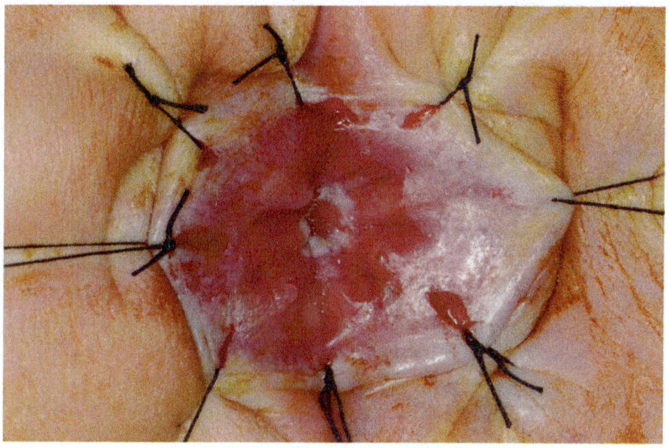

Fig. 29.15 After identifying the dentate line, the surgeon marks a circumference on the mucosa 5 mm more proximal

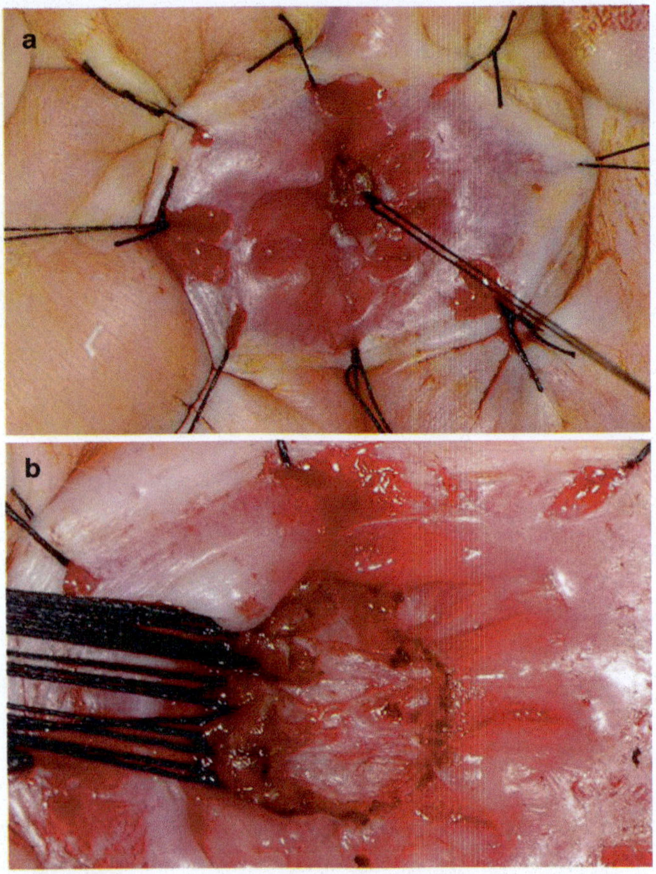

Fig. 29.16 (**a**, **b**) Using the needle-tipped diathermy, the mucosa is incised and interrupted silk sutures are placed from side to side in the mucosal layer and tied, to seal the mucosal tube

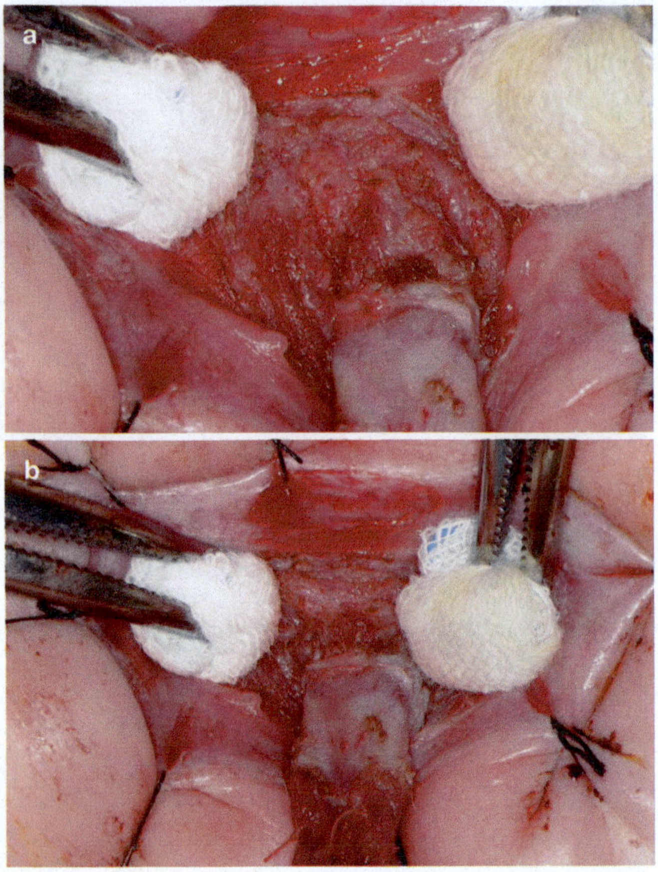

Fig. 29.17 (**a**, **b**) The plane between the submucosa and the white circular smooth muscle is developed inwards, with point diathermy controlling bleeding points

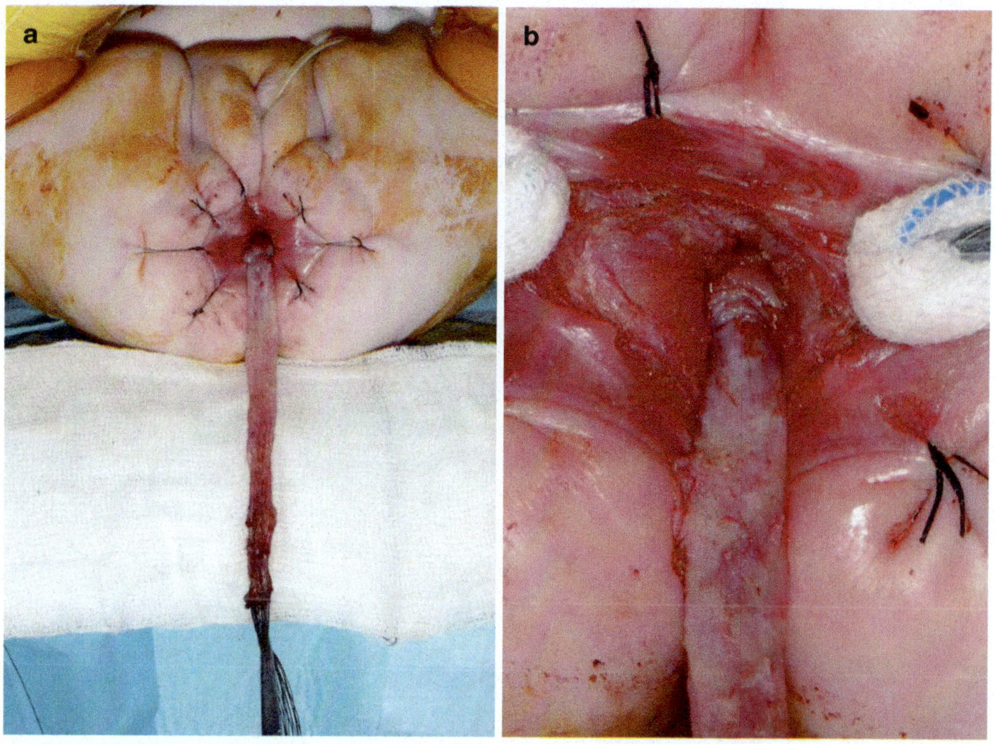

Fig. 29.18 (**a**, **b**) The mucosal tube is developed and the rolled circular muscle cuff is identified

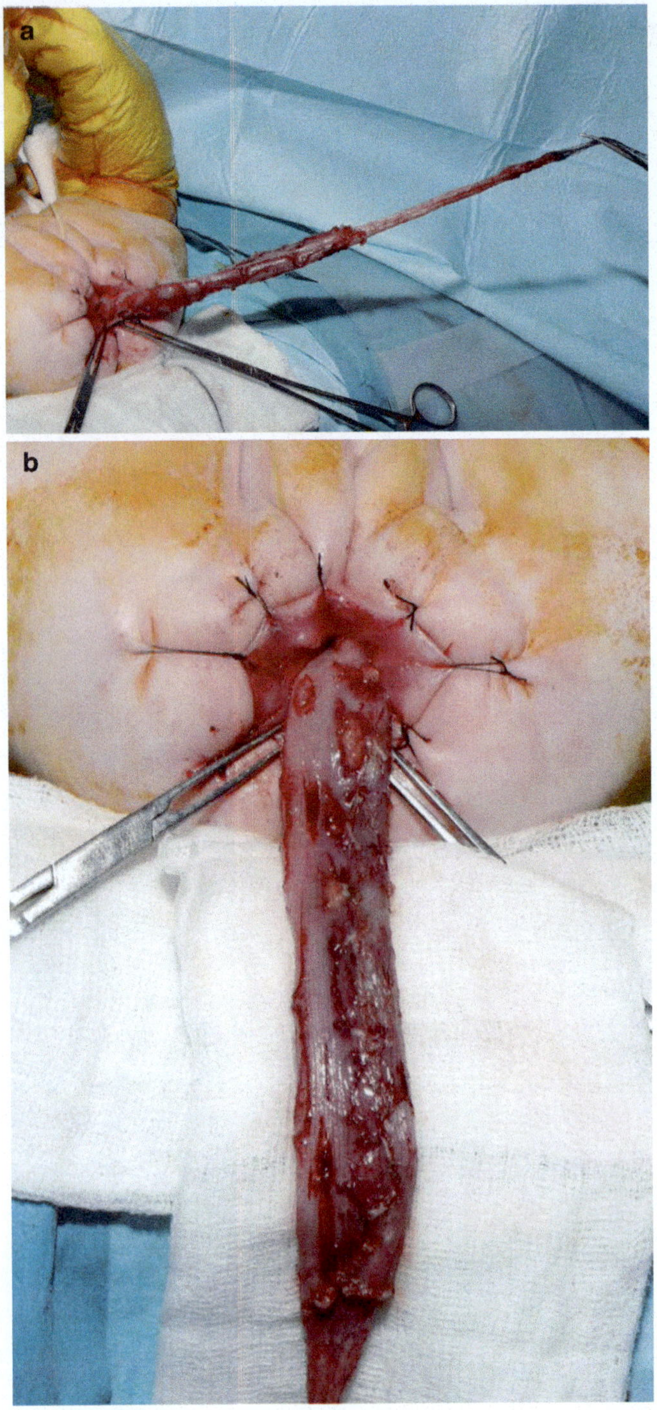

Fig. 29.19 (**a**) The cuff is grasped posteriorly with two Allis clamps. It is incised circumferentially, the bowel is prolapsed, and the biopsy sites are noted as they come through (**b**)

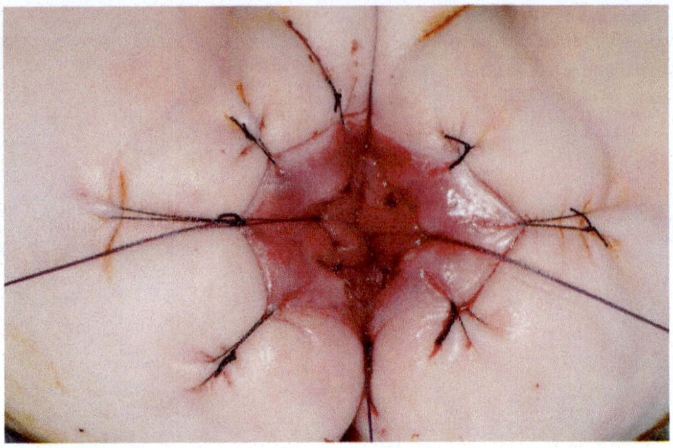

Fig. 29.20 An anastomosis is fashioned between the neorectum and the circumference just proximal to the dentate line

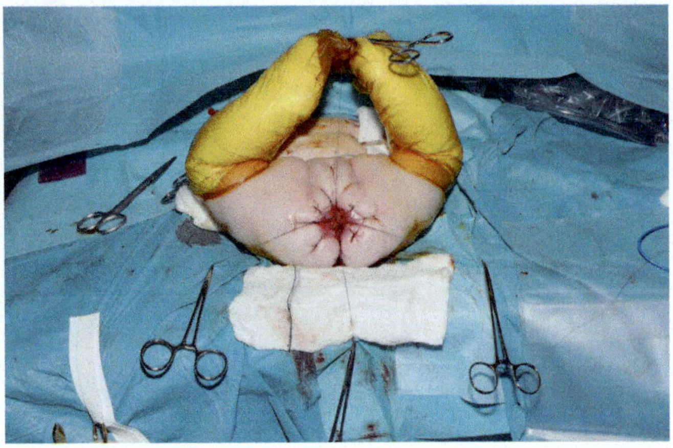

Fig. 29.21 Absorbable interrupted sutures are first placed at 3, 6, 9, and 12 o'clock. They are left long for retraction

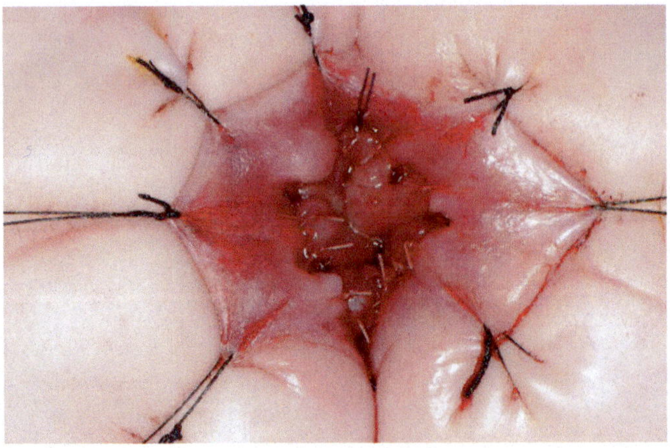

Fig. 29.22 Three further absorbable sutures are placed in each quadrant

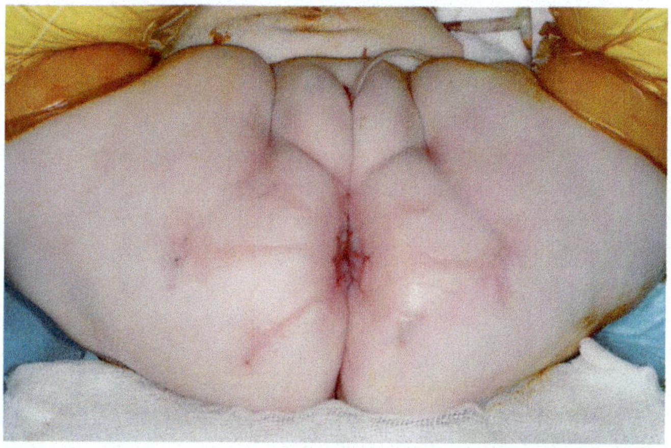

Fig. 29.23 The stay sutures used for eversion of the anus are then removed, allowing the tissue to gently retract inwards

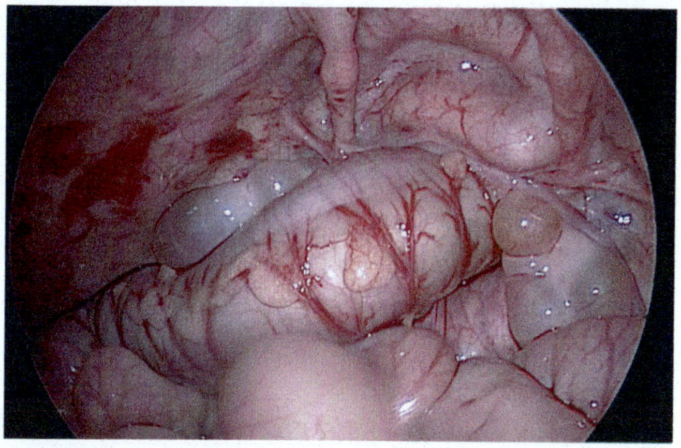

Fig. 29.24 The pulled-through portion of bowel is viewed laparoscopically to ensure that it is not twisted and it is lying without tension

29.6 Highlights and Pitfalls

- An on-table washout may be necessary if the sigmoid colon has not decompressed sufficiently (Fig. 29.25).
- Dissection for the initial 5-mm port must be kept perpendicular to the skin to aid precise dissection through the layers of the body wall.
- If the level of ganglionated bowel is beyond the midtransverse colon, it is best to wait for permanent section biopsies, consider placing a distal ileal stoma, permanent section biopsies of the more proximal colon, and consider placing a distal stoma. Do remember to send ilieal biopsies too.
- If the biopsy causes mucosal breach, the defect is closed with an absorbable suture.
- When the biopsies are sent, the gas and light source are turned off. Care is taken to protect the patient's skin from the pressure effect of resting ports during submucosal dissection (Fig. 29.26).

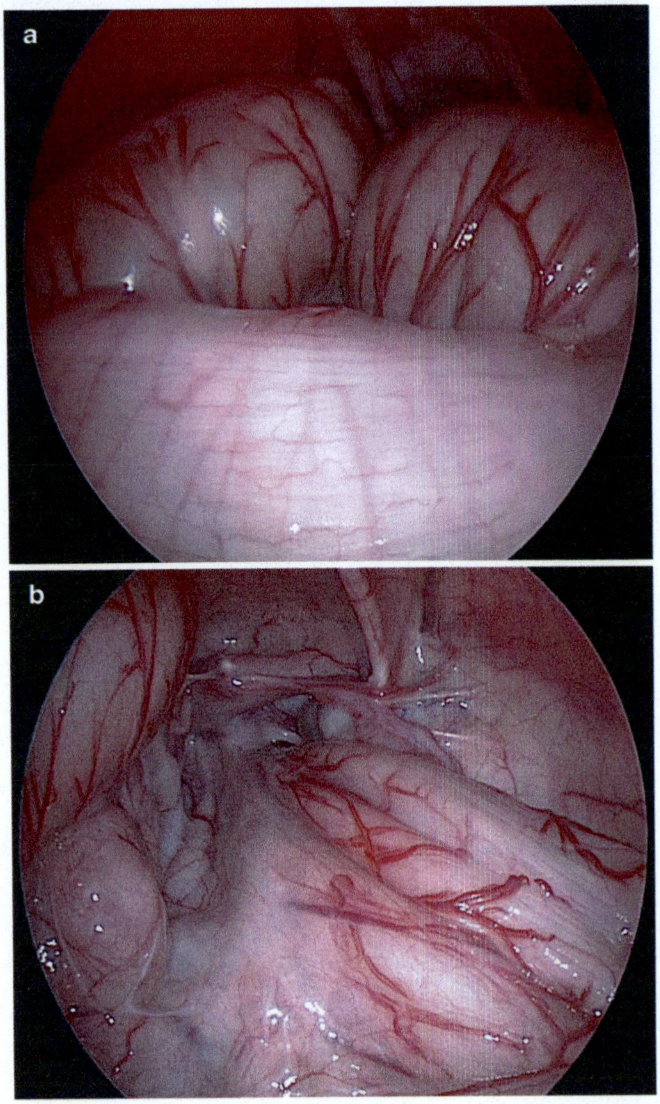

Fig. 29.25 (**a**, **b**) An on-table washout may be necessary if the sigmoid colon has not decompressed sufficiently

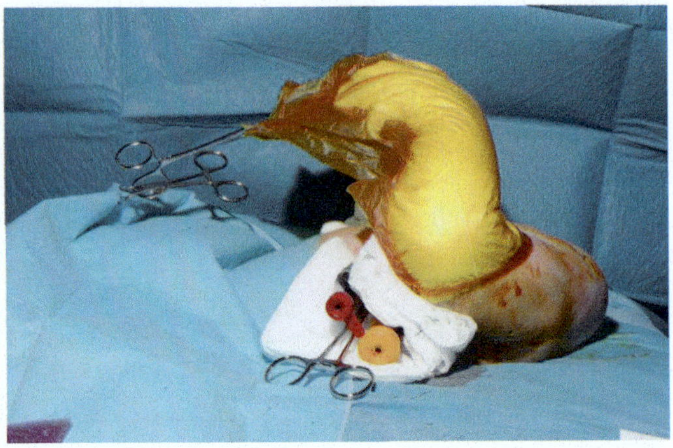

Fig. 29.26 The patient's skin should be protected from the pressure effect of resting ports during submucosal dissection

Laparoscopic-Assisted Swenson-Like Transanal Pullthrough for Hirschsprung Disease

30

Michael Stanton, Bala Eradi, and Marc A. Levitt

Abstract

Laparoscopic-assistance for Hirschsprung disease (HD) pullthrough procedures has increased in popularity. This change has been associated in the UK, for example, with a shift from the Duhamel procedure, which had been performed most frequently, to an endorectal pull-through (ERPT) procedure that is being practiced more widely [1]. The concept of the transanal only approach was put forth by Langer and de la Torre, but this was using a Soave-like dissection, a problem noted by Swenson himself who originated this concept with the original repair for HD [2]. The use of the Swenson-like technique for ERPT has been popularized by Levitt and colleagues [3], who reported a series of 67 patients in 2013. The advantages of this approach are its simplicity, preservation of the pelvic nerves, and avoidance of the obstructing muscle cuff that may occur after the Soave-Boley procedure [4]. The procedure can be undertaken as a purely transanal operation for cases in which rectosigmoid aganglionosis is obvious from the contrast study, but initial laparoscopic colonic mobilization and biopsy are preferred in most cases.

Keywords

Hirschsprung disease • Swenson-like pull-through • Laparoscopy

30.1 General Information

Laparoscopic-assistance for Hirschsprung disease (HD) pullthrough procedures has increased in popularity. This change has been associated in the UK, for example, with a shift from the Duhamel procedure, which had been performed most frequently, to an endorectal pull-through (ERPT) procedure that is being practiced more widely [1]. The concept of the transanal only approach was put forth by Langer and de la Torre, but this was using a Soave-like dissection, a problem noted by Swenson himself who originated this concept with the original repair for HD [2]. The use of the Swenson-like technique for ERPT has been popularized by Levitt and colleagues [3], who reported a series of 67 patients in 2013. The advantages of this approach are its simplicity, preservation of the pelvic nerves, and avoidance of the obstructing muscle cuff that may occur after the Soave-Boley procedure [4]. The procedure can be undertaken as a purely transanal operation for cases in which rectosigmoid

M. Stanton (✉)
Department of Paediatric Surgery, Southampton Children's Hospital, Southampton, UK

B. Eradi
Department of Paediatric Surgery, Leicester Royal Infirmary, Leicester, UK

M.A. Levitt, MD
Nationwide Children's Hospital, The Ohio State University, Columbus, OH, USA

M. McHoney et al. (eds.), *Color Atlas of Pediatric Anatomy, Laparoscopy, and Thoracoscopy*,
DOI 10.1007/978-3-662-53085-6_30

aganglionosis is obvious from the contrast study, but initial laparoscopic colonic mobilization and biopsy are preferred in most cases.

30.2 Working Instruments

- 3-mm Ports and instruments: hook diathermy, graspers, needle holders.
- 30° camera (5 or 3 mm)
- Lone Star retractor and needle-point (Lone Star Medical Products, Houston, TX); hand-held monopolar diathermy is used for the transanal approach.

30.3 Positioning, Port Siting and Ergonomic Considerations

For laparoscopic mobilization, the patient is positioned at the foot of the table and turned 90°. This allows the operating surgeon to stand at the patient's right shoulder facing the pelvis, and the assistant can stand on the patient's left side. Full skin preparation of the lower limbs and abdomen is undertaken, and the legs and feet are wrapped in crepe bandage and bio-occlusive dressing sheets to allow repositioning intraoperatively. The laparoscopic stack is positioned at the feet end of the patient. Urethral catheterization is required.

For the transanal approach, the prone position is preferable because this facilitates dissection between the rectum and urethra in males (the rectum and vagina in females). For laparoscopic dissection/biopsy and the transanal approach, the patient can be positioned supine with the legs elevated.

30.4 Relevant Anatomy

Dissection is done close to the rectal wall to avoid damage to the pelvic nerves. If laparoscopic abdominal mobilization is undertaken, both ureters (and vasa deferens in males) should be identified and preserved.

30.5 Surgical Technique

30.5.1 Laparoscopic Colonic Mobilization and Biopsy

The patient is initially positioned supine, as described above. Three or four ports are used. In small infants it is helpful to place these all above the umbilicus to allow more space for the instruments. The camera port is placed high in the epigastrium, to the right of the midline (and falciform ligament). Two lateral ports (3 mm) are placed. A fourth port placed in the left upper quadrant can be helpful to grasp the sigmoid colon. Insufflation pressure of 8–10 mmHg and a flow of 2–3 L/min are used. The patient is positioned head-down so that the small bowel can be displaced out of the pelvis. The rectosigmoid colon is evaluated, and the likely transition zone is identified. There are two options for confirmatory colonic biopsy with frozen section histologic evaluation. A seromuscular colonic biopsy can be taken from the taenia coli (Figs. 30.1, 30.2, 30.3, and 30.4) with a later full-thickness biopsy at the end of the pull-through. The alternative is to take an initial full-thickness colonic biopsy and suture the enterotomy closed (either laparoscopically or by exteriorizing the colon through one of the port sites). This latter technique avoids the pitfall of ganglion cells being noted in the muscular layer when there are hypertrophic nerves in the submucosal layer.

Colonic mobilization is started by elevating the sigmoid colon by grasping the mesenteric edge of the bowel. Hook monopolar diathermy is used to fashion a window in the mesentery (Fig. 30.5). Further mesenteric division is continued proximally and distally (Fig. 30.6). Great care must be taken to preserve the sigmoid mesenteric arcade and to ligate the inferior mesenteric artery high near the aorta. This will allow mobility of the colonic pull-through segment without compromising its blood supply. The lateral colonic peritoneal attachments are displayed by drawing the colon medially and are again divided using hook diathermy (Fig. 30.6). The left ureter should be identified at this stage. If need be, the splenic flexure is taken down. Dissection is continued (medially and laterally) distally to the peritoneal reflection (Figs. 30.7 and 30.8), together with identification of the right ureter. In males, the vas deferens should be identified at the level of the peritoneal reflection. Mobilization of the rectum beneath the peritoneal reflection should be to the deep pelvis, making the transanal part required relatively minimal. For a transition zone proximal to the mid-transverse colon, an open approach is performed to carefully delineate the mesentery and to derotate the colon if necessary.

30.5.1.1 Transanal Approach

If a transanal-only approach is used, the patient is positioned prone, with the buttocks elevated. The Lone Star retractor pins are placed initially at the anal mucocutaneous junction (Fig. 30.9). The pins are then replaced deeper, so that the dentate line is now buried and thus preserved (as is the distal 1.5 cm of the anal canal) (Fig. 30.10). The intended line of stay sutures can be marked on the rectal mucosa if necessary (Fig. 30.11). Interrupted stay sutures (e.g., silk) are placed in the rectal mucosa 1 cm above the dentate line circumferentially. The sutures are placed on a single artery clip, which is then used to provide uniform traction (Fig. 30.12).

A full-thickness rectal incision is made starting in the posterior midline using needle-point monopolar diathermy. Once the circumferential full-thickness Swenson plane is established, the rectum is drawn outwards using traction on the stay sutures (Fig. 30.13). Dissection is continued close to the rectal wall with coagulation of extrinsic vessels in the same fashion as employed in the posterior sagittal anorectoplasty. Dissection must be within the whitish fasciae that envelops the rectum. The dissection is continued up to and through the peritoneal reflection (Fig. 30.14). If a laparoscopic colonic mobilization was used, that plane of dissection is quickly reached. In such a case, supine with legs elevated is appropriate and the rectosigmoid can be withdrawn easily (Fig. 30.15). It is important to avoid twisting of the pull-through segment, and the antemesenteric border can be marked with a suture (Fig. 30.16).

The coloanal anastomosis is fashioned in two layers using absorbable sutures. The first layer is placed between the seromuscular layer of the pull-through colon and the proximal (internal) incised rectal layer. The second layer sutures the distal incised rectal tissue to the colon (mucosa to mucosa) (Fig. 30.17). If the approach has been transanal only, a Hegar dilator or large-bore tube is passed well into the pull-through in the colon to ensure that there has been no twist.

The port site wounds are closed with absorbable sutures under laparoscopic visualization and skin glue applied.

30.5.2 Highlights and Pitfalls

One of the key errors to avoid is inadvertent twisting of the pull-through segment. Some surgeons advocate laparoscopy to avoid this and passage of a Hegar dilator or large-bore tube at the end of the procedure.

As for all HD cases irrespective of operative technique, accurate intraoperative histologic confirmation of normal ganglion cells and the absence of hypertrophic nerves (defined as >40 μm) are essential. Some surgeons pause once the initial laparoscopic biopsy is taken until confirmation of the level is established; others continue but only complete the coloanal anastomosis once a full-thickness specimen has been evaluated. The pitfalls to avoid here are co-localized ganglion cells in the intermuscular layer with hypertrophic nerves in the submucosal plexus and spiral configuration of the transition zone.

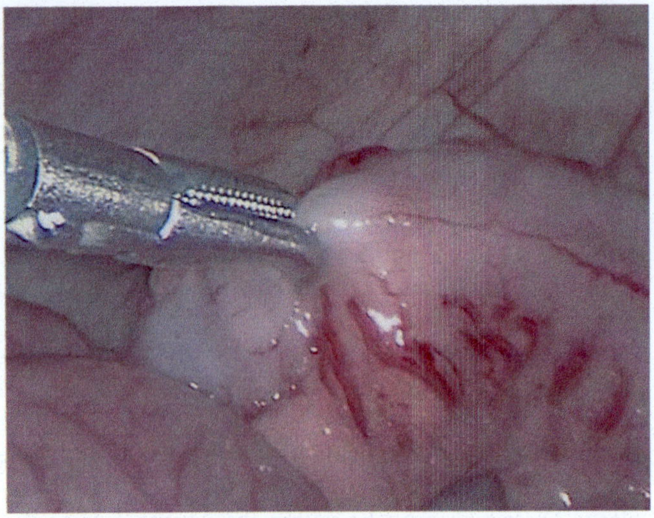

Fig. 30.1 Laparoscopic colonic biopsies are taken for histologic assessment, initially by grasping and elevating the taenia coli

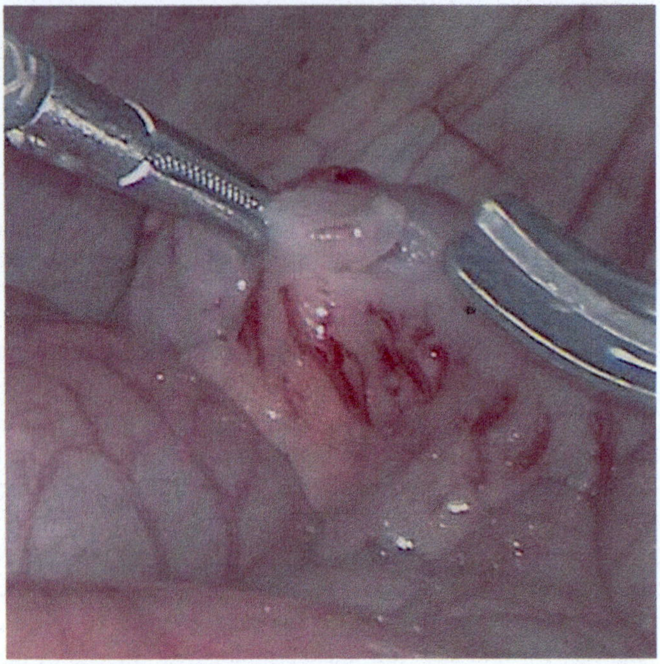

Fig. 30.3 The biopsy instrument is grasped again and a seromuscular or full-thickness biopsy is taken (see text)

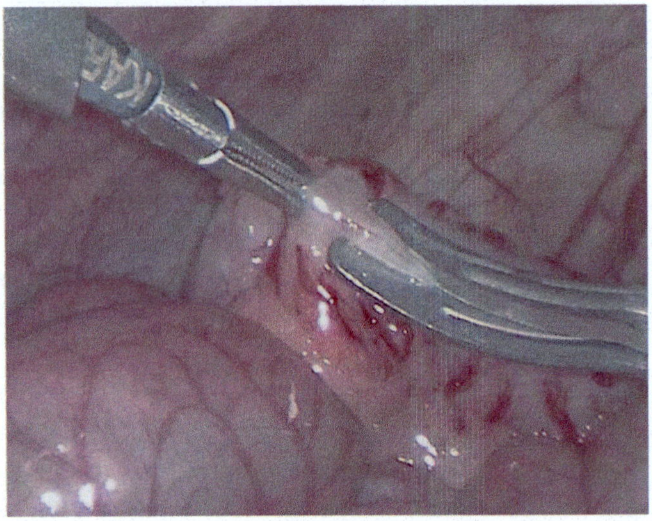

Fig. 30.2 Dissecting scissors are used to cut the taenia coli

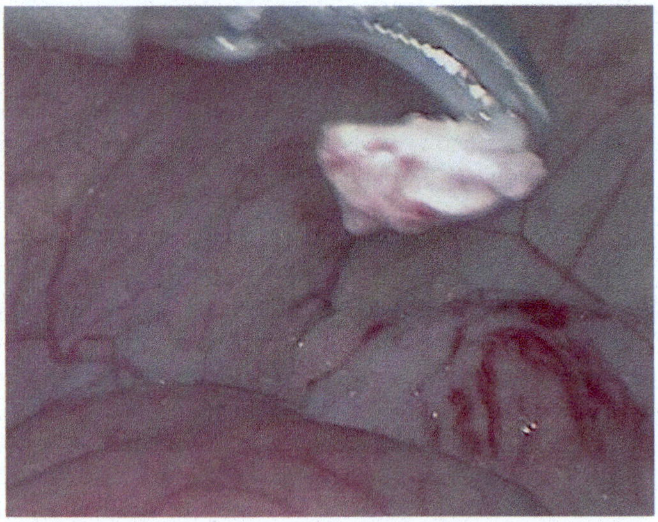

Fig. 30.4 The biopsy is withdrawn from the abdomen via one of the ports under direct vision

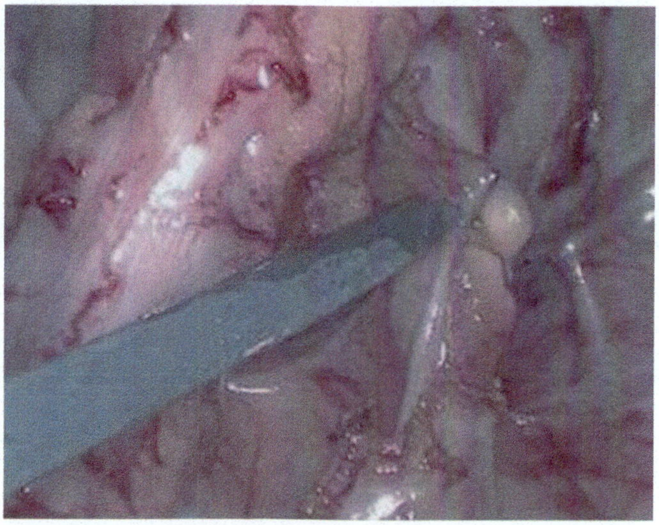

Fig. 30.5 A window in the sigmoid mesentery is made using monopolar hook diathermy

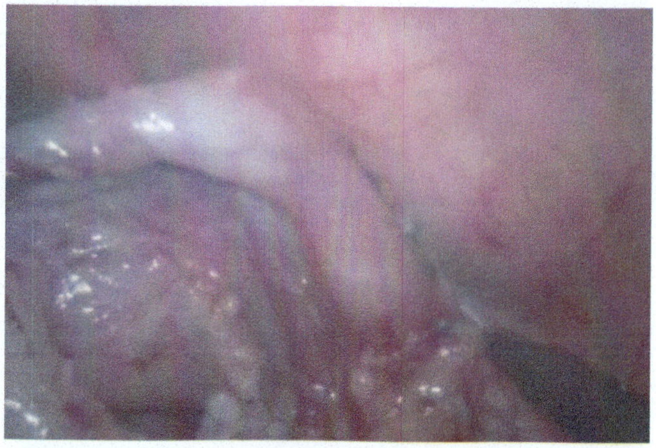

Fig. 30.7 Dissection continues to the peritoneal reflection, taking care to preserve the ureters and vasa

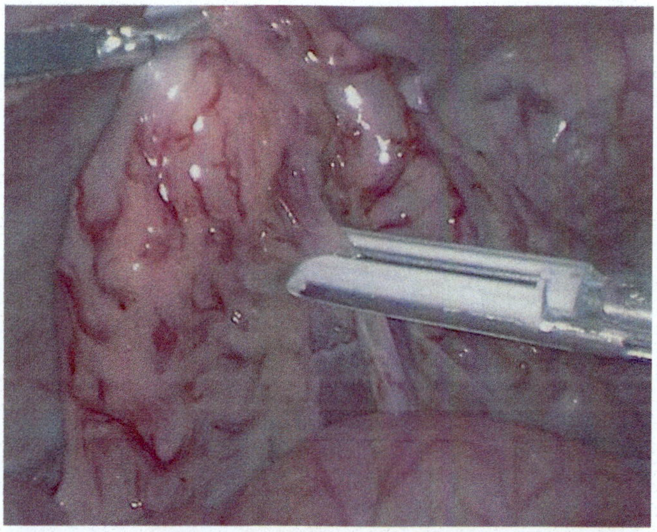

Fig. 30.6 The mesentery is divided proximally and distally staying close to the bowel wall and using an electrocautery (or monopolar hook) device

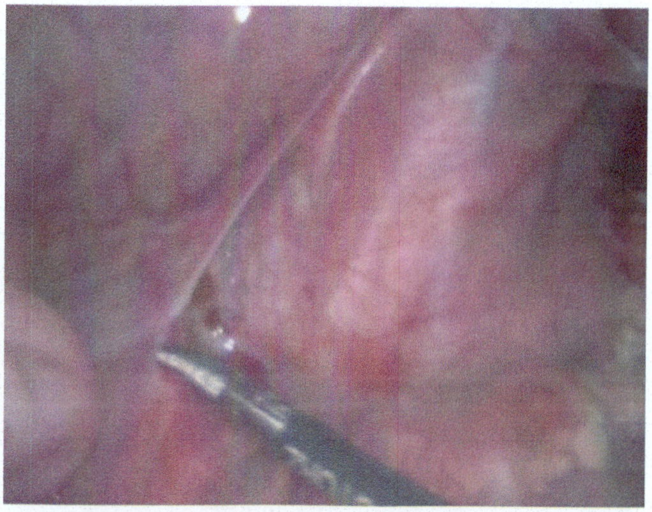

Fig. 30.8 The lateral colonic peritoneal attachment is divided to further mobilize the pull-through segment. The left ureter is identified and preserved

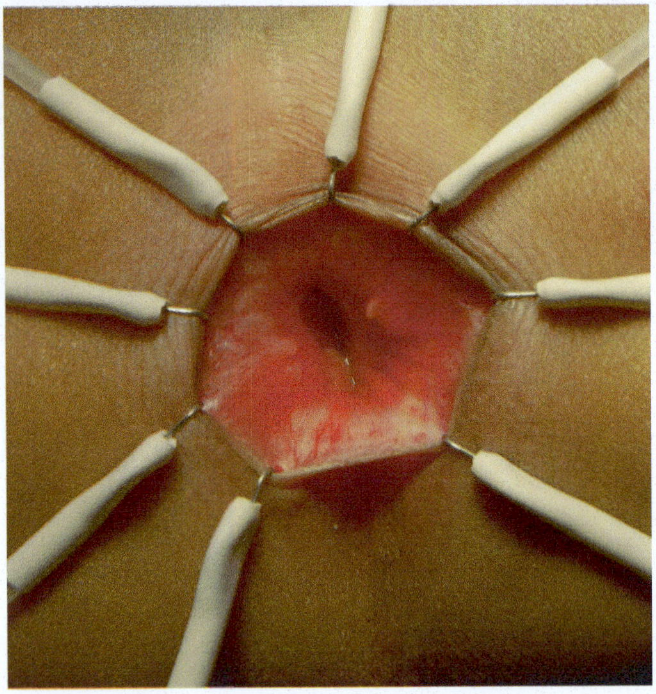

Fig. 30.9 Transanal approach: the Lone Star retractor in place, with pins intitially at the dentate line

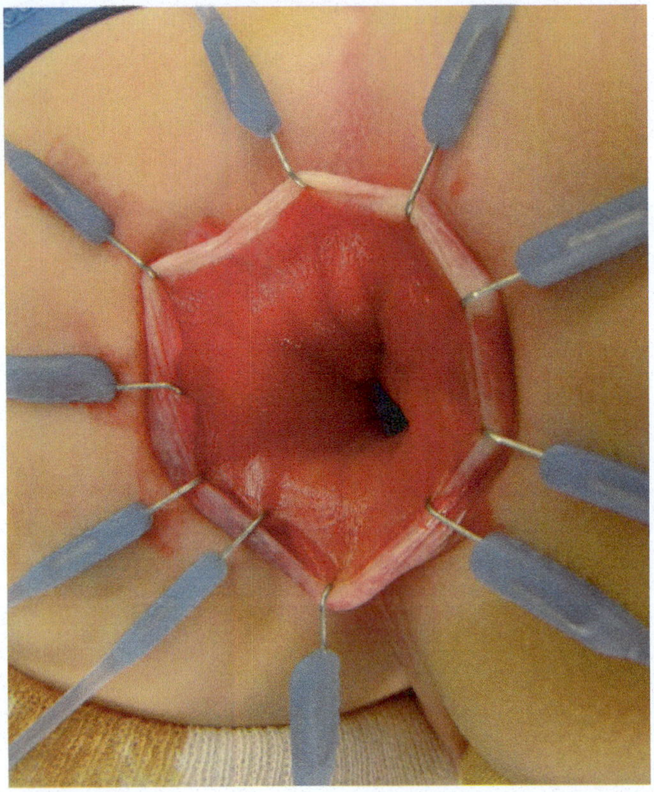

Fig. 30.10 The Lone Star pins are replaced sequentially so that the dentate line is now hidden

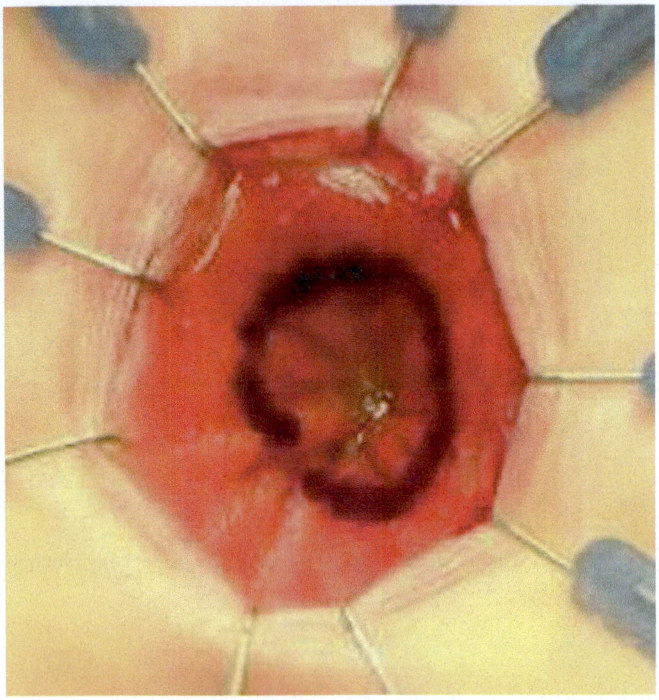

Fig. 30.11 A marking line is made 1–1.5 cm above the dentate line

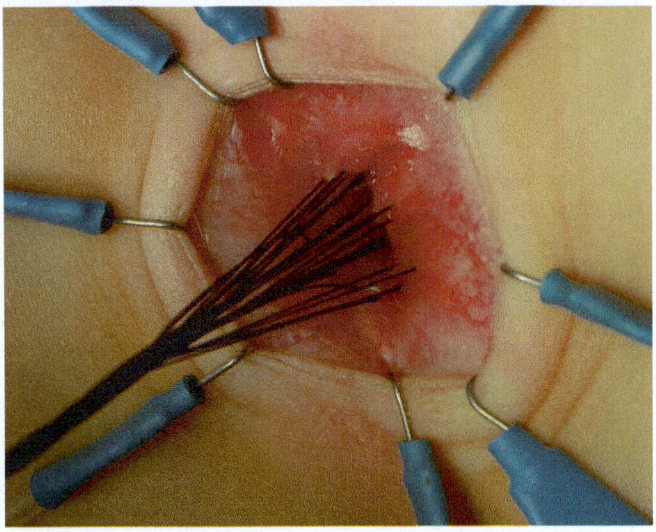

Fig. 30.12 A circumferential row of sutures (e.g., 5/0 silk) is placed 1.0 cm above the dentate line

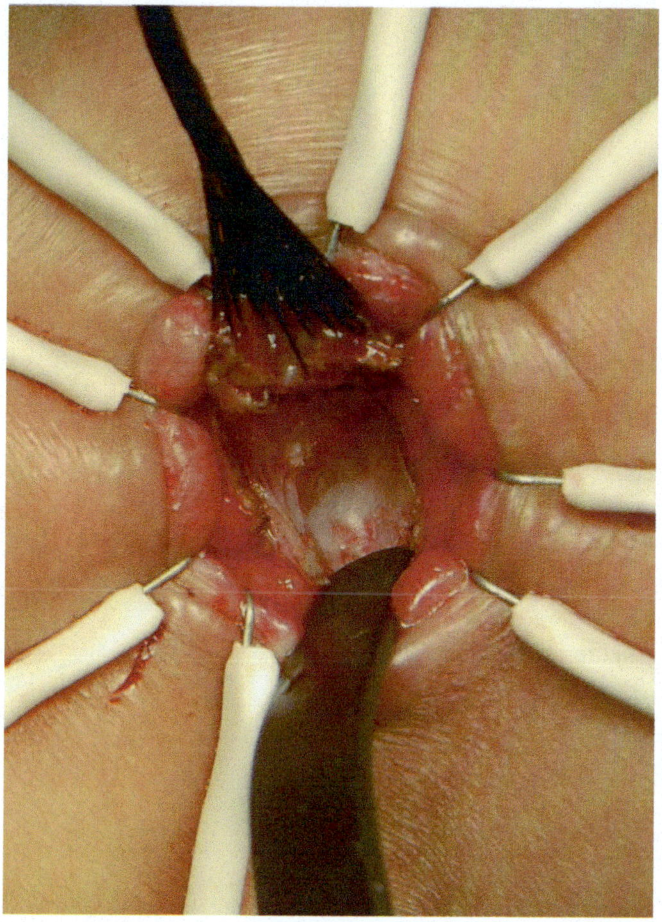

Fig. 30.13 Full-thickness dissection is started posteriorly and continues close to the rectal wall

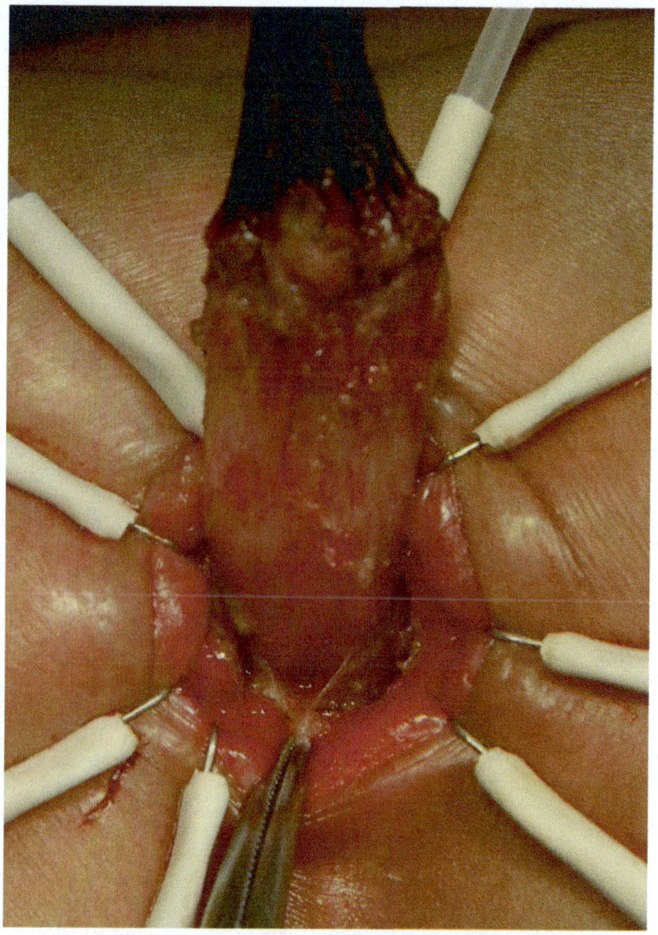

Fig. 30.14 Dissection continues to the peritoneal reflection (clip applied)

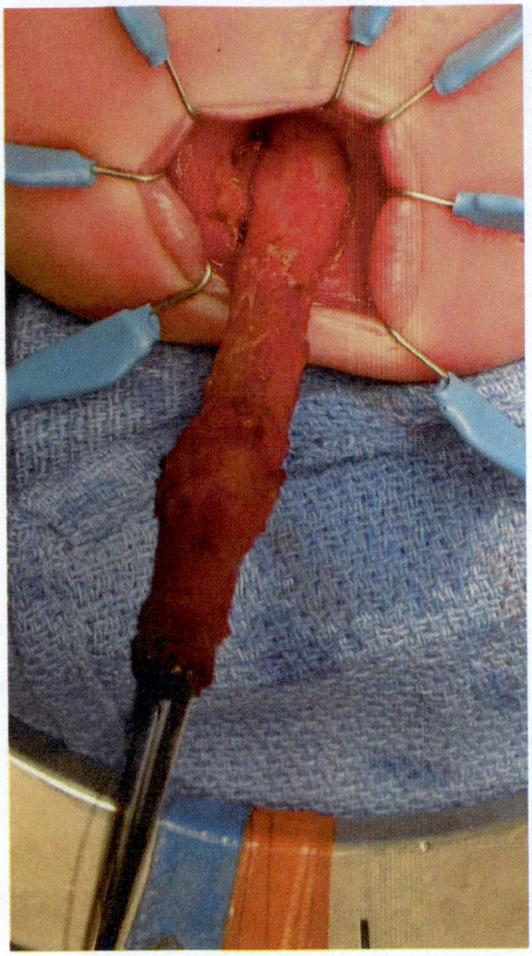

Fig. 30.15 The pull-through segment is now easily drawn down

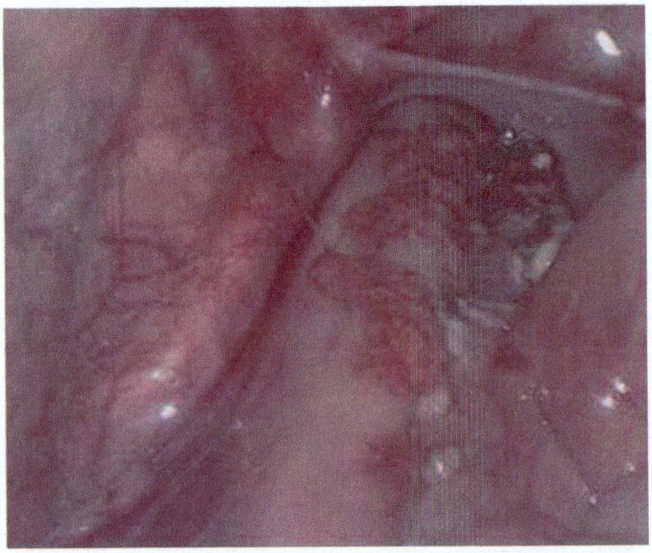

Fig. 30.16 A two-layer coloanal anastomosis is completed

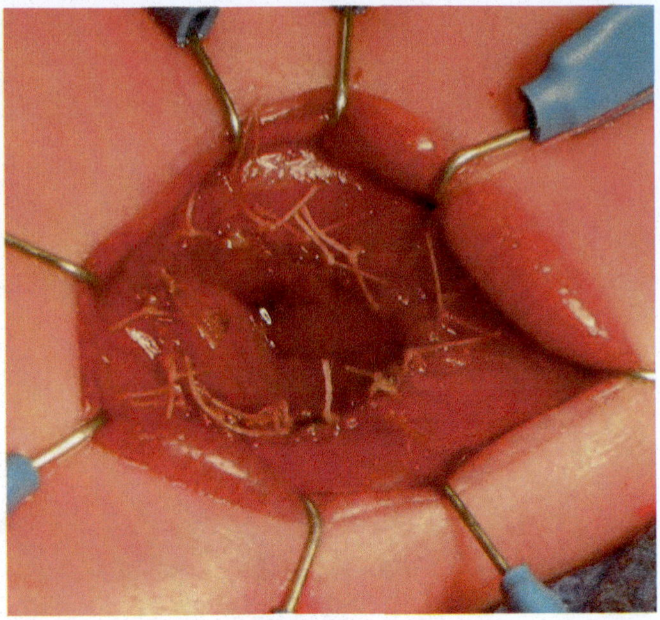

Fig. 30.17 Laparoscopic view confirming the pull-through is not twisted (note intact left ureter)

References

1. Bradnock TJ, Walker GM. Evolution in the management of Hirschsprung's disease in the UK and Ireland: a national survey of practice revisited. Ann R Coll Surg Engl. 2011;93:34–8.
2. Langer JC, Durrant AC, de la Torre L, Teitelbaum DH, Minkes RK, Caty MG, Wildhaber BE, et al. One-stage transanal Soave pullthrough for Hirschsprung disease: a multicenter experience with 141 children. Ann Surg. 2003;238(4):569–83.
3. Levitt MA, Hamrick MC, Eradi B, Bischoff A, Hall J, Peña A. Transanal, full-thickness, Swenson-like approach for Hirschsprung disease. J Pediatr Surg. 2013;48:2289–95.
4. Dickie BH, Webb KM, Eradi B, Levitt M. The problematic soave cuff in Hirschsprung disease: manifestations and treatment. J Pediatr Surg. 2014;49:77–81.

Inguinal Hernia Repair

31

Merrill McHoney

Abstract

Potential advantages of the laparoscopic approach to inguinal hernias in children include avoiding handling of the vas and vessels in boys, inspection of the internal genital organs in girls, and diagnosis of asymptomatic contralateral hernia. In experienced hands, the recurrence rate is minimal and approaches that of open surgery. It may be an advantageous approach to incarcerated hernias.

Keywords

Laparoscopic hernia repair • Paediatrics • Inguinal hernia • Patent processus vaginalis

31.1 General Information

Potential advantages of the laparoscopic approach to inguinal hernias in children include avoiding handling of the vas and vessels in boys, inspection of the internal genital organs in girls, and diagnosis of asymptomatic contralateral hernia. In experienced hands, the recurrence rate is minimal and approaches that of open surgery. It may be an advantageous approach to incarcerated hernias.

31.2 Working Instruments

- 5-mm Hasson port
- 30° (preferable) or 0° telescope
- 3-mm needle holders
- 3-mm scissors
- 3-mm Maryland forceps

M. McHoney
Department of Paediatric Surgery,
Royal Hospital for Sick Children,
9 Sciennes Road, Edinburgh EH9 1LF, UK

M. McHoney et al. (eds.), *Color Atlas of Pediatric Anatomy, Laparoscopy, and Thoracoscopy*,
DOI 10.1007/978-3-662-53085-6_31

31.3 Positioning, Port Siting, and Ergonomic Considerations

A 5-mm umbilical port is used as the primary port. Other working instruments are placed as shown in Fig. 31.1. In bilateral cases, the two working ports are placed in each of the lumbar regions. A direct puncture of the abdominal wall, without port placement, is used for the working instruments, as there is very little exchange of instruments during the procedure. A needle holder is placed in the dominant hand and a grasper in the opposite. In small infants and neonates, shorter instruments allow more ergonomic movements.

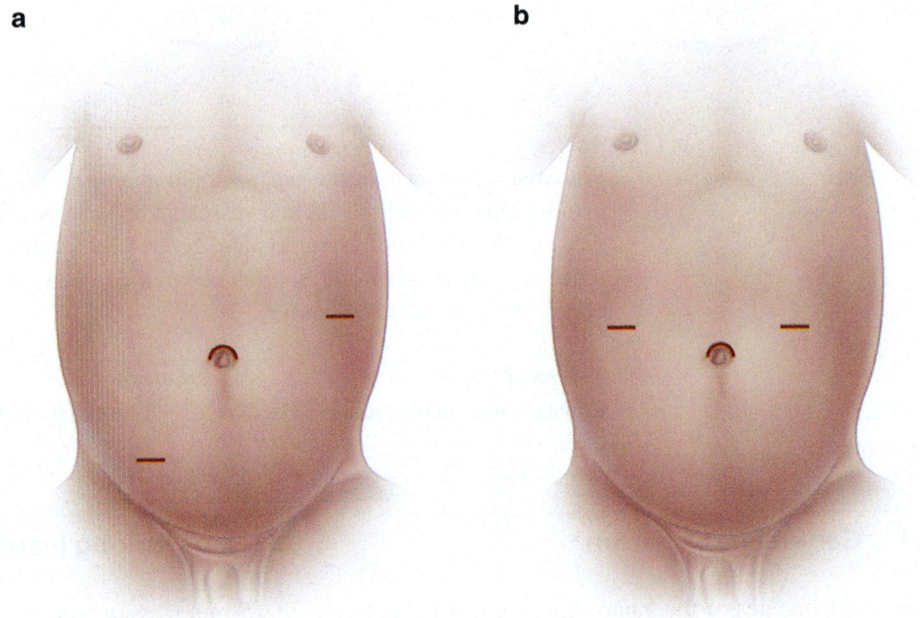

Fig. 31.1 Positioning of ports for left (**a**) and bilateral (**b**) inguinal hernia in an infant

31.4 Relevant Anatomy (Fig. 31.2)

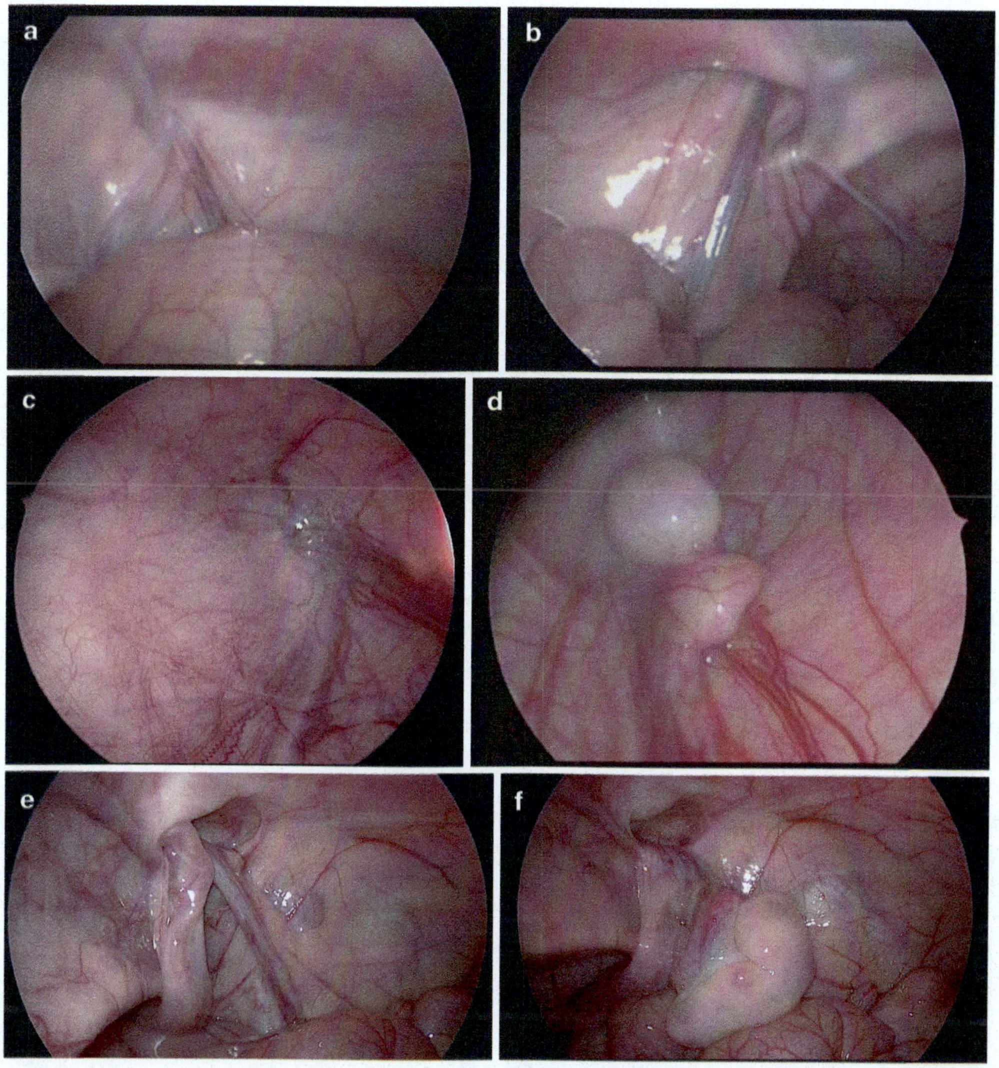

Fig. 31.2 (**a**) Closed right inguinal canal in a boy. The ring is closed, and the cord structures may be seen entering the closed deep inguinal ring. (**b**) Open left inguinal ring in a boy. The ring is seen here to be widely opened, and it has a horseshoe shape. (**c**) Inside of an inguinal hernia in a boy. In this instance, the patent processus vaginalis (hernia sac) reaches the scrotal level but stops short of the testis itself. The vas (*lower middle*) and vessels (from *right lateral* side) may be seen travelling beyond the hernia sac to the testis. (**d**) Inside of an inguinal hernia with the testis in the sac. In this instance, the patent hernia sac reaches the testis itself, which may be seen with the laparoscope. (**e**) Open inguinal canal and right inguinal hernia in a girl. The inguinal ligament is seen clearly from inside the abdomen in this photograph. (**f**) Reduced ovary. The ovary is shown in this second picture after being reduced from the inguinal canal

31.5 Surgical Technique (Figs. 31.1, 31.2, 31.3, 31.4, 31.5, 31.6, 31.7, and 31.8)

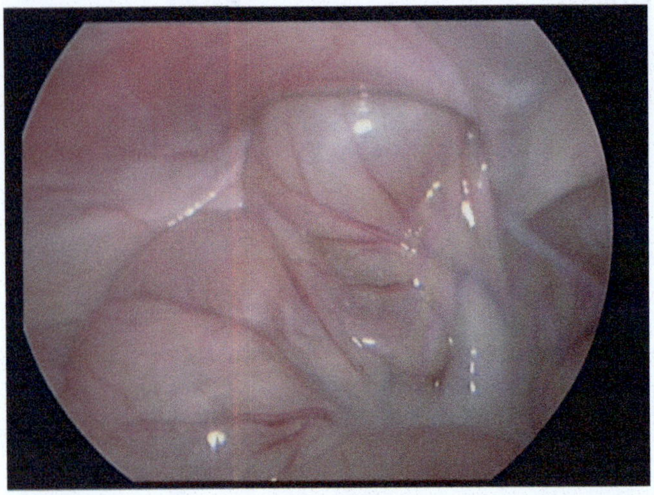

Fig. 31.3 Reduction of herniated contents. If any herniated bowel is present, it may be reduced with a combination of external pressure on the groin and gentle pulling of the viscera from the abdomen. A nonabsorbable suture of the surgeon's choice is introduced into the abdomen, either through the abdominal wall under vision or through a port, if possible. The needle may be introduced into the abdomen with the needle holder if no ports are being used

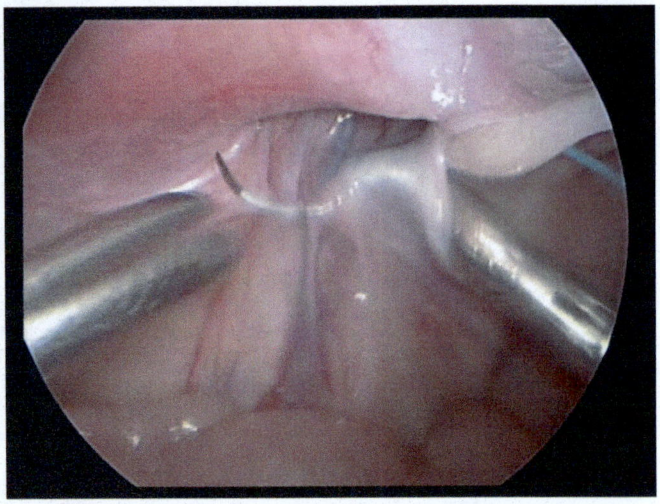

Fig. 31.5 Avoiding the vas and vessels. The suture then is used to incorporate the peritoneum at the internal ring, but avoiding the vas and vessels, by picking up the peritoneum on either side and between these structures

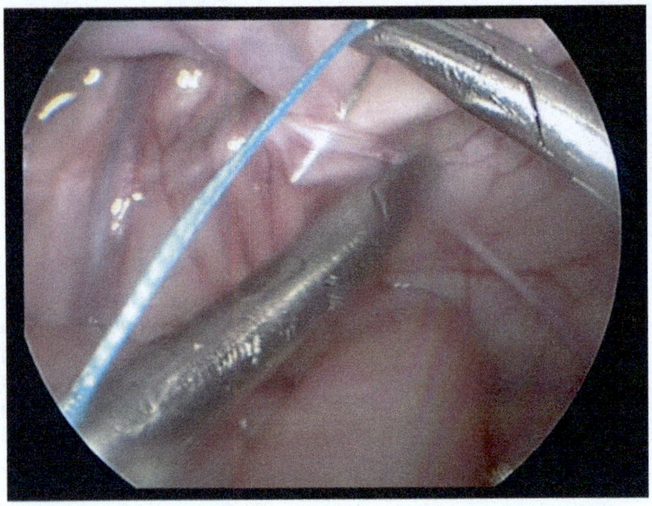

Fig. 31.4 Suture of the internal ring. The first stitch may be placed at the most ergonomic point or preference for the surgeon. The stitch incorporates the peritoneum at the level of the ring while avoiding the vas and vessels. Suture ligation in this patient began medially at the internal ring, incorporating the peritoneum medial to the vas

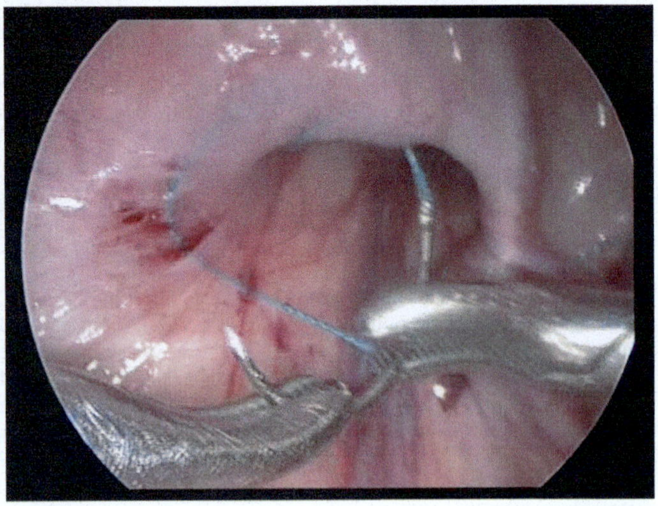

Fig. 31.6 Lateral incorporation of the peritoneum and abdominal wall. The suture then is continued laterally, and a small amount of abdominal wall muscle may be included in the stitch. Care is taken to avoid the epigastric vessels while suturing the ring at that level

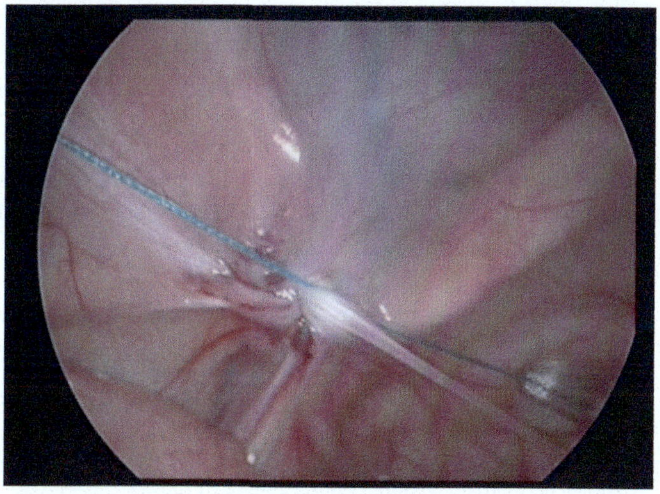

Fig. 31.7 Closure of the ring. On completion of the purse-string suture at the starting point, confirmation of the integrity of the closure may be confirmed by pulling the suture tight

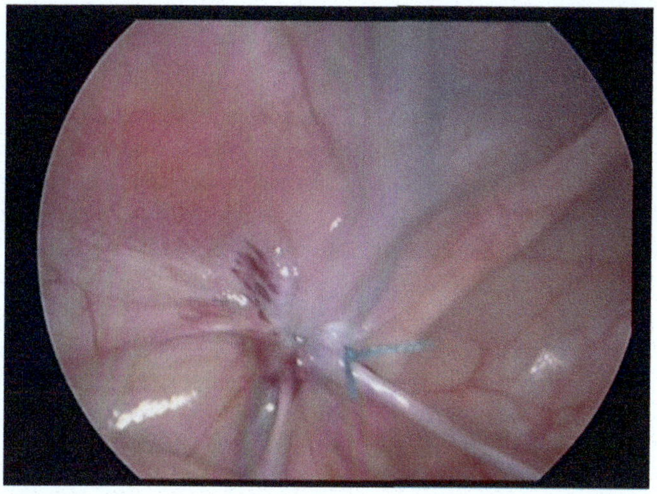

Fig. 31.8 Intracorporeal knot. The suture then is tied intracorporeally while gentle pressure on the scrotum evacuates the CO_2 within it, and the suture is cut. The instruments are removed under vision. The wounds are closed with an absorbable suture to muscle and with subcuticular suture or skin glue

31.6 Alternatives

If preferred, a left- or right-handed needle holder may be used (instead of a grasper in the nondominant hand).

Other techniques for repairing these hernias are described. They include dividing the hernia sac before suturing and dissection of the vas and vessels of the sac before placing the purse-string suture. Other techniques include a combined percutaneous and endoscopic approaches such as the subcutaneous endoscopically assisted ligation or SEAL technique. Laparoscopic extraperitoneal repair also is described. These techniques are said to lessen the incidence of recurrence but may have to be balanced with the increased risks associated with handling the vas and vessels.

31.7 Highlights and Pitfalls

- The first knot may require tension to maintain closure before the second knot is done. If maintaining tension is difficult, three throws of the first knot may help overcome this.
- Vigilant care must be taken to steer away from the epigastric vessels to avoid bleeding and obscuring the view.
- In small infants, positioning across the table may be advantageous, especially if bilateral access is required, as the ergonomics are optimal in that position.
- Studies suggest that a nonabsorbable braided suture is associated with a lower incidence of recurrence compared with absorbable and monofilament sutures.

Suggested Reading

Endo M, Watanbe T, Nakano M, Yoshida F, Ukiyama E. Laparoscopic completely extraperitoneal repair of inguinal hernia in children: a single-institution experience with 1,257 repairs compared with cut-down herniorrhaphy. Surg Endosc. 2009;23:1706–12.

Kelly KB, Krpata DM, Blatnik JA, Ponsky TA. Suture choice matters in rabbit model of laparoscopic, pre-peritoneal, inguinal hernia repair. J Laparoendosc Adv Surg Tech A. 2014;24:428–31.

Montupet P, Esposito C. Fifteen years experience in laparoscopic inguinal hernia repair in pediatric patients. Results and considerations on a debated procedure. Surg Endosc. 2011;25:450–3.

Nah SA, Giacomello L, Eaton S, de Coppi P, Curry JI, Drake DP, et al. Surgical repair of incarcerated inguinal hernia in children: laparoscopic or open? Eur J Pediatr Surg. 2011;21:8–11.

Undescended Testis: Laparoscopic Fowler-Stephens Orchidopexy

32

Francisca Yankovic and Naima Smeulders

Abstract

Undescended testis is the most frequent congenital genitourinary anomaly, affecting 1–2 % of boys [1]. About 20 % of them will have an impalpable testicle (IPT) [1, 2], situated intra-abdominally in nearly half. In the remainder, the testis may be absent, may have atrophied, or may be hidden in the inguinal canal or fat [2, 3]. For patients with IPT, laparoscopy is the gold standard for diagnosis, and it allows the surgical treatment to be performed in the same setting [4]. The laparoscopic surgical options for the intra-abdominal testis include single-stage orchidopexy, single-stage Fowler-Stephens procedure (FSP), and two-stage FSP. Intra-abdominal testes with sufficient vessel length may be mobilised into the scrotum in one procedure, but the gonadal vessels are too short in the vast majority of patients. In 1959, Fowler and Stephens [5] described division of the testicular vasculature to aid mobilisation, thereby leaving the testes to rely on collateral blood supply along the vas deferens. Twenty-five years later, Ransley *et al.* [6] advised a two-stage procedure with an interval between vessel ligation and testicular mobilisation to allow time for enhancement of the collateral blood supply along the vas deferens. Laparoscopy for division of the vessels was introduced 20 years ago, and more recently, it has been used for mobilisation during the second stage [7–9]. The FSP has been demonstrated to be an effective and safe technique with reported success rates consistently above 80 %; the complications of testicular atrophy and ascent are documented in about 10 % and 5 %, respectively. A systematic review comparing single-stage and two-stage FSP concluded that the staged approach has a higher rate of success, with better testicular viability [10]. This chapter describes the surgical technique of first- and second-stage Fowler-Stephens orchidopexy.

Keywords

Cryptorchidism • Undescended testis • Impalpable testis • Intra-abdominal testis • Laparoscopy • Orchidopexy • Fowler-Stephens procedure

32.1 General Information

Undescended testis is the most frequent congenital genitourinary anomaly, affecting 1–2 % of boys [1]. About 20 % of them will have an impalpable testicle (IPT) [1, 2], situated intra-abdominally in nearly half. In the remainder, the testis may be absent, may have atrophied, or may be hidden in the inguinal canal or fat [2, 3]. For patients with IPT, laparoscopy is the gold standard for diagnosis, and it allows the surgical treatment to be performed in the same setting [4]. The

F. Yankovic
Department of Paediatric Urology,
Hospital Exequiel Gonzales Cortes and Clinica Santa Maria,
Santiago, Chile

N. Smeulders (✉)
Great Ormond Street Hospital for Children
NHS Foundation Trust, London, UK

M. McHoney et al. (eds.), *Color Atlas of Pediatric Anatomy, Laparoscopy, and Thoracoscopy*,
DOI 10.1007/978-3-662-53085-6_32

laparoscopic surgical options for the intra-abdominal testis include single-stage orchidopexy, single-stage Fowler-Stephens procedure (FSP), and two-stage FSP. Intra-abdominal testes with sufficient vessel length may be mobilised into the scrotum in one procedure, but the gonadal vessels are too short in the vast majority of patients. In 1959, Fowler and Stephens [5] described division of the testicular vasculature to aid mobilisation, thereby leaving the testes to rely on collateral blood supply along the vas deferens. Twenty-five years later, Ransley *et al.* [6] advised a two-stage procedure with an interval between vessel ligation and testicular mobilisation to allow time for enhancement of the collateral blood supply along the vas deferens. Laparoscopy for division of the vessels was introduced 20 years ago, and more recently, it has been used for mobilisation during the second stage [7–9]. The FSP has been demonstrated to be an effective and safe technique with reported success rates consistently above 80 %; the complications of testicular atrophy and ascent are documented in about 10 % and 5 %, respectively. A systematic review comparing single-stage and two-stage FSP concluded that the staged approach has a higher rate of success, with better testicular viability [10]. This chapter describes the surgical technique of first- and second-stage Fowler-Stephens orchidopexy.

32.2 Working Instruments

- 5-mm Hasson port
- Two ports of 5 mm or 3 mm
- 5-mm, 30° telescope
- 5-mm or 3-mm Kelly forceps
- 5-mm or 3-mm scissors
- 5-mm endoclips or bipolar coagulation forceps
- Laparoscopic diathermy lead

32.3 Positioning, Port Siting, and Ergonomic Considerations

The patient is positioned supine and the IPT is confirmed on examination under anaesthesia. A 5-mm umbilical port is used as the primary port. Two additional 5-mm working ports are placed as shown in Fig. 32.1a. In bilateral cases, the working ports are placed in each flank, as illustrated in Fig. 32.1b. For small patients, 3-mm working ports may be used, but a 5-mm port will be necessary for clip placement.

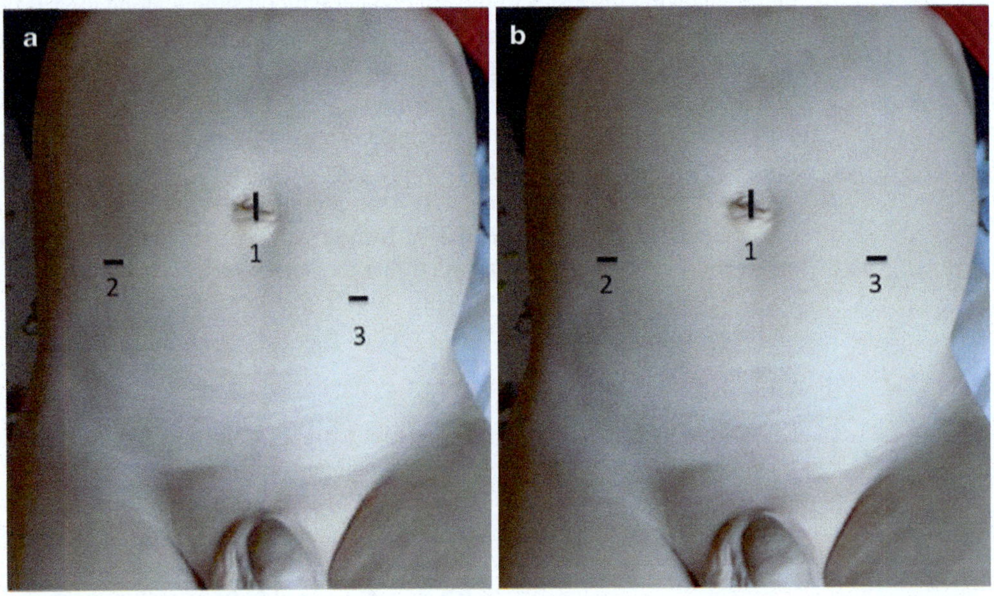

Fig. 32.1 (**a**) Port placement for right impalpable testicle (IPT). (**b**) Port placement for bilateral IPT. Shown are a 5-mm umbilical port (*1*) and the two additional 5-mm working ports (*2* and *3*)

32.4 Relevant Anatomy and Diagnostic Laparoscopy Findings

After placement of the umbilical port, a diagnostic laparoscopy is carried out. The normal anatomy of the contralateral side is shown in Fig. 32.2. The laparoscopic assessment of the IPT has a number of possible outcomes:

- The finding of an intra-abdominal testicle (Fig. 32.3)
- The presence of a vas deferens and spermatic vessels running into the inguinal canal with the internal/deep ring closed or open (Fig. 32.4)
- An atrophic intra-abdominal testicle or nubbin (Fig. 32.5)
- A blind-ending vas deferens and spermatic vessels (Fig. 32.6).

When the vas deferens and vessels are observed to run into the deep ring, we recommend an open inguinal exploration for a testis hidden within the inguinal canal/fat or a testicular nubbin or vanishing testis. An atrophic intra-abdominal testicle or nubbin should be removed and sent for histological analysis. If a blind-ending vas deferens is found, the spermatic vessels must be identified (Fig. 32.6), to ensure that a gonadal structure dissociated from the vas deferens-epididymis (for instance, located adjacent to the lower pole of the kidney) is not missed. When both the vas deferens and vessels are absent, testicular agenesis is diagnosed. No additional intraoperative procedure is indicated, but an ultrasound of the urinary tract should be undertaken subsequently, in view of the association with urinary tract anomalies, particularly renal agenesis.

Finding an intra-abdominal testis will prompt a brief assessment of the length of the testicular vessels, using the following guide: If an intra-abdominal testis can be brought to the contralateral deep inguinal ring without tension, the vessel length is likely to be sufficient to enable mobilisation to a scrotal position in a single procedure. For the vast majority of intra-abdominal testes, this is not the case, however, and a staged Fowler-Stephens procedure is carried out.

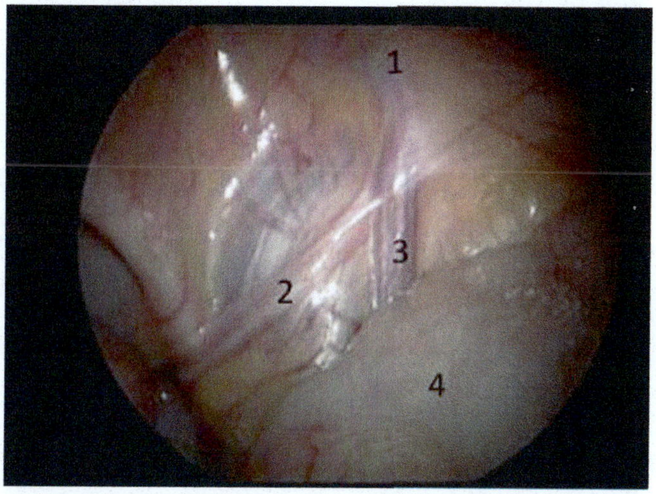

Fig. 32.2 Contralateral normal side (*right*): the vas deferens (*2*) and spermatic vessels (*3*) exit through a closed deep/internal inguinal ring (*1*); *4* designates the caecum

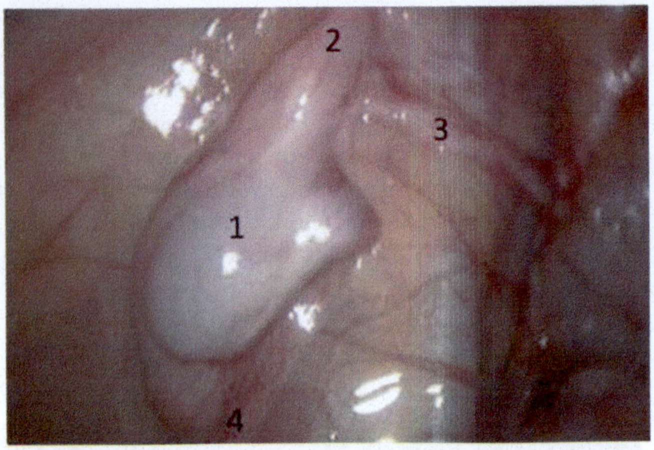

Fig. 32.3 Intra-abdominal left testicle (*1*). Also shown are the gubernaculum (*2*), vas deferens (*3*), and spermatic vessels (*4*)

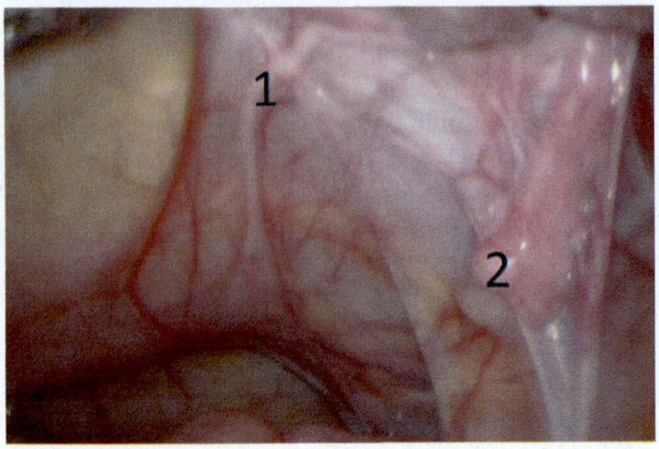

Fig. 32.5 Right atrophic intra-abdominal testicular nubbin (*2*); also present is a vas deferens (*1*)

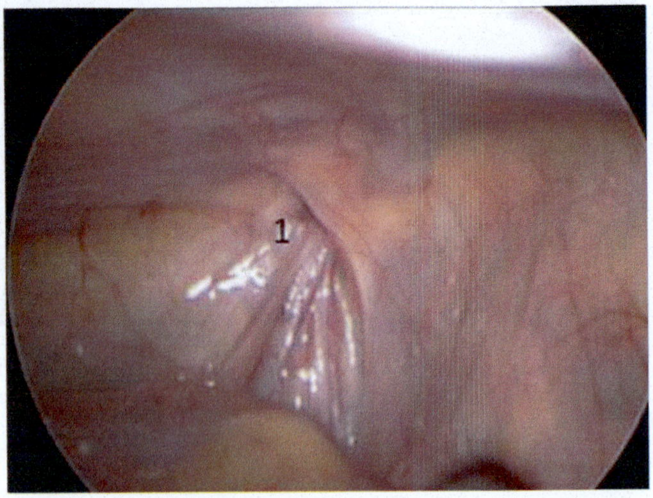

Fig. 32.4 Vas deferens and vessels exit through an open left deep ring into the inguinal canal (*1*)

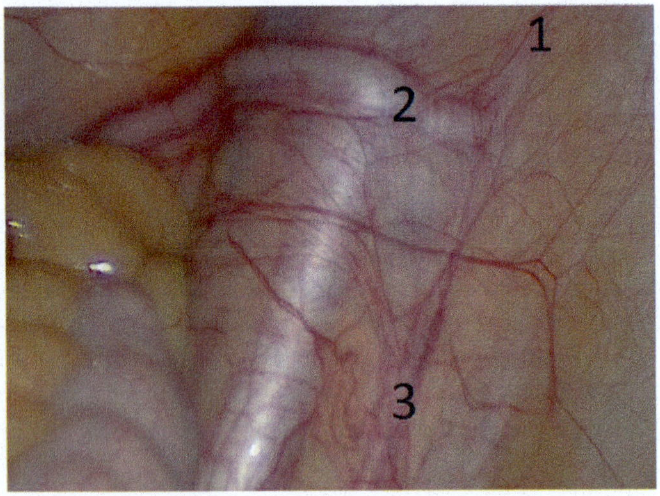

Fig. 32.6 Right blind-ending vas deferens (*2*) and atretic spermatic vessels (*3*), with a closed deep inguinal ring (*1*)

32.5 Surgical Technique: First-Stage Fowler-Stephens Procedure

1. The patient is positioned supine; the laparoscopic insufflator is set to 10 mmHg of pressure and 2 L/min of flow. The umbilical port is placed using an open Hasson technique. A 5-mm 30° camera is used and two additional 5-mm ports are placed as shown in Fig. 32.1a (or Fig. 32.1b if bilateral) under direct vision.
2. The intra-abdominal testicle is then identified. A Kelly forceps is used to explore the relevant anatomy. If the bowel obscures the view, it can be gently swept towards the upper abdomen using the length of a closed Kelly forceps. A Trendelenburg position will help keep the bowel in the upper abdomen. Good exposure of the spermatic vessels is essential to avoid subsequent diathermy damage to the testicle or neighbouring structures.
3. Once good exposure of the vessels has been achieved, with a Kelly forceps in the nondominant hand and a laparoscopic scissor with the diathermy lead attached in the dominant hand, the surgeon creates a window in the peritoneum over the spermatic vessels to the retroperitoneum (Fig. 32.7). This window should be proximal on the spermatic vessels well away from the testis, in order to avoid compromising it during subsequent diathermy and division, and so as to maintain a broad base for the testis, preventing torsion.
4. The Kelly forceps can be passed under the vessels to enable the proximal aspect of the vasculature to be lifted up into the operative field for clip placement or coagulation well away from the testis. The vessels are then occluded en masse, either by placement of endoclips or by coagulation using bipolar forceps (Fig. 32.8). The Kelly forceps is maintained in position while the endoclip or bipolar instrument is exchanged for the scissors for division of the vessels between the endoclips or over the area of coagulation (Figs. 32.8 and 32.9).
5. After the vessels are divided, the abdominal cavity is inspected to exclude bleeding or inadvertent injury to a neighbouring structure (Fig. 32.10). The ports are removed and the pneumoperitoneum is carefully allowed to deflate. The port sites are closed with a dissolvable suture (*eg*, 4-0 or 3-0 vicryl) and sealed with skin glue.
6. The collateral blood supply along the vas deferens is allowed to strengthen for an interval (for instance, 6 months) before proceeding to the second stage for the mobilisation of the testis on the vas deferens and its vascular collaterals.

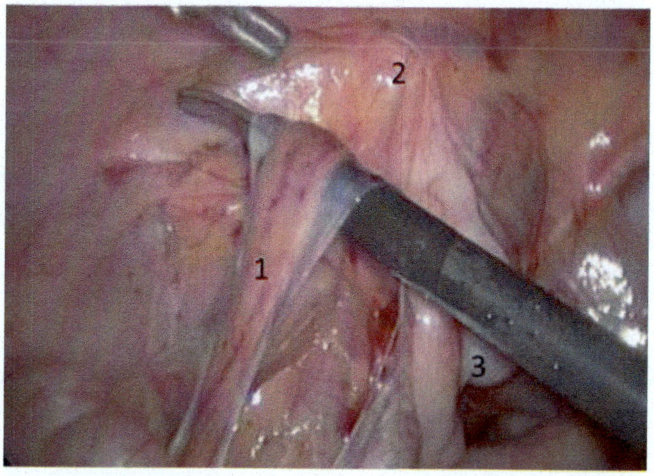

Fig. 32.7 Retroperitoneal window over the left spermatic vessels (*1*); also visible are the deep inguinal ring (*2*) and the intra-abdominal testicle (*3*)

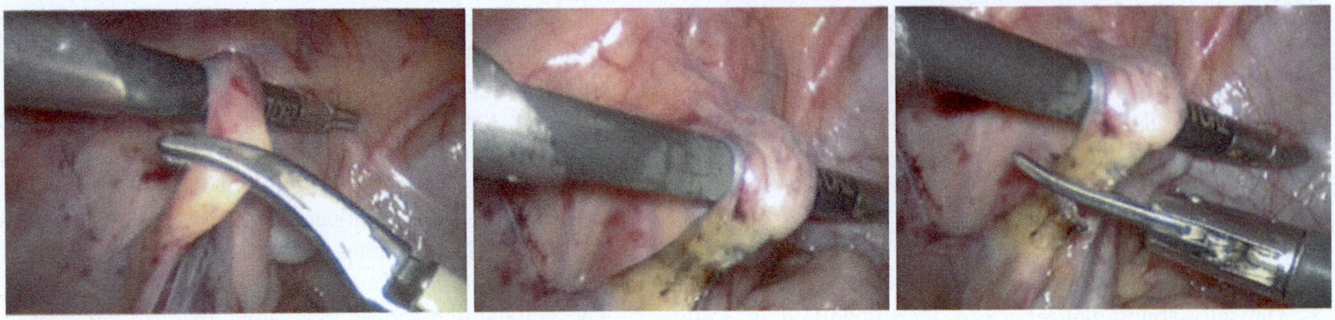

Fig. 32.8 Bipolar diathermy forceps coagulation and scissor division of left spermatic vessels

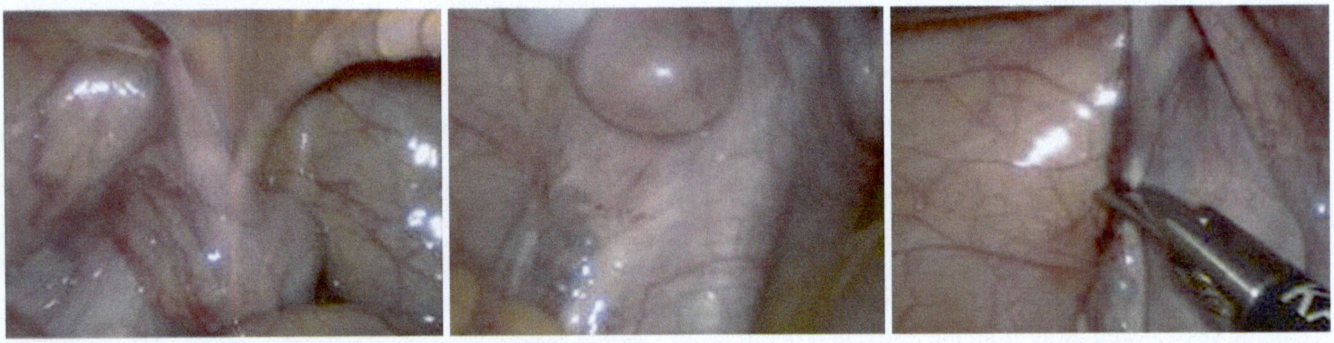

Fig. 32.9 Scissor division of left spermatic vessels between endoclips

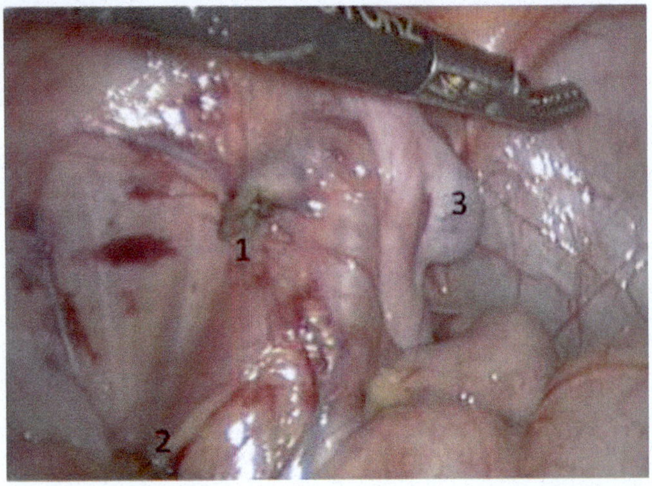

Fig. 32.10 Final appearance after a left first-stage Fowler-Stephens procedure (FSP). Visible are the intra-abdominal testicle (*3*) and the distal end (*1*) and proximal end (*2*) of the left spermatic vessels

32.6 Surgical Technique: Second-Stage Fowler-Stephens Procedure

1. The procedure commences with an examination of the groin and scrotum under anaesthesia, as the testis may have descended to an inguinal position following the division of the short spermatic vessels in the first stage. Nevertheless, a high inguinal or "peeping" testis is still best approached laparoscopically, at least initially, for maximal mobilisation.
2. Again, the patient is positioned supine; the laparoscopic insufflator is set to 10 mmHg of pressure and 2 L/min of flow. The umbilical port is placed using an open Hassan technique. A 5-mm 30° camera is used, and two additional 5-mm ports are positioned under direct vision, using the previous scars from the first stage.
3. An initial laparoscopy is performed (Fig. 32.11). In our experience, testicular atrophy after the first stage is extremely infrequent. The appearance and size of the testis and its connection to the epididymis should be recorded. If the bladder is distended (as seen in the figure), subsequent passage of the testis through the conjoint tendon will be facilitated by using a urethral catheter to empty the bladder.
4. With a Kelly forceps in the nondominant hand and a laparoscopic scissor with the diathermy lead attached in the dominant hand, the surgeon mobilises the testicle. The first step is the division of the gubernaculum. To avoid dividing the vas and the collateral blood supply to the testis, care must be taken to identify anatomical variants such as a looping vas deferens (Fig. 32.12).

   A short loop of vas deferens usually can be mobilised laparoscopically from within the abdomen, but a long-looping vas may require additional groin incision for its preservation. The gubernaculum is retracted into the peritoneal cavity and divided caudal to the Kelly and the vas deferens (Fig. 32.13). It has been suggested that by preserving the gubernaculum and routing the testis via the inguinal canal, a further blood supply to the testis via the gubernaculum can be maintained [11]. It is our experience, however, that the gubernacular attachments are usually abnormal and require division for optimal mobilisation, without detriment to outcome.
5. The mobilization of the testicle continues first laterally, then across the divided spermatic vessels (Fig. 32.14), and finally medially, releasing the vas deferens on a strip of peritoneum supporting the deferential vessels. As the dissection continues medially over the medial umbilical ligament, care must be taken to maintain the width of the peritoneal strip (Fig. 32.15). Once the mobilisation enables the testis to reach to the contralateral deep ring without tension, sufficient length has been obtained to place the testis in the scrotum (Fig. 32.16).
6. To reduce the distance to the scrotum, a more direct path than that afforded by the inguinal canal can be forged through the conjoint tendon. We recommend careful blunt dissection with a 5-mm laparoscopic Kelly forceps between the medial border of the deep ring, marked by the inferior epigastric vessels, and the medial umbilical ligament (Fig. 32.17). If the length of the strip of peritoneum carrying the deferential vessels to the mobilised intra-abdominal testis is too short to reach the scrotum via this route, an even more direct path can be created between the medial umbilical ligament and the bladder, as illustrated in Fig. 32.21. However, extreme care is needed to avoid inadvertent damage to the bladder, so emptying the bladder via a urethral catheter can be advantageous in this situation.
7. The next step is to expose the empty scrotum and create a subdartos pouch (Fig. 32.18). When the pouch is completed, an artery forceps is directed through the passage created in the previous steps into the abdominal cavity (Fig. 32.19). Alternatively, the dissecting Kelly forceps can be advanced from within the abdomen through the passage and scrotal incision, enabling a further port (5 or 10 mm) to be railroaded over the Kelly forceps into the abdominal cavity.
8. Attention to the orientation of the testis as it is grasped through the additional port by either the artery forceps or a toothed laparoscopic grasping forceps avoids torsion of the mobilised strip of peritoneum and testis as they are brought to the scrotum. By bringing the testis into the port, it slips more easily down the passage created through the conjoint tendon (Fig. 32.20). Furthermore, the Kelly forceps is used to lift the peritoneum away from the passage, thereby preventing the testis from being caught on the peritoneum and pulled out of the forceps.
9. Finally, the testicle is placed into the subdartos pouch (Fig. 32.21). If present, the hydatid of Morgagni is removed. Again, note is made of the testicular size and appearance, as well as its epididymal association. We recommend stitching the testis within the pouch with one or two dissolvable sutures (Vicryl or Monocryl) to secure the position of the testicle while it becomes adherent within the pouch.

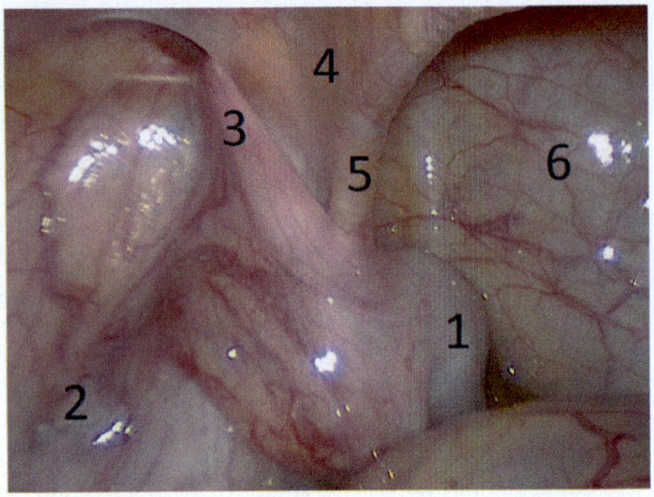

Fig. 32.11 Laparoscopy for left second-stage Fowler-Stephens procedure. Visible are the left intra-abdominal testis (*1*), an endoclip on a distal section of divided spermatic vessels (*2*), the gubernaculum extending through the open deep inguinal ring (*3*), the conjoint tendon (*4*), the medial umbilical ligament (*5*), and a distended bladder (*6*)

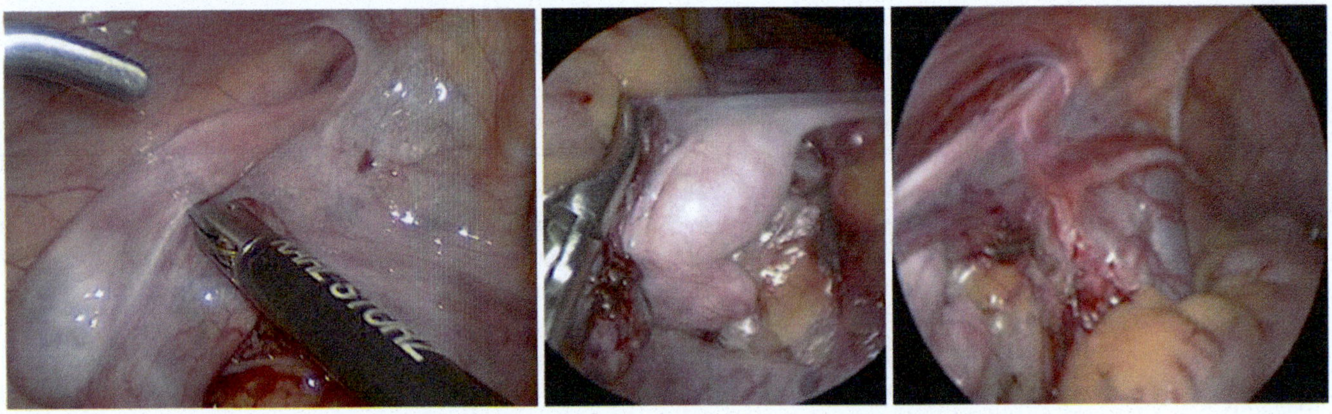

Fig. 32.12 Careful assessment for anatomical variations: long-looping vas deferens to left intra-abdominal testis

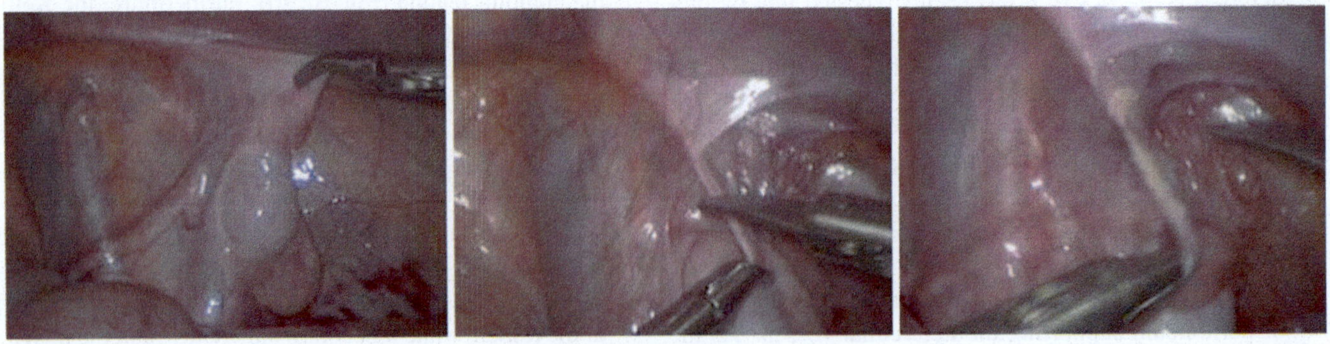

Fig. 32.13 Division of the gubernaculum (right intra-abdominal testis)

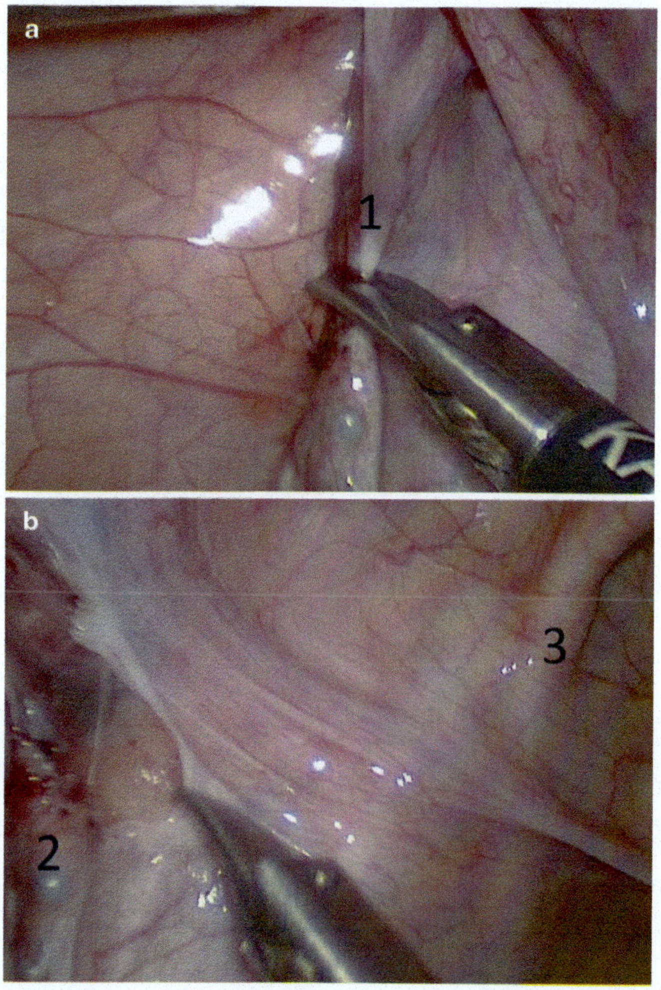

Fig. 32.14 (**a, b**) Mobilisation of the left testis by dissection between the endoclips placed on the spermatic vessels divided at the first stage (*1* distal clip; *2* proximal clip) and medially towards the medial umbilical ligament (*3*)

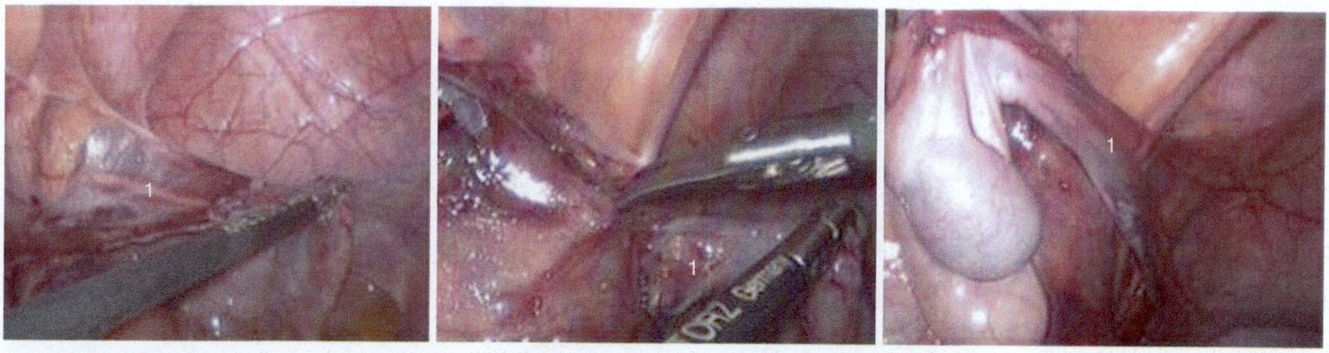

Fig. 32.15 Mobilisation of the vas deferens and collateral vessels, maintaining a wide strip of peritoneum 1 as the dissection continues over the medial umbilical ligament (left FSP)

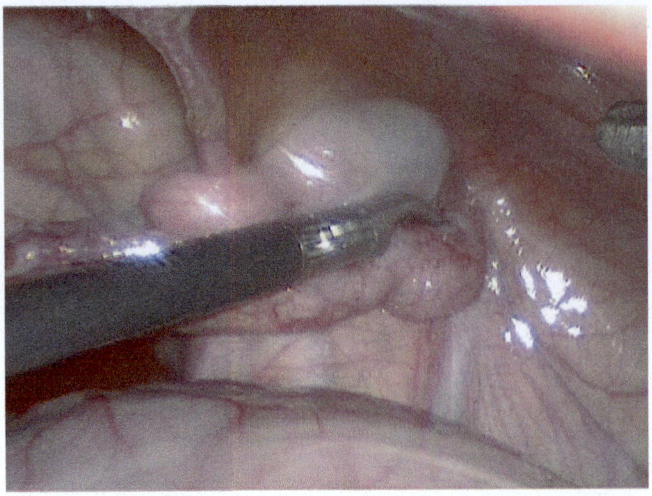

Fig. 32.16 The mobilised left testis reaches contralateral (*right*) deep inguinal ring without tension. The right deep inguinal ring is indicated by the right vas deferens and vessels, and can be seen behind the left testis

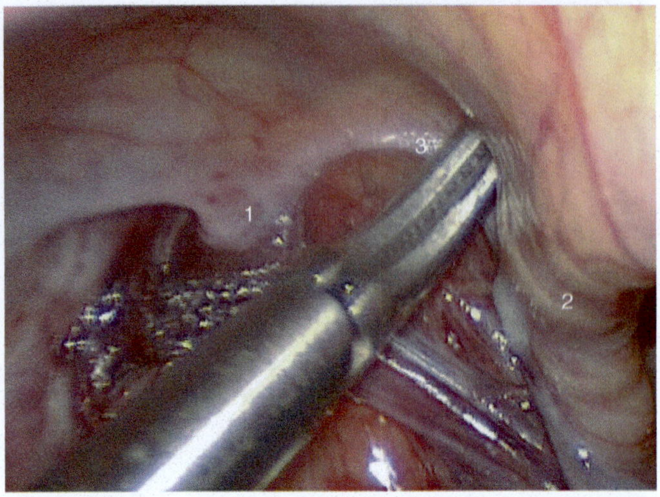

Fig. 32.17 In this left FSP, blunt dissection is used to create the passage through the conjoint tendon (*3*). Visible are the inferior epigastric vessels and medial border of the deep inguinal ring (*1*) and the medial umbilical ligament (*2*)

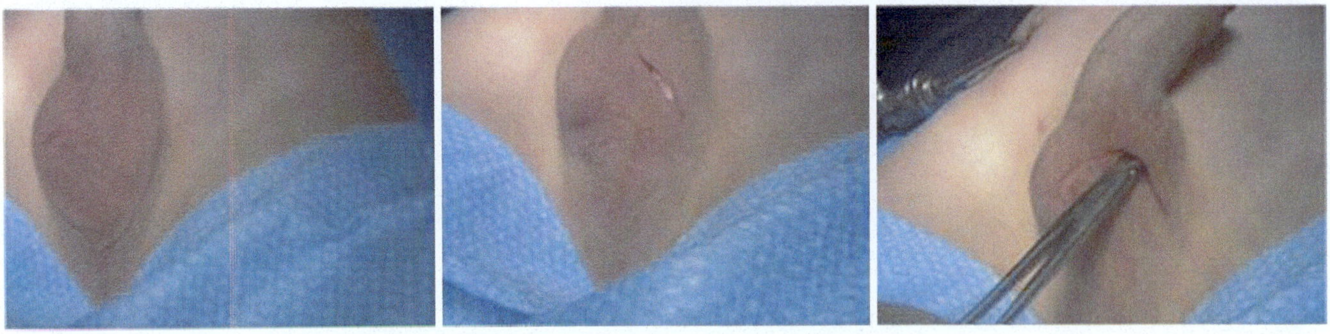

Fig. 32.18 A subdartos pouch is created and an artery forceps is used to reach the abdominal cavity (left FSP)

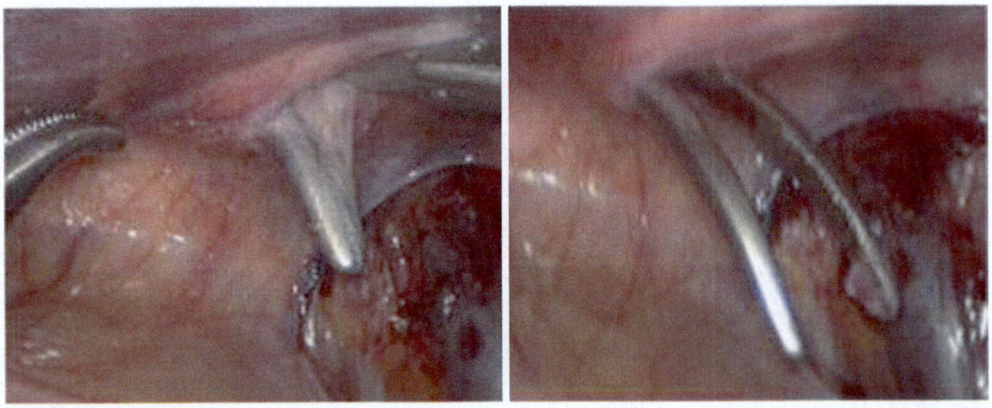

Fig. 32.19 Laparoscopic view of the artery forceps inside the abdominal cavity (right FSP)

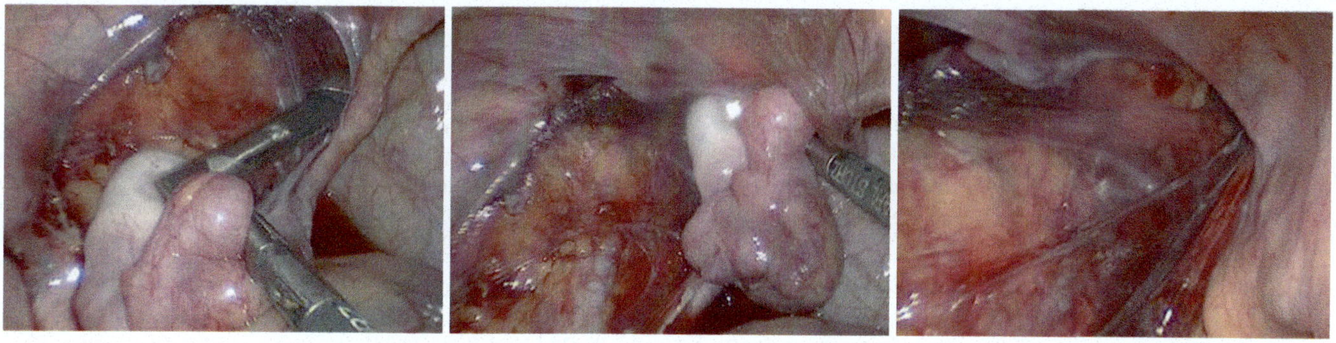

Fig. 32.20 Laparoscopic view of the left testis held in a toothed laparoscopic grasping forceps. The testis is guided along the passage by retraction into the additional railroaded port and by using a Kelly forceps, held in the surgeon's right hand, to lift the peritoneum over the testis

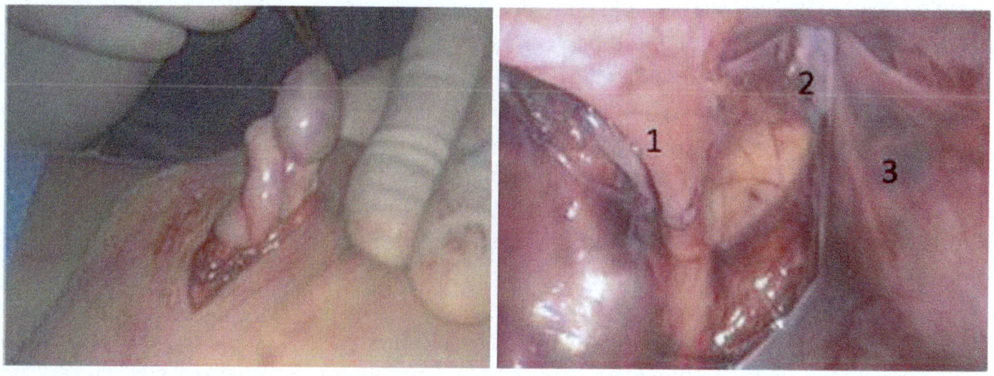

Fig. 32.21 Left testis reaches scrotum without tension. Note that the left testis has been routed medial to the medial umbilical ligament (*1*) via an exit site through the conjoint tendon alongside the bladder (*2*). Also seen is the vas deferens with the strip of peritoneum (*3*)

32.7 Highlights and Pitfalls

- After achieving good exposure of the spermatic vessels in the first-stage FSP, lifting them over a Kelly forceps (Fig. 32.7) allows the vessels to be clipped or coagulated and divided well away from the testis, maintaining a broad base and thereby avoiding heat damage and loss from torsion.
- Assessment for variations in vasal anatomy prevents inadvertent transection of a looping vas deferens (Fig. 32.12).
- During the mobilisation of the intra-abdominal testis and the deferential vessels in the second-stage procedure, care must be taken to maintain the width of the peritoneal strip as the dissection continues medially over the medial umbilical ligament (Figs. 32.14 and 32.15).
- Attention to the orientation of the mobilised testis and its vasculature as it is grasped will prevent its torsion as the testis is brought to the scrotum (Fig. 32.20).
- Passage of the testis through the conjoint tendon is facilitated in two ways: (1) by retraction into a further port, railroaded over the Kelly forceps used for blunt dissection of the passage from the abdominal cavity, and (2) using the Kelly forceps to lift the peritoneum over the testis as it slips down the passage (Fig. 32.20).
- When the peritoneal strip is exceptionally short, an even more direct route to the scrotum can be achieved medial to the medial umbilical ligament (Fig. 32.21). Emptying the bladder by placement of a urethral catheter will help avoid inadvertent injury to the bladder.

References

1. Daher P, Nabbout P, Feghali J, Riachy E. Is the Fowler-Stephens procedure still indicated for the treatment of nonpalpable intraabdominal testis? J Pediatr Surg. 2009;44:1999–2003.
2. Essam EM, Gamal W, Hussein MM. Laparoscopic orchidopexy for non-palpable testes: outcome of two techniques. J Pediatr Urol. 2011;7:178–81.
3. O'Brien MF, Hegarty PK, DeFrietas D, Bredin HC. One-stage Fowler-Stephens orchidopexy for impalpable undescended testis. Ir J Med Sci. 2004;173:18–9.
4. Taran I, Elder JS. Results of orchiopexy for the undescended testis. World J Urol. 2006;24:231–9.
5. Fowler R, Stephens FD. The role of testicular vascular anatomy in the salvage of high undescended testes. Aust N Z J Surg. 1959;29:92–106.
6. Ransley PG, Vordermark JS, Caldamone AA, Berllinger MF. Preliminary ligation of the gonadal vessels prior to orchidopexy for the intra-abdominal testicle. A staged Fowler-Stephens procedure. World J Urol. 1984;2:266–8.
7. Bloom DA, Ayers JW, McGuire EJ. The role of laparoscopy in management of nonpalpable testes. J Urol. 1988;94:465–70.
8. Jordan GH, Winslow BH. Laparoscopic single stage and staged orchidopexy. J Urol. 1994;152:1249–52.
9. Diamond DA, Caldamone AA. The value of laparoscopy for 106 impalpable testes relative to clinical presentation. J Urol. 1992;148:632–4.
10. Elyas R, Guerra LA, Pike J, DeCarli C, Betolli M, Bass J, et al. Is staging beneficial for Fowler-Stephens orchiopexy? a systematic review. J Urol. 2010;183:2012–9.
11. Robertson SA, Munro FD, Mackinlay GA. Two-stage Fowler-Stephens orchidopexy preserving the gubernacular vessels and a purely laparoscopic second stage. J Laparoendosc Adv Surg Tech A. 2007;17:101–7.

Laparoscopic Varicocelectomy

33

Vassilis J. Siomos, Cole Wiedel, and Duncan T. Wilcox

Abstract

Varicoceles are thought to alter testicular development and semen parameters in some men. Whom to treat and the ideal time to treat varicoceles in adolescents have not been demonstrated, but surgical treatment often is recommended. We review the indications and describe the steps of a laparoscopic varicocelectomy.

Keywords

Varicocele • Pediatrics • Laparoscopy • Varicocelectomy • Surgical treatment

Varicoceles are thought to alter testicular development and semen parameters in some men. Whom to treat and the ideal time to treat varicoceles in adolescents have not been demonstrated, but surgical treatment often is recommended. We review the indications and describe the steps of a laparoscopic varicocelectomy.

33.1 General Information

Scrotal varicoceles are similar in nature to varicose veins of the lower extremity. They are caused by abnormal dilatation and tortuosity of the veins due to valvular dysfunction within the pampiniform plexus. The prevalence is 15–20 % but may be as high as 40 % in the subgroup of infertile men. Over 90 % of varicoceles occur on the left side [1]. The peak age for varicocele presentation is in adolescence, during the time of rapid testicular growth [2]. In addition to an undesirable appearance, often causing distress, it may be the cause of scrotal pain in up to 10 % of patients [3]. More importantly, several studies show that varicoceles may lead to abnormal semen parameters and testicular growth [4]. Surgical correction of the varicocele has been shown to improve semen parameters and inspire catch-up growth of the affected testicle [5–7].

Because approximately 90 % of males with a varicocele appear fertile and men with abnormal semen parameters have achieved pregnancy and resolution of a varicocele, there has been considerable debate regarding if and when surgical correction should occur. This debate is compounded by many pediatric studies that do not include data on semen parameters.

33.2 Evaluation

Usually, patients describe the gradual onset of a scrotal lump, commonly described as a "bag of worms." On examination, the volume of the testes should be similar. Studies have demonstrated that the affected side may have a smaller volume or even be atrophic [8]. Varicoceles are graded on a scale of

V.J. Siomos, MD (✉)
Department of Surgery-Urology, University of Colorado Hospital Children's Hospital Colorado, Aurora, CO, USA

C. Wiedel
Department of Surgery-Urology, University of Colorado Denver School of Medicine, University of Colorado Hospital, Children's Hospital Colorado, Aurora, CO, USA

D.T. Wilcox, MD
Department of Pediatric Urology, Children's Hospital Colorado, Aurora, CO, USA

M. McHoney et al. (eds.), *Color Atlas of Pediatric Anatomy, Laparoscopy, and Thoracoscopy*,
DOI 10.1007/978-3-662-53085-6_33

severity, designated I, II, or III. Grade I implies a small varicocele, palpable during Valsalva. Grade II implies a medium-sized varicocele, palpable at rest. A grade III varicocele is large and visualized at rest. Isolated right-sided varicoceles are rare and should be investigated further for causes of extrinsic compression or mass effect. Varicoceles also should decrease in size in a supine position versus standing. There is debate over the necessity of testicular ultrasound in the initial workup.

33.3 Indications for Treatment

Several studies demonstrated a benefit in adolescent varicocelectomy, whereas others advocate watchful waiting [9]. A recent Cochrane review on the subject could not identify a significant benefit to early repair [10]. Thus far, optimal timing for surgery has not been established, and there are limited data suggesting that postponement of surgery affects outcomes. Initial treatment with nonsteroidal anti-inflammatory medications and scrotal support is recommended for patients who have mild discomfort. At this point, the accepted treatment indications in adolescents include pain, discomfort, and a testicular size discrepancy greater than 15–20% on more than one occasion, although these often are debated [11].

33.4 Surgical Technique

There are several approaches to performing a varicocelectomy, including open inguinal, microscopic inguinal, laparoscopic, and open high (Palomo) ligation, and sclerotherapy. This chapter focuses on the laparoscopic Palomo technique. The goal of this procedure is to take advantage of the good collateral blood flow to the testes and to target the internal spermatic vessels before their descent into the deep inguinal ring.

The patient is asked to void before proceeding to the operative suite to avoid intraoperative catheterization and to keep the bladder out of the surgical field. The patient is placed under general anaesthesia, then prepared and draped in the supine position. An incision is made in or around the umbilicus along a natural skin line. We prefer to use an open Hasson technique and dissect down to the peritoneum using opposing hemostats to present each layer. Once the peritoneum is entered, a 5-mm port is placed in the abdomen and pneumoperitoneal pressure is kept between 10 and 15 mmHg. The lens is placed into the abdomen, and the abdomen is evaluated thoroughly for possible bowel or vessel injury, as well as to assess the anatomy of both inguinal canals (Figs. 33.1, 33.2, 33.3, 33.4, 33.5 and 33.6).

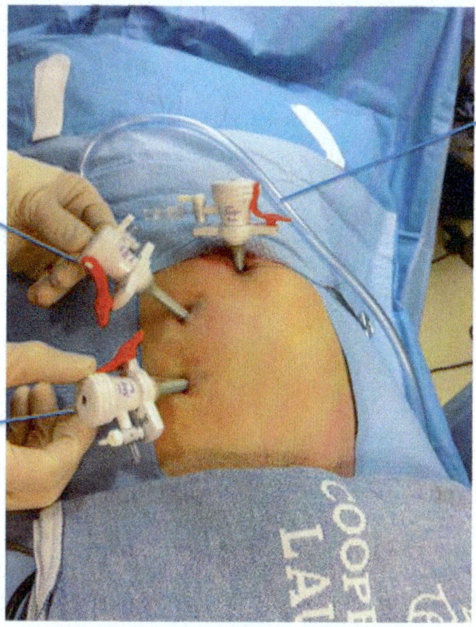

Fig. 33.1 The port incisions are marked in an attempt to spatially triangulate the affected side. The epigastric vessels are identified and avoided. Bupivicaine is used to infiltrate the incisions, and two more 5-mm ports are placed under direct vision. Port placement is shown here for a left-sided procedure. Ultimately, the camera is repositioned from port I and moved to the middle, which is labeled "camera port." We have chosen this arrangement to make it more ergonomic for the surgeon versus symmetric port placement around the umbilicus

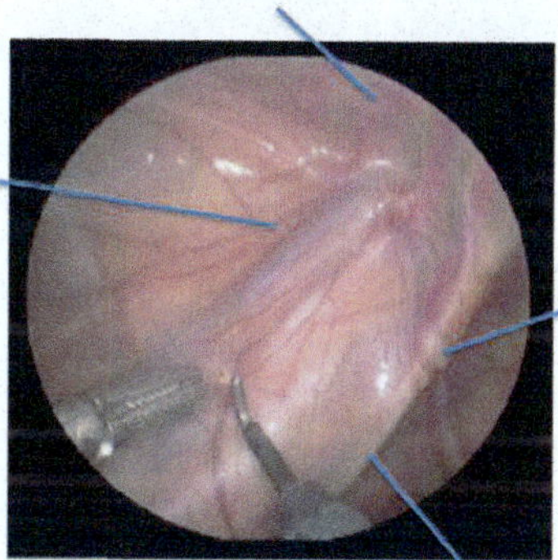

Fig. 33.2 The internal spermatic vessels and vas deferens are identified and are seen joining at the entrance of the inguinal canal. Note the vas deferens obliquely traversing the lateral umbilical ligament

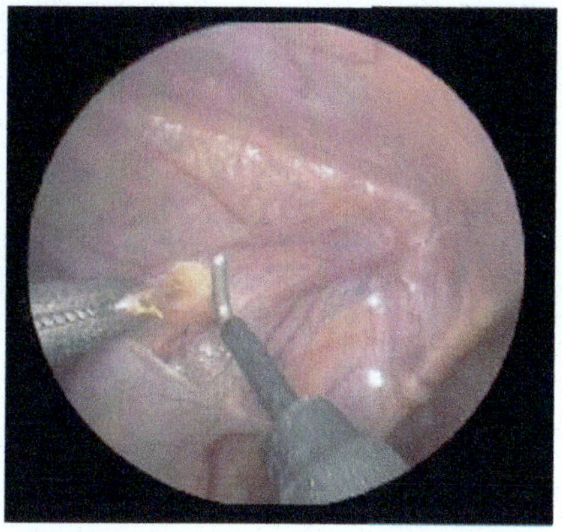

Fig. 33.3 A Maryland grasper and a hook with electrocautery are used to separate the posterior peritoneum from the vessels at least 5 cm from the canal to avoid injury to the vas deferens

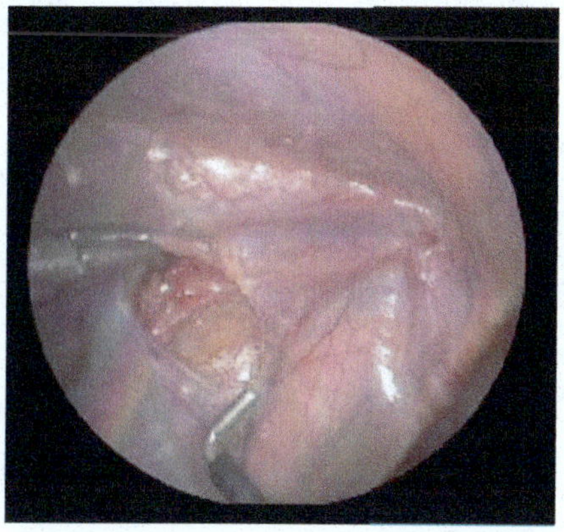

Fig. 33.4 Once a spacious peritoneal window has been made, the internal spermatic vessels are lifted anteriorly. Most of the time, the artery is fused with the veins and no attempts are made to spare the artery or the lymphatics

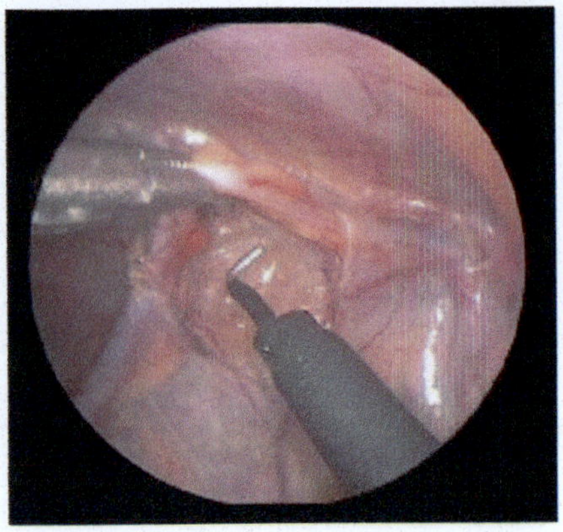

Fig. 33.5 Next, circumferential dissection is completed with the surgeon being mindful of the posteriorly located iliac vessels

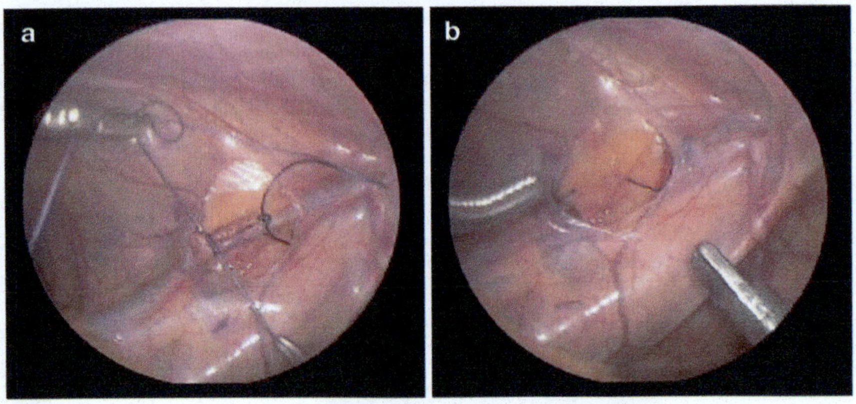

Fig. 33.6 (**a**, **b**) A suture then is passed at the proximal and distal ends of the vessels, leaving ample space between them. The vessels then are transected with laparoscopic scissors. A thermal sealing surgical device or hemostatic clips also may be used at this point in lieu of suture. Hemostasis is confirmed after decompressing the pneumoperitoneum. The ports then are removed under direct vision. The skin is closed in two layers, and dermabond (Ethicon; Somerville, NJ, USA) is applied. Intraoperative ketorolac is administered before extubation

33.5 Postoperative Follow-up

Patients are discharged the same day. We recommend scheduling and alternating ibuprofen and acetaminophen for the next 48 h and prescribing oxycodone as needed. We also recommend that ice be applied to the scrotum for the next 24 h and that the patient's activities be limited for the following week. Patients may return to sporting events and full activity in 6 weeks. We ask patients to return at 3 months for a wound check and at 1 year for a testicular ultrasound to evaluate testicular catch-up growth.

33.6 Discussion

Laparoscopic varicocelectomy is well tolerated in adolescents and has achieved good results. Because the internal spermatic vessels and lymphatics are closely associated, it often is difficult to separate them. We prefer taking the vessels en bloc, which makes subtotal ligation less likely. The two main complications relate to varicocele recurrence and hydrocele. When the spermatic vessels are taken en bloc, the recurrence rate ranges between 0 and 5 % [12–14]. In patients in whom the artery is spared, recurrence has been reported to range from 4 to 37 % [15]. Although there were initial concerns regarding ligation of the internal spermatic artery, several studies reported no evidence of testicular atrophy in adolescents treated with the Palomo approach. Good collateral flow via the external spermatic artery and artery of the vas deferens is maintained with the laparoscopic approach. This cannot be assumed in children who have had previous surgery on the ipsilateral side; thus, the artery should be preserved [16]. In the unfortunate event of a recurrence, it is possible to repeat the laparoscopic approach, which many authors recommend performing by radiologically guided sclerotherapy [17].

The reported range of hydrocele formation following en bloc vessel ligation is 10–30 %; artery-sparing laparoscopic procedures have been shown to lower the hydrocele rate to 6 % [18]. When the artery is spared, some of the lymphatics likely are spared as well. The theoretic cause of postoperative hydroceles is thought to be lymphatic disruption. In attempts to mitigate postoperative hydroceles, lymphatic-sparing procedures have been reported. Because the lymphatics are difficult to isolate, lymphatic-staining agents have been used. Four different studies reported a 0 % hydrocele rate with similar results in varicocelectomy efficacy [19–22].

Acknowledgments We thank Dr. David J. Chalmers for providing the video footage and Dr. A. Carson Baisden for photographic editing.

References

1. Witt MA, Lipshultz LI. Varicocele: a progressive or static lesion? Urology. 1993;42(5):541–3.
2. Oster J. Varicocele in children and adolescents. An investigation of the incidence among Danish school children. Scand J Urol Nephrol. 1971;5(1):27.
3. Chen S-S, Chen L-K. Risk factors for progressive deterioration of semen quality in patients with varicocele. Urology. 2012;79(1):128–33.
4. The influence of varicocele on parameters of fertility in a large group of men presenting to infertility clinics. World Health Organization. Fertil Steril. 1992;57:1289.
5. Thomas JC, Elder JS. Testicular growth arrest and adolescent varicocele: does varicocele size make a difference? J Urol. 2002;168(4):1689–91.
6. Paduch DA, Niedzielski J. Repair versus observation in adolescent varicocele: a prospective study. J Urol. 1997;158(3):1128–33.
7. Kass EJ, Chandra RS, Belman AB. Testicular histology in the adolescent with a varicocele. Pediatrics. 1987;79(6):996–8.
8. Korets R, Woldu SL, Nees SN, Spencer BA, Glassberg KI. Testicular symmetry and adolescent varicocele—does it need followup? J Urol. 2011;186(4):1614–9.
9. Preston MA, Carnat T, Flood T, Gaboury I, Leonard MP. Conservative management of adolescent varicoceles: a retrospective review. Urology. 2008;72(1):77–80.
10. Kroese A, de Lange N, Collins J, Evers J. Surgery or embolization for varicoceles in subfertile men. Cochrane Database Syst Rev. 2012;(10):CD000479.
11. Koyle MA, Oottamasathien S, Barqawi A, Rajimwale A, Furness 3rd PD. Laparoscopic Palomo varicocele ligation in children and adolescents: results of 103 cases. J Urol. 2004;172(4):1749–52.
12. Atassi O, Kass EJ, Steinert BW. Testicular growth after successful varicocele correction in adolescents: comparison of artery sparing techniques with the Palomo procedure. J Urol. 1995;153(2):482–3.
13. Palmer LS, Maizels M, Kaplan WE, Stokes S, Firlit CF. The influence of surgical approach and intraoperative venography on successful varicocelectomy in adolescents. J Urol. 1997;158(3):1201–4.
14. Parrott T, Hewatt L. Ligation of the testicular artery and vein in adolescent varicocele. J Urol. 1994;152(2 Pt 2):791.
15. Kass E, Marcol B. Results of varicocele surgery in adolescents: a comparison of techniques. J Urol. 1992;148(2 Pt 2):694.
16. Iselin C, Almagbaly U, Borst F, et al. Safety and efficiency of laparoscopic varicocelectomy in one hundred consecutive cases. Urol Int. 2010;58(4):213–7.
17. Mazzoni G, Minucci S, Gentile V. Recurrent varicocele: role of antegrade sclerotherapy as first choice treatment. Eur Urol. 2002;41(6):614–8.
18. Diamond DA, Xuewu J, Cilento Jr BG, et al. Varicocele surgery: a decade's experience at a children's hospital. BJU Int. 2009;104(2):246–9.
19. Schwentner C, Radmayr C, Lunacek A, et al. Laparoscopic varicocele ligation in children and adolescents using isosulphan blue: a prospective randomized trial. BJU Int. 2006;98(4):861–5.
20. Golebiewski A, Krolak M, Komasara L, Czauderna P. Dye-assisted lymph vessels sparing laparoscopic varicocelectomy. J Laparoendosc Adv Surg Tech A. 2007;17(3):360–3.
21. Chiarenza SF, D'Agostino S, Scarpa M, Fabbro M, Costa L, Musi L. Lymphography prior to laparoscopic Palomo varicocelectomy to prevent postoperative hydrocele. J Laparoendosc Adv Surg Tech A. 2006;16(4):394–6.
22. Capolicchio J-P, El-Sherbiny M, Brzezinski A, Eassa W, Jednak R. Dye-assisted lymphatic-sparing laparoscopic varicocelectomy in children. J Pediatr Urol. 2013;9(1):33–7.

Retroperitoneoscopic Adrenalectomy

34

Imran Mushtaq and Francisca Yankovic

Abstract

Laparoscopic adrenalectomy is the standard of care for the surgical excision of the adrenal gland, adrenal tumours, and adrenal biopsy. The choice between laparoscopic and retroperitoneoscopic approaches is dictated by the experience of the surgeon [1]. Patients requiring surgery for a phaeochromocytoma, adrenal adenoma, adrenal adenocarcinoma, Cushing's syndrome, neuroblastoma, or an incidentaloma are all candidates for this approach. There are no absolute contraindications to the minimally invasive (MI) approach, but for neuroblastomas and other adrenal neoplasms, care must be taken to maintain the principles of cancer surgery. Relative contraindications include previous surgery of the liver or kidney, large tumours (>8–10 cm in diameter), or coagulation disorders.

Keywords

Retroperitoneoscopic • Adenoma • Adrenal • Laparoscopic

34.1 General Information

Laparoscopic adrenalectomy is the standard of care for the surgical excision of the adrenal gland, adrenal tumours, and adrenal biopsy. The choice between laparoscopic and retroperitoneoscopic approaches is dictated by the experience of the surgeon [1]. Patients requiring surgery for a phaeochromocytoma, adrenal adenoma, adrenal adenocarcinoma, Cushing's syndrome, neuroblastoma, or an incidentaloma are all candidates for this approach. There are no absolute contraindications to the minimally invasive (MI) approach, but for neuroblastomas and other adrenal neoplasms, care must be taken to maintain the principles of cancer surgery. Relative contraindications include previous surgery of the liver or kidney, large tumours (>8–10 cm in diameter), or coagulation disorders.

A large body of literature supports MI adrenalectomy in adults, but the experience in children is relatively sparse. The transperitoneal route is used more widely and offers a larger working space [2]. The retroperitonscopic approach provides direct access to the adrenal gland and more direct visualization of the adrenal vein. It also avoids colonic mobilization and minimizes the risk of injury to hollow viscera and the potential risk of adhesion formation. However, the reversed orientation of the kidney and hilum, combined with a significantly smaller working space, may make this approach difficult to master [3]. This chapter describes the retroperitoneoscopic MI adrenalectomy.

I. Mushtaq, MD (✉) • F. Yankovic
Department of Paediatric Urology,
Great Ormond Street Hospital NHS Foundation Trust,
Great Ormond Street, London WC1N 3JH, UK

M. McHoney et al. (eds.), *Color Atlas of Pediatric Anatomy, Laparoscopy, and Thoracoscopy*,
DOI 10.1007/978-3-662-53085-6_34

34.2 Working Instruments

- Primary camera port: 6-mm Hasson
- Secondary 5-mm ports (×2)
- 30° 5-mm laparoscope
- Kelly forceps (×2).
- Metzenbaum scissors
- Laparoscopic hook
- LigaSure™ (Covidien) or 5-mm endoclips
- Endopouch® (Ethicon Endo-Surgery) for specimen retrieval

34.3 Relevant Anatomy

The left adrenal gland usually is smaller than the right and lies at the medial aspect of the upper pole of the left kidney. The arterial supply is derived from three adrenal arteries: superior (left inferior phrenic artery), middle (aorta), and inferior (left renal artery). The main left adrenal vein joins with the left inferior phrenic vein to drain into the left renal vein.

The right adrenal gland is located at the medial aspect of the upper pole of the right kidney, behind the vena cava in a very deep and high position. The arterial supply derives from the superior (inferior phrenic artery), middle (aorta), and inferior (right renal artery) adrenal arteries. The main right adrenal vein drains into the posterior lateral aspect of the vena cava after a short horizontal course. Variant venous anatomy is more frequent in the right adrenal gland, being encountered in about 10–15 % of patients [4].

34.4 Positioning, Port Siting, and Ergonomic Considerations

The patient is positioned fully prone with the chest and hips raised, allowing the abdomen to be free (Fig. 34.1a). It is important to place the patient at the lateral edge of the table, to allow unrestricted movements of the laparoscopic instruments. Once the position is ready, the patient should not be moved. Before skin preparation, the landmarks are drawn (Fig. 34.1b).

The working ports are placed using the landmarks: The camera port is placed lateral to the sacrospinalis muscle, midway between the 12th rib and the iliac crest; the first instrument port is placed 2–3 cm lateral to the tip of the 12th rib, and the second instrument port is placed medial to the camera port (Fig. 34.2a). The camera port is placed using the technique described by Gaur [5], with a balloon made from the middle finger of a size 8 surgical glove tied to a Fr 12 Jacques catheter (Fig. 34.2b).

After incising the skin (5–10 mm), an artery forceps is advanced through two fascial layers to reach the plane just outside Gerota's fascia (Fig. 34.3a). Once in position, the balloon is inflated with 150–250 mL of air; a lateral bulge confirms the extraperitoneal position of the balloon (Fig. 34.3b).

The 6-mm Hasson port is placed and secured to the skin by a suture, followed by insufflation of the retroperitoneum with carbon dioxide to a pressure of 10 mmHg (Fig. 34.4a). The two instrument ports are positioned under direct vision, with the first port under the 12th rib and the second port through the paravertebral muscle (Fig. 34.4b).

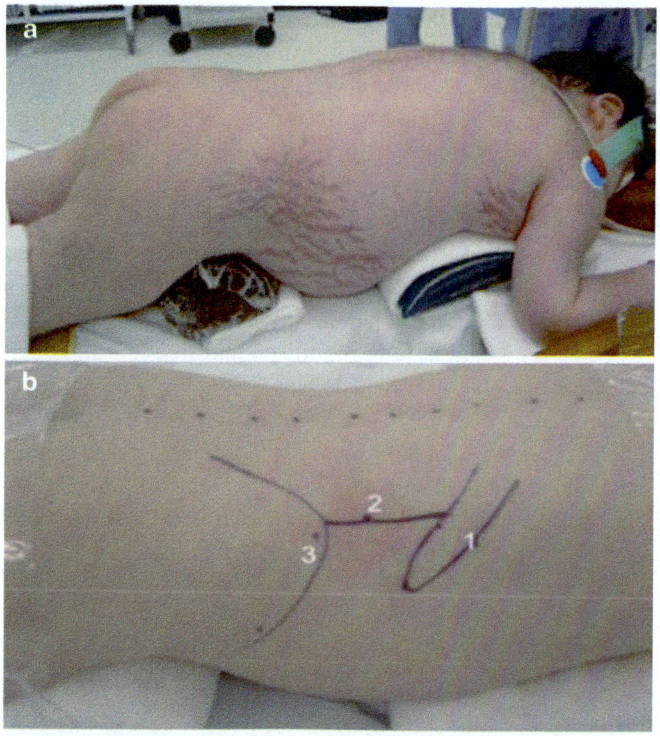

Fig. 34.1 (**a**) Position for right adrenalectomy. (**b**) Landmarks: 12th rib (*1*), sacrospinalis muscle (*2*), iliac crest (*3*)

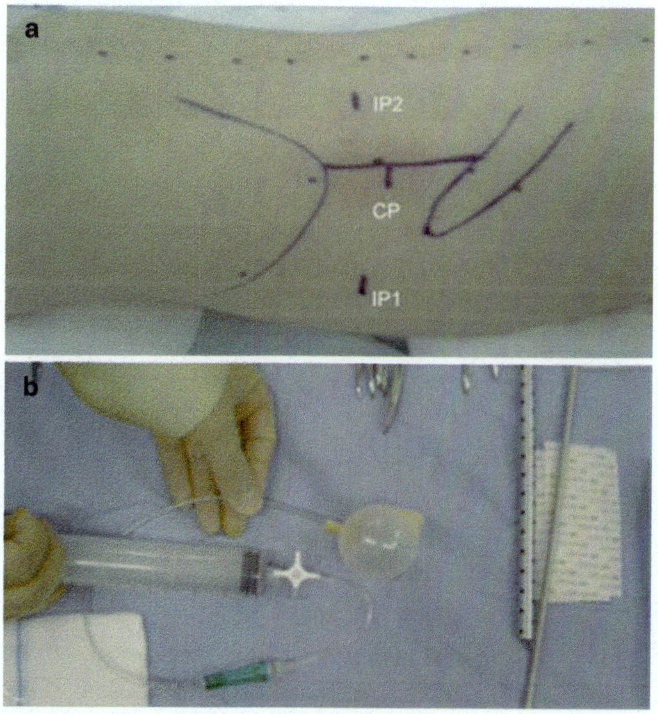

Fig. 34.2 (**a**) Port siting: camera port (*CP*), instrument port 1 (*IP1*), instrument port 2 (*IP2*). (**b**) Homemade balloon

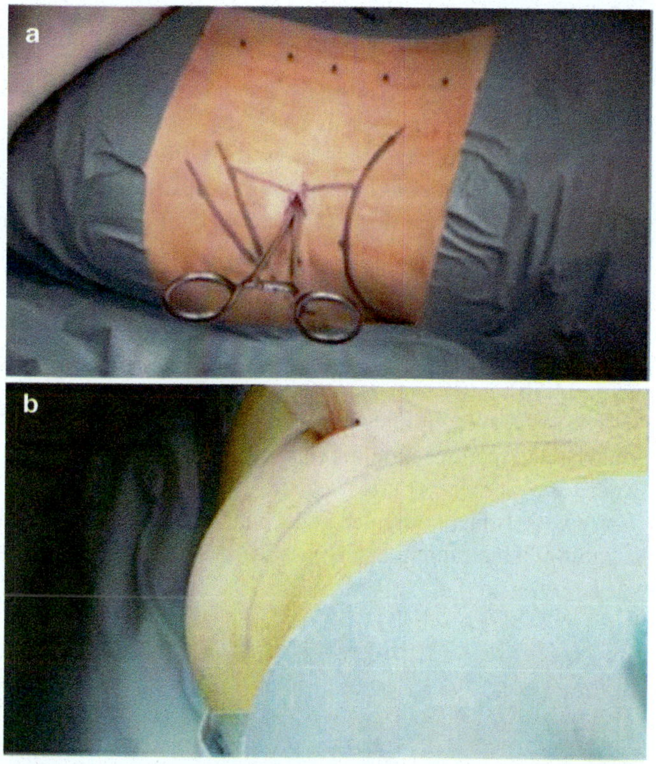

Fig. 34.3 (**a**) Artery forceps in the retroperitoneal space. (**b**) Lateral bulge confirms adequate positioning of the balloon

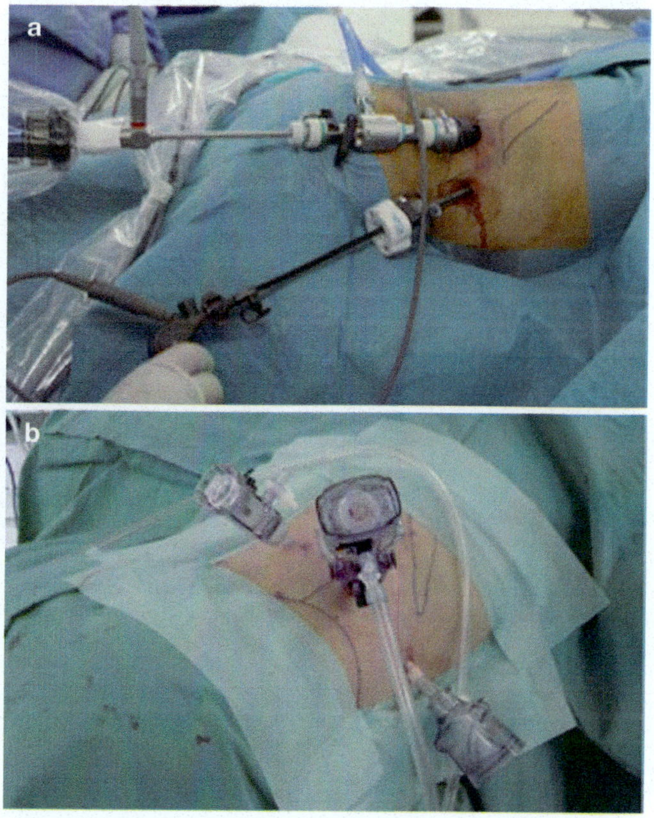

Fig. 34.4 (**a**) Camera port and instrument port 1. (**b**) Port siting

34.5 Surgical Technique

The first step in retroperitoneoscopic adrenalectomy is the exposure of the kidney. Using scissors connected to the diathermy, Gerota's fascia is incised. With a combination of blunt dissection and diathermy, the posterior surface of the kidney is exposed (Fig. 34.5a). The lateral and inferior attachments are not divided, as they anchor the kidney in position and aid in exposure of the upper pole. The inferior margin of the adrenal gland or tumour can then be visualised at the superomedial border of the kidney (Fig. 34.5b).

The surgery continues with the division of the adrenal vessels. Dissection should proceed on the medial aspect, where the adrenal arteries can be identified (Fig. 34.6a). These vessels can be ligated and divided between clips, or a LigaSure™ instrument can be used [6] (Fig. 34.6b).

The next step is the division of the adrenal vein (Fig. 34.7). Reflecting the adrenal gland posteromedially facilitates the exposure. The left adrenal vein is less prominent than the right and therefore is more difficult to identify. On the right side, the vein is short and enters directly into the vena cava; sometimes there can be more than one vein.

Once all the vessels to the tumour have been divided, it can be separated from the remaining attachments. This can be achieved with a hook instrument (Fig. 34.8a) and/or Kelly forceps connected to the diathermy lead (Fig. 34.8b).

Once the gland is fully mobilised and freed of all attachments, it can be removed. The gland is then placed within an Endopouch® device (Fig. 34.9a) and removed through the camera port incision, which often needs to be slightly enlarged to facilitate an intact specimen removal (Fig. 34.9b). The wounds are closed with absorbable suture and infiltrated with local anaesthetic.

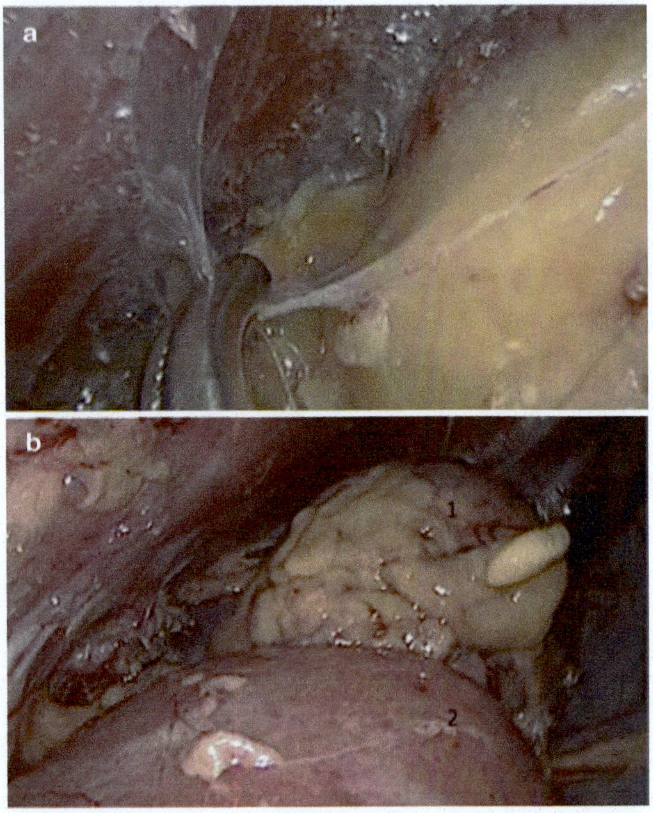

Fig. 34.5 (**a**) Blunt dissection with Kelly forceps. (**b**) Exposure of the adrenal mass (*1*) and the superomedial border of the kidney (*2*)

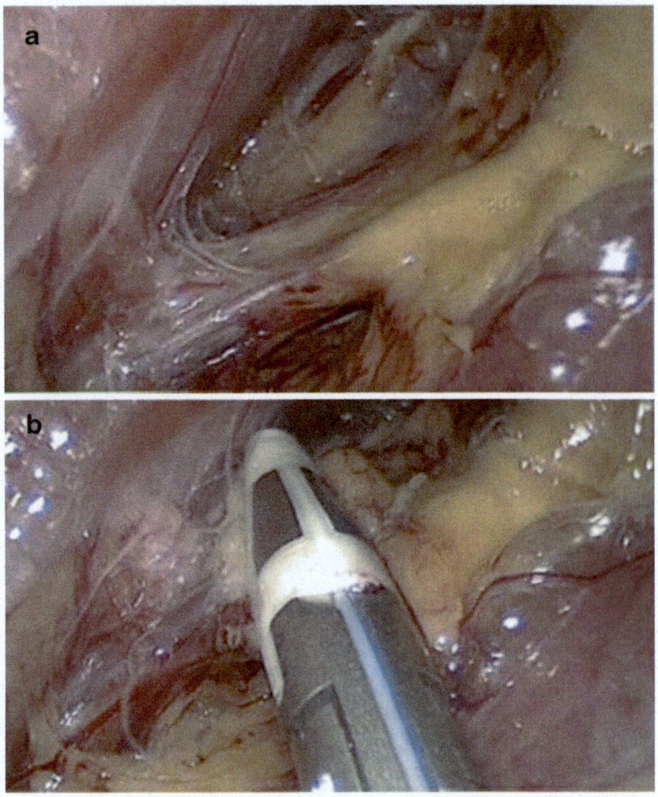

Fig. 34.6 (**a**) Adrenal arteries. (**b**) LigaSure coagulation of vessels

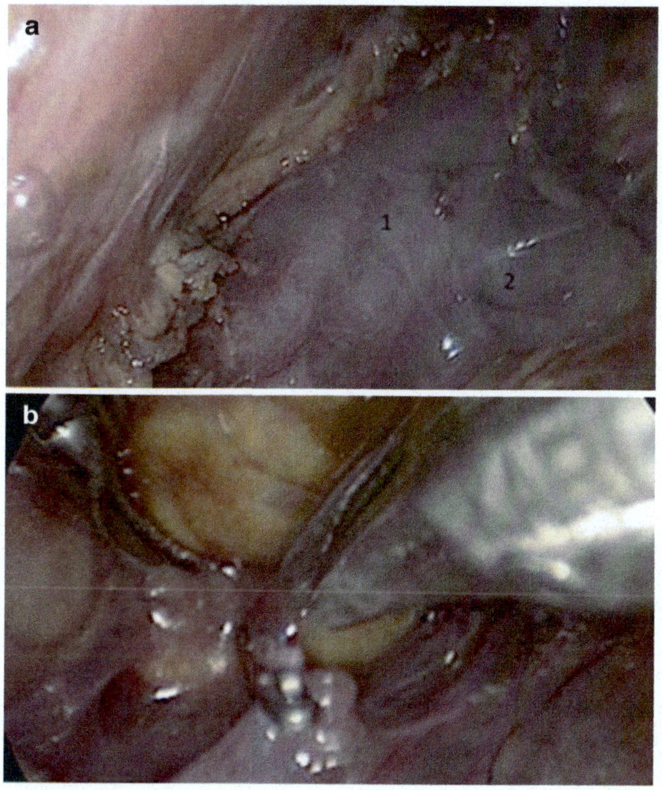

Fig. 34.7 (**a**) Vena cava (*1*) and adrenal vein (*2*). (**b**) Division of the adrenal vein with clips

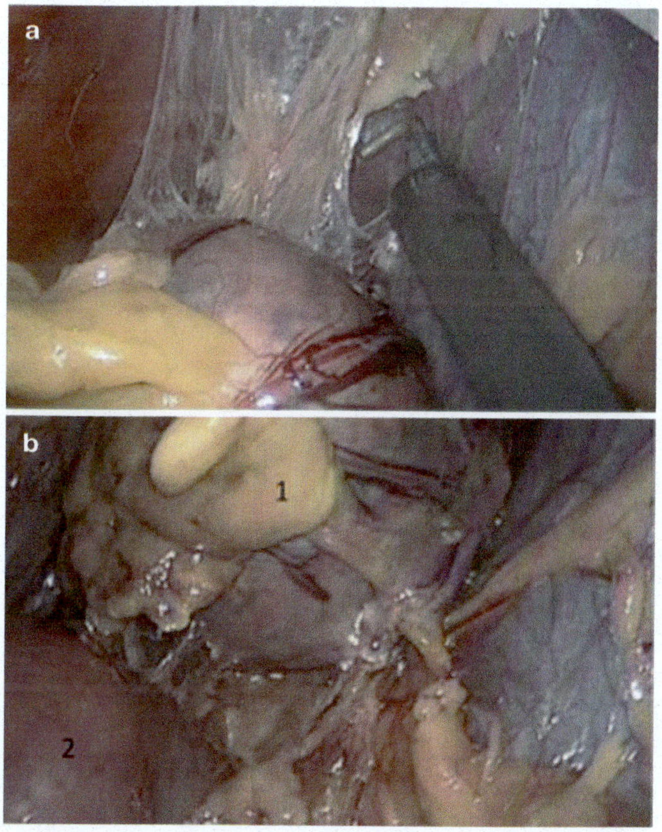

Fig. 34.8 (**a**) Division of superior attachments with hook. (**b**) Division of inferior attachments of tumour (*1*) from upper pole of kidney (*2*)

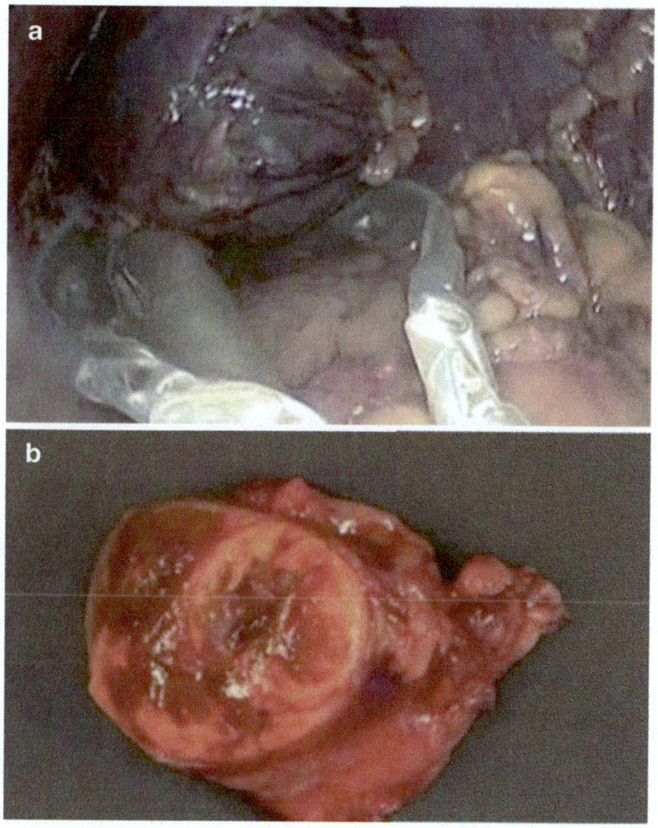

Fig. 34.9 (**a**) Specimen retrieval with Endopouch. (**b**) Adrenal gland and tumour completely removed

34.6 Highlights and Pitfalls

- The position of the patient is very important. Inadequate positioning may lead to technical difficulties and surgical failure. Once the position is ready, the patient should not be moved; the landmarks are drawn before skin preparation.
- To maximize the working area, the retroperitoneal space must be created outside Gerota's fascia.
- The adrenal gland is often surrounded by a large amount of fat. The characteristic yellow or bright orange appearance of the adrenal gland allows it to be differentiated from the surrounding fat.
- Dissection should advance outside the fatty layer. Dissection within this tissue generates excessive bleeding that interferes with the surgeon's vision.
- Care must be taken to identify variations in the vascular supply to the adrenal gland. Vessels that are more than 3 mm in diameter are more safely controlled with endoclips. For the right adrenal vein, which is characteristically short, it is also safer to use clips.
- It is imperative to remove an adrenal tumour as an intact specimen, so use of an Endopouch® device is highly recommended.

References

1. International Pediatric Endosurgery Group. IPEG guidelines for the surgical treatment of adrenal masses in children. J Laparoendosc Adv Surg Tech A. 2010;20:vii–ix.
2. Romano P, Avolio L, Martucciello G, Steyaert H, Valla JS. Adrenal masses in children: the role of minimally invasive surgery. Surg Laparosc Endosc Percutan Tech. 2007;17:504–7.
3. Heloury Y, Muthucumaru M, Panabokke G, Cheng W, Kimber C, Leclair MD. Minimally invasive adrenalectomy in children. J Pediatr Surg. 2012;47:415–21.
4. Scholten A, Cisco RM, Vriens MR, Shen WT, Duh QY. Variant adrenal venous anatomy in 546 laparoscopic adrenalectomies. JAMA Surg. 2013;148:378–83.
5. Gaur DD. Retroperitoneoscopy: the balloon technique. Ann R Coll Surg Engl. 1994;76:259–63.
6. Surgit O. Clipless and sutureless laparoscopic adrenalectomy carried out with the LigaSure device in 32 patients. Surg Laparosc Endosc Percutan Tech. 2010;20:109–13.

Laparoscopic Nephrectomy

35

Pankaj Kumar Mishra and Abraham Cherian

Abstract

Laparoscopic (transperitoneal) nephrectomy is a well-established and versatile technique in pediatric practice [1, 2]. It allows surgery to be carried out by a minimally invasive approach in the familiar intraperitoneal compartment. It provides ample working space and has a short learning curve. It facilitates a one stop solution for anatomic variations, is helpful in complete excision of the ureter when needed (avoiding the need for multiple separate incisions) [1, 2], and it also allows the surgeon to manage other intraabdominal pathologies at the same time (e.g., inguinal hernia, undescended testis).

Keywords

Laparoscopy • Transperitoneal • Nephrectomy • Pediatric urology • Renal hilum

35.1 General Information

Laparoscopic (transperitoneal) nephrectomy is a well-established and versatile technique in pediatric practice [1, 2]. It allows surgery to be carried out by a minimally invasive approach in the familiar intraperitoneal compartment. It provides ample working space and has a short learning curve. It facilitates a one stop solution for anatomic variations, is helpful in complete excision of the ureter when needed (avoiding the need for multiple separate incisions) [1, 2], and it also allows the surgeon to manage other intraabdominal pathologies at the same time (e.g., inguinal hernia, undescended testis).

35.2 Working Instruments

- 5-mm Hasson port
- 5-mm 30° Telescope
- 5-mm Working instruments (Kelly forceps, Johan forceps, and monopolar hook)
- 5-mm Ligaclips (Ethicon-Medline Industries, Mundelein, IL)/Harmonic scalpel/LigaSure (Valleylab Inc., Boulder CO)/needle holders.
- Organ retrieval bag if the kidney is very large.

P.K. Mishra (✉) • A. Cherian
Department of Paediatric Urology, Great Ormond Street Hospital, London, UK

M. McHoney et al. (eds.), *Color Atlas of Pediatric Anatomy, Laparoscopy, and Thoracoscopy*,
DOI 10.1007/978-3-662-53085-6_35

35.3 Positioning, Port Siting, and Ergonomic Considerations

The patient is placed in lateral position, with pathologic side up, and at the edge of the table close to the operating surgeon. The laparoscopy stack and screen are on the other side of the operating table, facing the surgeon (Fig. 35.1).

A 5-mm trans-umbilical port is placed under direct vision by open cut down. A second 5-mm port is placed in the epigastric region midway between the xiphisternum and the umbilicus. The third 5-mm port is placed in the ipsilateral flank [2], maintaining the principle of triangulation (Fig. 35.2). The positions of the working ports have to be adjusted according to the size of patient, the size and position of the kidney, and the need for ureterectomy. The ports are secured in place with 3–0 suture to prevent dislodgement. Intra-abdominal pressure of 10–12 mmHg and a CO_2 flow rate of 2L/min are generally sufficient.

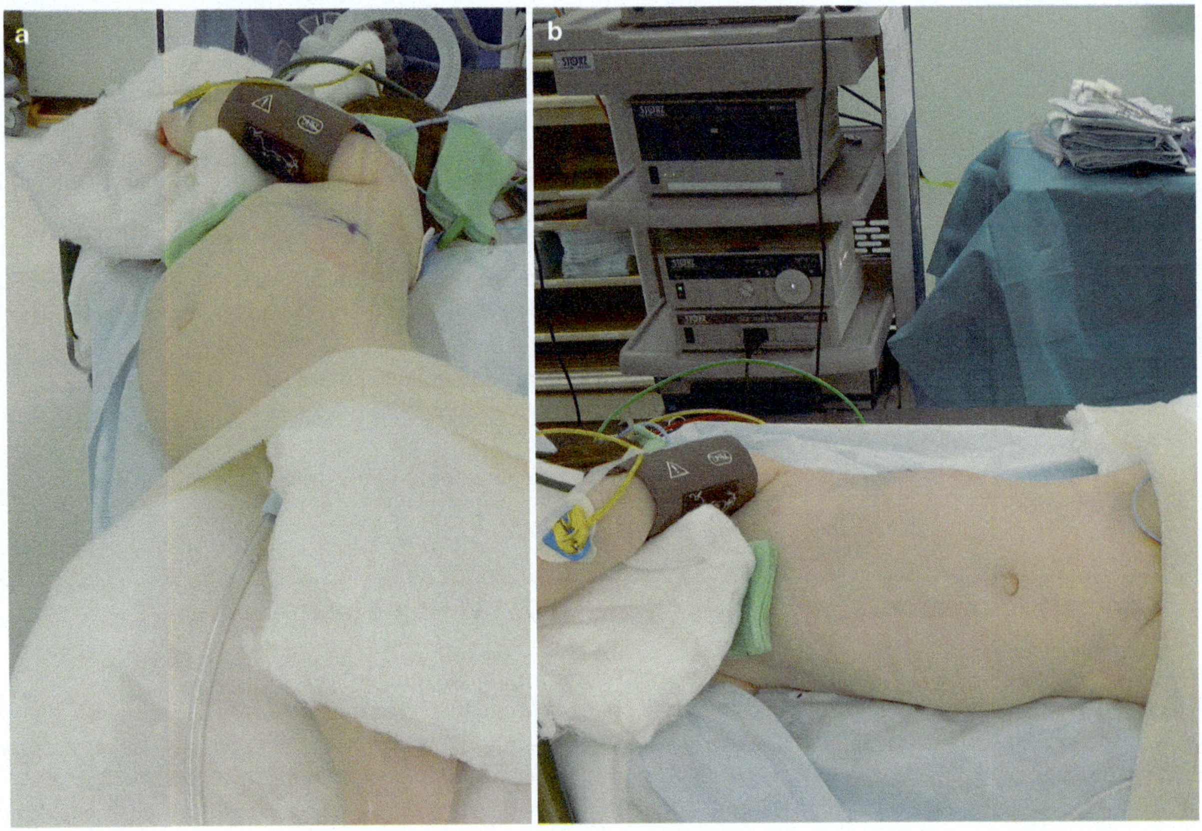

Fig. 35.1 (**a**, **b**) Patient position

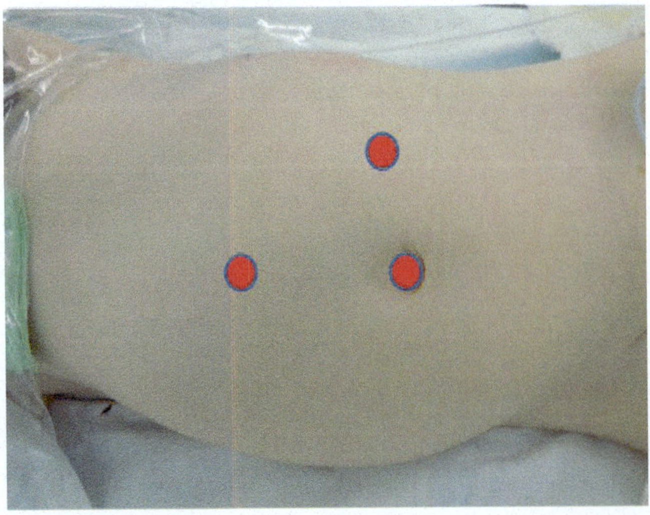

Fig. 35.2 Port positions

35.4 Relevant Anatomy

The kidney is a retroperitoneal organ with an arterial blood supply from the abdominal aorta and venous drainage into the inferior vena cava. The most relevant structures in close anterior relation to it are the colon, liver, duodenum, small intestine, spleen, suprarenal, and pancreas (Fig. 35.3).

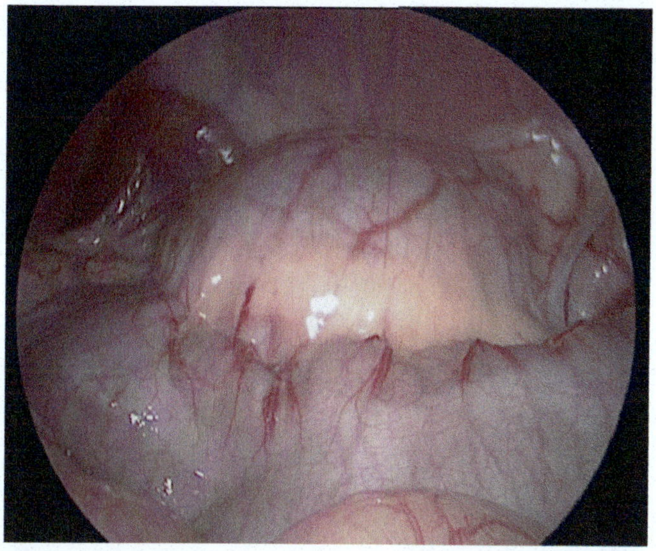

Fig. 35.3 Left kidney anatomic relations: left colon is displaced more medially in this patient

35.5 Surgical Technique

After diagnostic laparoscopy the colon is mobilized and reflected medially (Fig. 35.4). The perinephric tissue is dissected with monopolar diathermy to avoid bleeding and maintain good vision. The ureter is identified early at the level of the lower pole of the kidney (Fig. 35.5) and used for traction to faciliate further dissection at and around the renal hilum (Fig. 35.6) [1, 2]. The renal artery and vein are clearly and separately dissected, clipped, and divided (Fig. 35.7). Presence of accessory or aberrant renal vessels is kept in mind while dissecting the kidney from the surrounding structures and dealt with accordingly. The kidney is retrieved by slightly enlarging the umbilical port. The fasciae at the port sites are closed with interrupted 3–0 sutures and the skin with 5–0 subcuticular sutures.

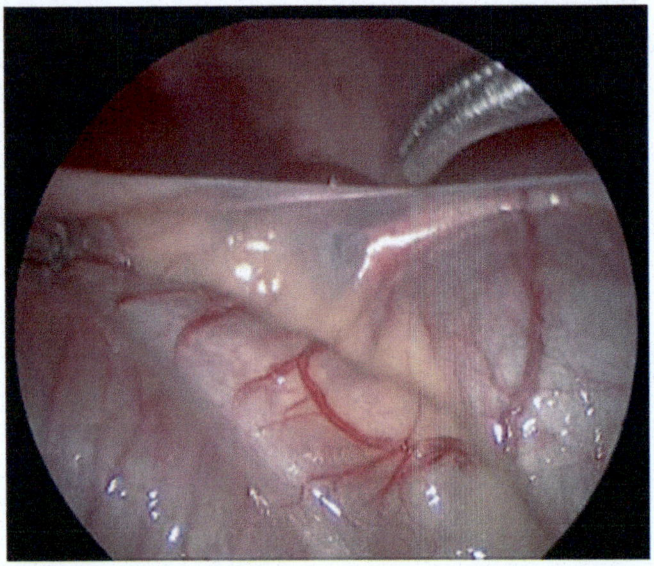

Fig. 35.4 Reflecting the colon

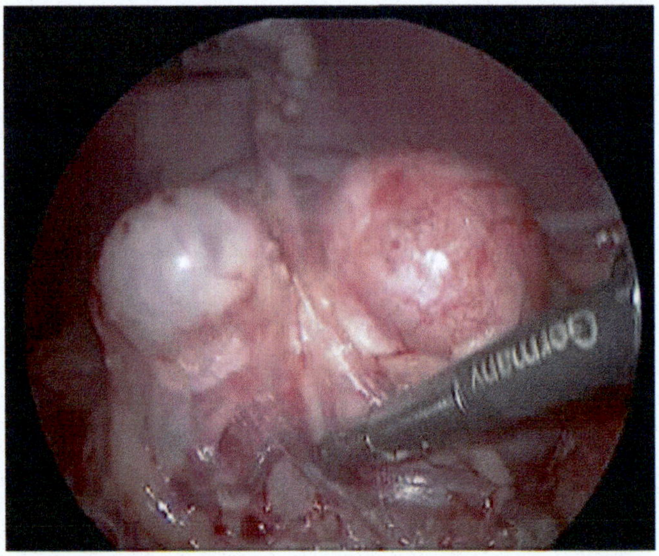

Fig. 35.6 Ureter used as traction for dissection

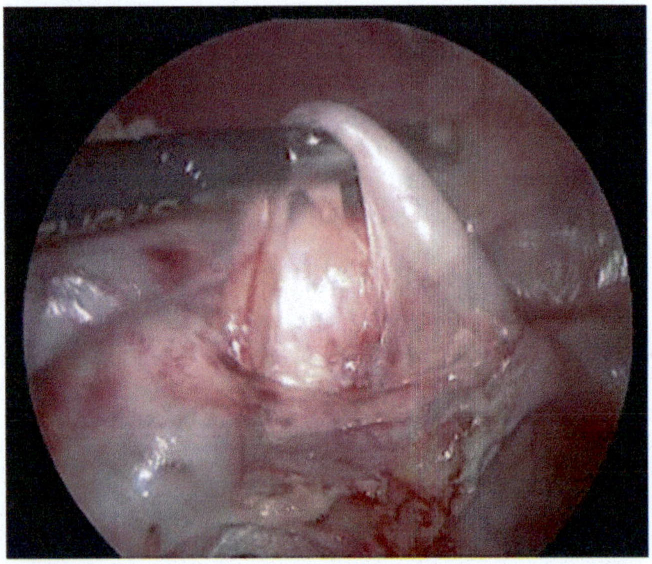

Fig. 35.5 Isolating the ureter at the the lower pole of the kidney

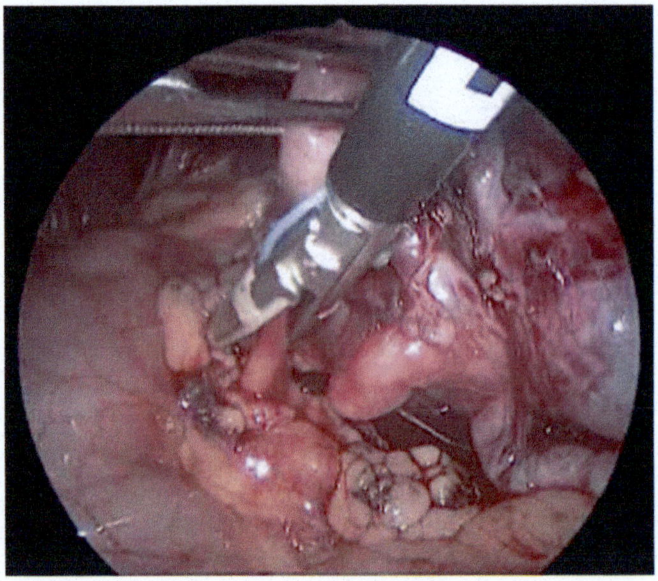

Fig. 35.7 Dividing renal vessel with LigaSure

35.6 Alternatives

Sometimes the colon is very mobile and falls off the kidney easily, avoiding the need for colonic mobilization. When the colon is well fixed to the lateral abdominal wall, the surgery can be carried out through a mesocolic window conveniently. (Transmesocolic approach). This is especially true for the left side. Renal vessels can be secured with Ligaclips, a harmonic scalpel, LigaSure, or intracorporeal ties, depending on the size of the vessels and the availability of instruments.

The lower end of the ureter can be left as it is or ligated with endoclip or endo-loop, depending on its size and reflux status (Fig. 35.8). Refluxing ureters are generally ligated. Nonrefluxing ureters can be left unligated. However, it is advisable to put in a bladder catheter overnight.

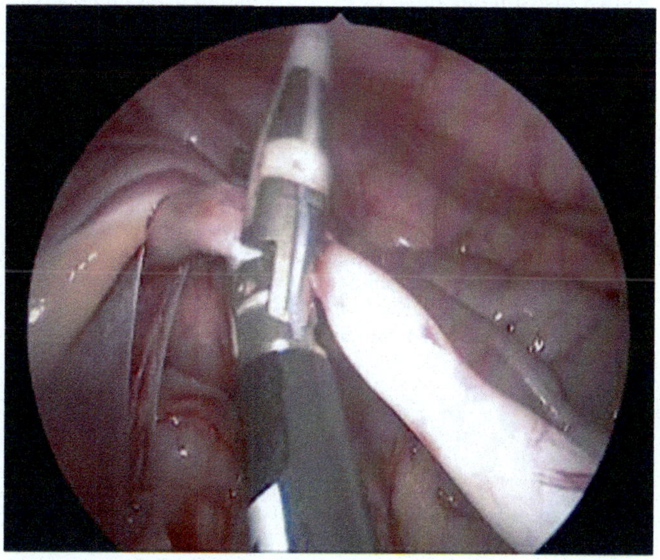

Fig. 35.8 Dividing lower ureter

35.7 Highlights and Pitfalls

- The open access (cutdown) technique is safer but not without hazard, including inadvertent injury to bowel [1].
- The advantage of the transperitoneal approach is that the kidney, ureter, and bladder can be approached in the same procedure by just changing the camera position and working instruments without changing the port sites [1].
- If the kidney is very large as a result of hydronephrosis or cyst, it might need decompression for better handling or space. However, this should be delayed as much as possible as it is easier to dissect around a tense distended kidney.
- A fourth port might occasionally be required for liver retraction for a right nephrectomy.
- Complete mobilization of the colon or a separate peritoneal window might have to be created in the pelvis for complete excision of the ureter down to the bladder.
- If dissection and complete excision of the ureter is done, then a urethral catheter is left in situ drain the bladder overnight.
- If the kidney is very large, following dissection and detachment it is positioned in an endobag and morcellated with sponge-holding forceps and removed piecemeal.
- The contraindications to laparoscopic nephrectomy are severe acute coagulopathy, uncontrolled sepsis, severe adhesions from previous laparotomy [1, 2], and distended small bowel loops with very little working space.
- Distention of bowel loops is more commonly seen in small babies who have been crying and hence swallowing air or because of bagging during induction of anesthesia. A nasogastric tube is usually helpful to decompress a distended stomach.
- All instruments must be exchanged and used under vision to minimize inadvertent visceral injury.

References

1. Yeung CK, Thakre AA. Laparoscopy in pediatric urology. In: Gearhart JP, Rink RC, Mouriquand PDE, editors. Pediatric urology. 2nd ed. Philadelphia: Saunders Elsevier; 2009. p. 92–4.
2. Peters CA. Logical laparoscopy. In: Holcomb III GW, Murphy JP, editors. Aschraft's pediatric surgery. 5th ed. Philadelphia: Saunders Elsevier; 2009. p. 821–5.

Retroperitoneoscopic Nephrectomy/ Heminephrectomy

36

Jimmy Lam

Abstract

The retroperitoneoscopic approach to renal surgery is a well-established technique that allows direct access to the kidney without transgression of the peritoneal cavity, as occurs in the laparoscopic approach. This has the advantage of having no intestinal loops obscuring the view, minimal risk of ileus, reduced postoperative pain, and avoidance of the risk of intestinal adhesions. The majority of children successfully undergoing this procedure are therefore fit for discharge within 24 h of surgery. It can be undertaken in patients who have had previous transperitoneal surgery and also preserves the integrity of the peritoneal cavity for those in end-stage renal failure who require immediate postoperative peritoneal dialysis.

Keywords

Retroperitoneoscopy • Peritoneal cavity

36.1 General Information

The retroperitoneoscopic approach to renal surgery is a well-established technique that allows direct access to the kidney without transgression of the peritoneal cavity, as occurs in the laparoscopic approach. This has the advantage of having no intestinal loops obscuring the view, minimal risk of ileus, reduced postoperative pain, and avoidance of the risk of intestinal adhesions. The majority of children successfully undergoing this procedure are therefore fit for discharge within 24 h of surgery. It can be undertaken in patients who have had previous transperitoneal surgery and also preserves the integrity of the peritoneal cavity for those in end-stage renal failure who require immediate postoperative peritoneal dialysis.

36.2 Working Instruments

- 5-mm ports (balloon blunt tip preferable to maximize working space and avoid gas leakage or accidental displacement)
- 5-mm 30° telescope
- 5-mm Endopledgets
- 5-mm Johan forceps
- 5-mm Maryland forceps
- 5-mm right-angled dissector
- 5-mm hook diathermy
- 5-mm Metzenbaum scissors
- Bipolar or ultrasonic forceps (essential for heminephrectomy)

J. Lam
Department of Paediatric Surgery, Royal Hospital for Sick Children, Edinburgh, UK

M. McHoney et al. (eds.), *Color Atlas of Pediatric Anatomy, Laparoscopy, and Thoracoscopy*,
DOI 10.1007/978-3-662-53085-6_36

36.3 Positioning, Port Siting, and Ergonomic Considerations

Retroperitoneoscopy can be performed in either the lateral or prone position, according to the surgeon's preference. My personal preference is for the lateral approach (as illustrated in this chapter) because it is more versatile and allows access to the lower ureter for complete ureterectomy when required. The lateral approach is also used in retroperitoneoscopic pyeloplasty, and therefore familiarity with this approach will provide a consistent orientation to the kidney for a variety of procedures.

The patient is placed in a full lateral position with flexion of the operating table to increase the distance between the costal margin and the iliac crest (Fig. 36.1). Stabilization of the patient using a vacuum bag and strapping is advised to allow for rolling of the operating table if required. The surgeon and assistant stand at the patient's back with the video screen on the opposite side.

The primary port is inserted in the midaxillary line below the last rib (11th or 12th depending on rib length) using open cut-down to the retroperitoneal space (Fig. 36.2). A space is then created using blunt dissection just anterior to the quadratus lumborum with a wet pledget to accommodate a homemade balloon; the balloon is inflated with air to expand the working space within the retroperitoneum (Fig. 36.3). Following insertion of the primary port, the space is further expanded with CO_2 insufflation to 10 mmHg pressure and blunt dissection using the tip of the telescope. The working ports are then inserted under direct vision posteriorly just lateral to the erector spinae in the costal angle and superior to the iliac crest.

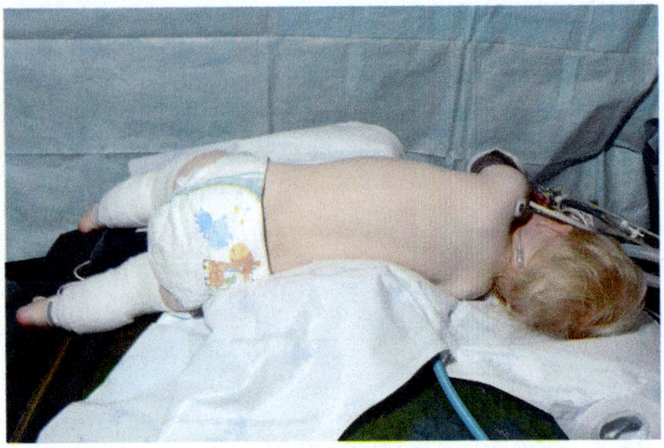

Fig. 36.1 The patient is placed in full lateral position with flexion of the operating table to increase the distance between the costal margin and iliac crest

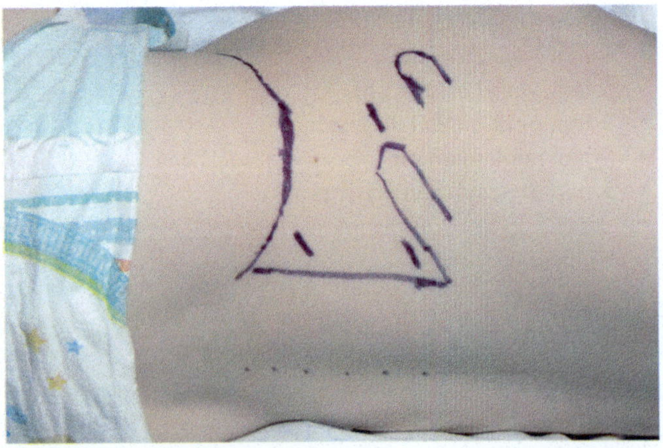

Fig. 36.2 The primary port is inserted in the mid-axillary line below the last rib (11th or 12th depending on rib length) using an open cut-down procedure to the retroperitoneal space

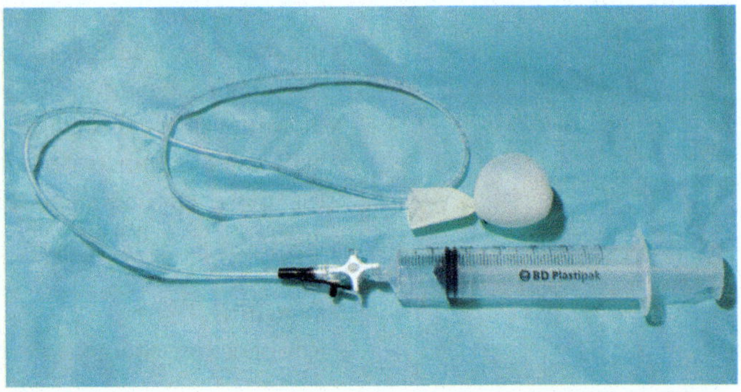

Fig. 36.3 A space is then created using blunt dissection just anterior to the quadratus lumborum muscle with a wet pledget to accommodate a homemade balloon, which is inflated with air to expand the working space within the retroperitoneum

36.4 Relevant Anatomy (Figs. 36.4, 36.5 and 36.6)

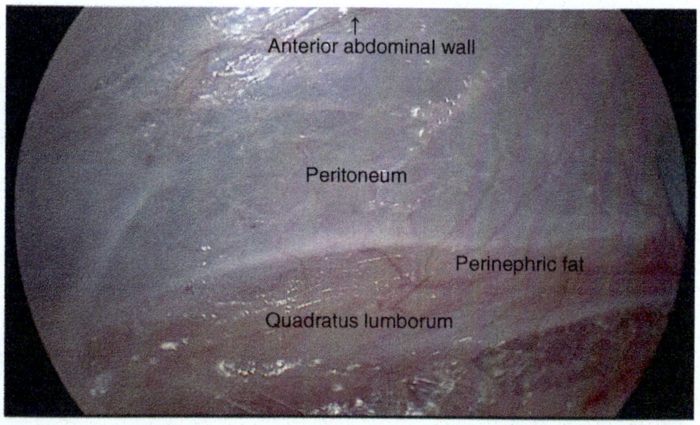

Fig. 36.4 Since the retroperitoneal space needs to be created, it is easy to become disorientated initially within such a confined space. It is best to orientate the telescope and camera to keep the peritoneum anterior and to develop the space by sweeping it off the quadratus lumborum to expose the perinephric fat. The anatomic landmarks are then gradually exposed by extending this working space on the posterior aspect of the kidney

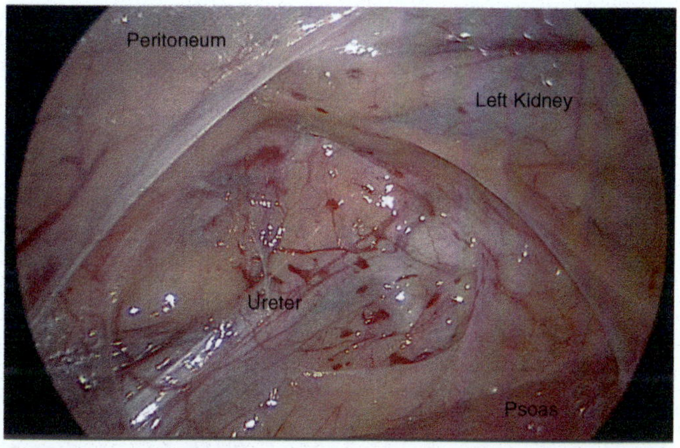

Fig. 36.5 With further exposure, the ureter and the posterior aspect of the kidney should come into view

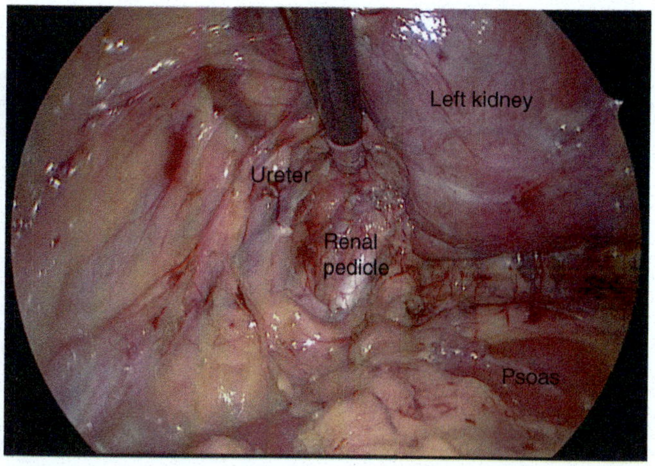

Fig. 36.6 Keeping the kidney attached anteriorly to the peritoneum, the renal hilum is exposed by dissection on the medial aspect of the kidney, aided by traction of the ureter laterally

36.5 Surgical Technique (Figs. 36.7, 36.8, 36.9, 36.10, 36.11, 36.12, 36.13, 36.14, 36.15, 36.16, 36.17, 36.18, 36.19 and 36.20)

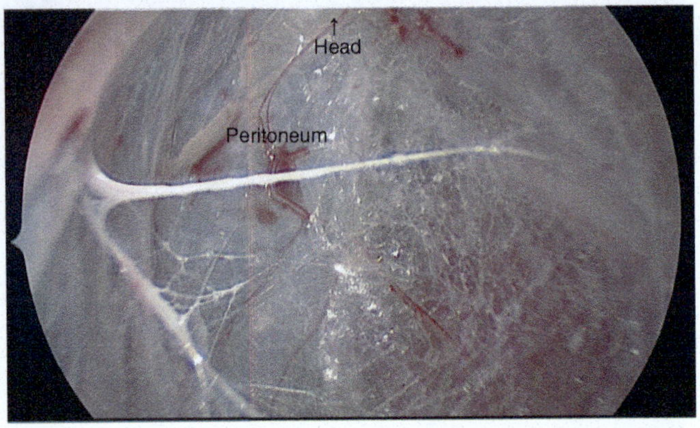

Fig. 36.7 Following insertion of the primary port, the retroperitoneal space is further expanded with CO_2 insufflation to 10 mmHg. This, along with blunt dissection using the tip of the telescope, creates the space between the peritoneum and the inner aspect of the quadratus lumborum allowing insertion of the working ports

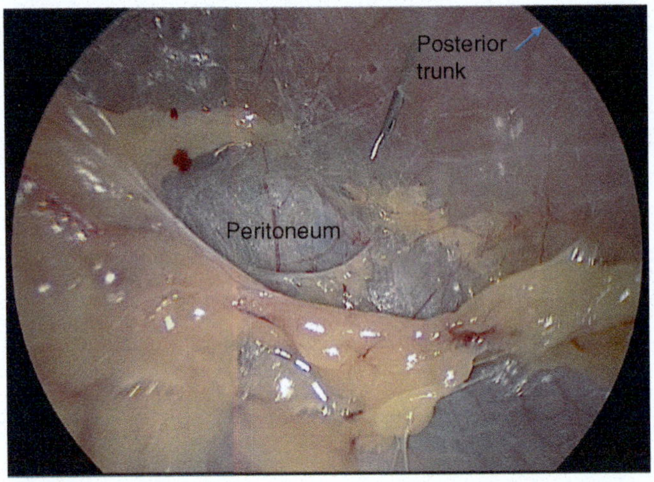

Fig. 36.8 The site of insertion of the port at the costal angle is confirmed with the needle used to infiltrate the local anesthetic. Ensure that there is sufficient space between the attachment of the peritoneum to the quadratus lumborum and port insertion site, thereby avoiding the risk of breaching the peritoneum and making possible further expansion of the retroperitoneal space by not fixing the peritoneum with the port

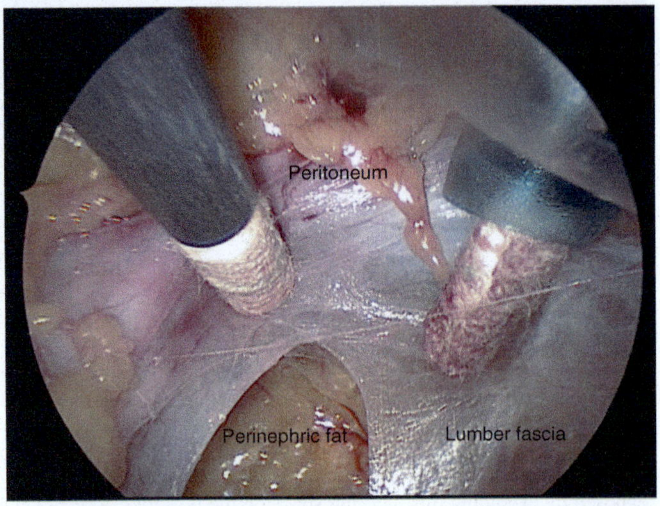

Fig. 36.9 Following insertion of the two working ports, the peritoneum is swept off the lumbar fasciae using endopledgets to expose the perinephric fat and the ureter and tracing the ureter cranially to the kidney. This dissection needs to be done extensively in both the cranial and caudal directions in order to create a sufficient operating space by allowing the peritoneum and the intra-abdominal contents to be displaced anteriorly by the CO_2 insufflation

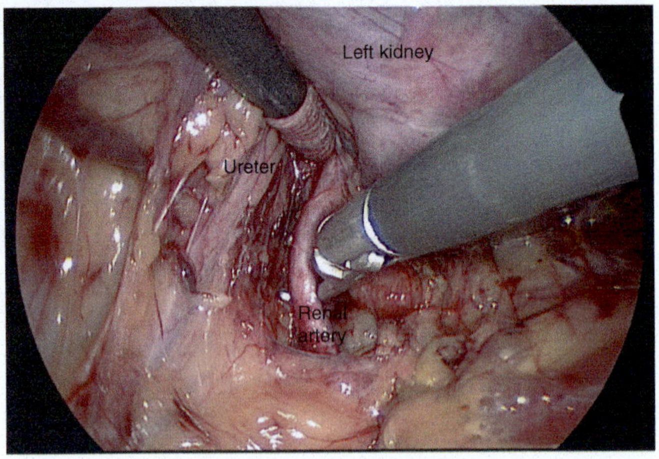

Fig. 36.10 In total nephrectomy, the ureter is mobilized to allow it to be grasped and used to traction the kidney laterally to expose the renal hilum. The renal vessels are then individually isolated using the Maryland or right-angled forceps before being sealed and divided using hook monopolar diathermy if they are small or a bipolar or ultrasonic forceps if they are larger

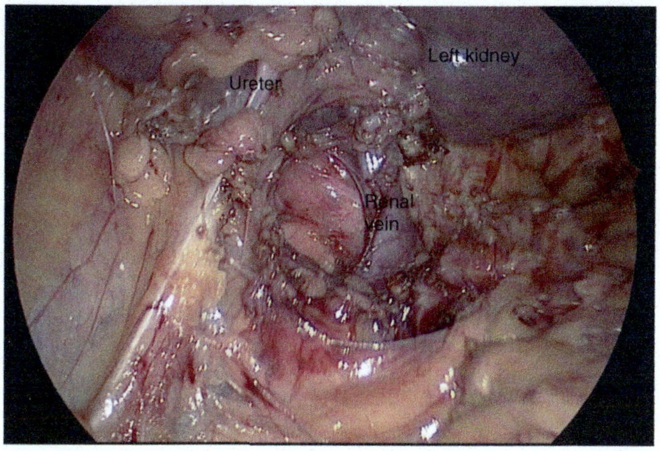

Fig. 36.11 The arterial branches as shown in Fig. 36.10 should ideally be divided first before similarly sealing and dividing the renal vein(s), which should then become less engorged as shown above once there is no blood flow through the kidney. The kidney is then freed by dividing the perinephric attachments to the adrenal gland and peritoneum using a combination of blunt dissection and hook diathermy. Care must be taken not to breach the peritoneum on the anterior aspect

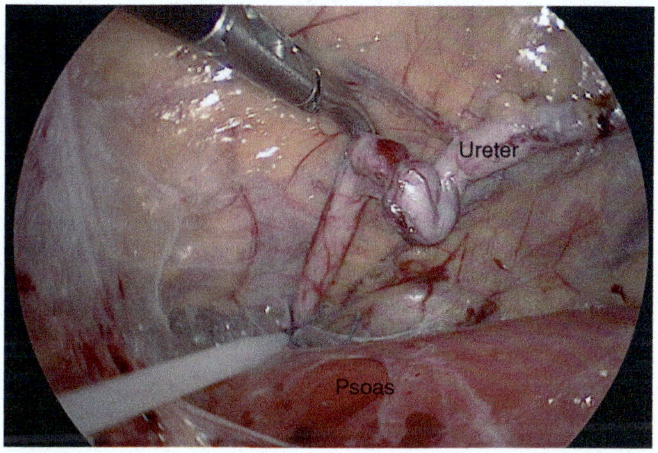

Fig. 36.12 The necessary amount of ureter can then be resected by dissection down to the pelvis aided by traction on the ureter cranially. It can then be ligated with an endoloop at the distal end if necessary before resection. The ureter will normally deliver through one of the working 5-mm ports and the kidney should be delivered through the primary port site. The site may need to be stretched or extended, depending on the size of the specimen

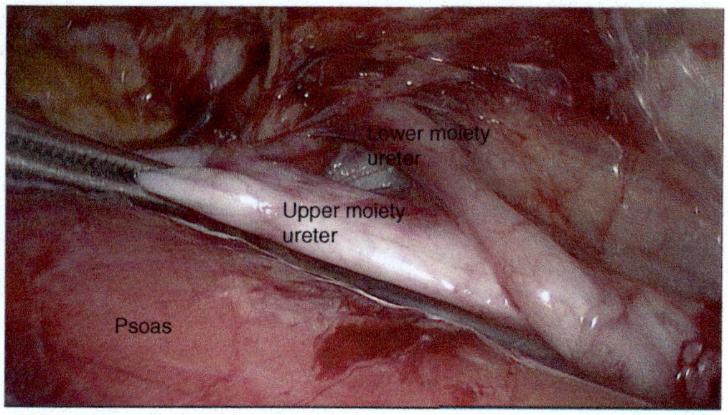

Fig. 36.13 For heminephrectomy, it is essential to identify the ureter of the moiety to be resected by tracing them up to the duplex kidney; this is usually the more dilated, dysplastic ureter as shown. A portion of the ureter is then skeletalized to preserve the blood supply to the retained ureter before it is transected using hook diathermy and/or scissors

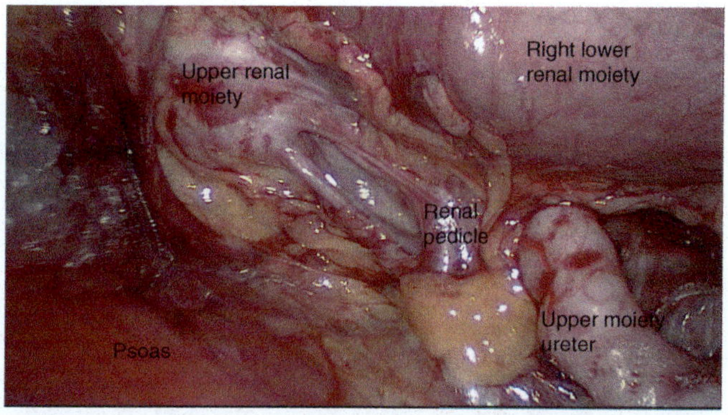

Fig. 36.14 The proximal ureter is then mobilized cranially. Often for an upper heminephrectomy (as shown), it may traverse between the main renal vessels and therefore needs to be carefully mobilized through these while preserving the vessels

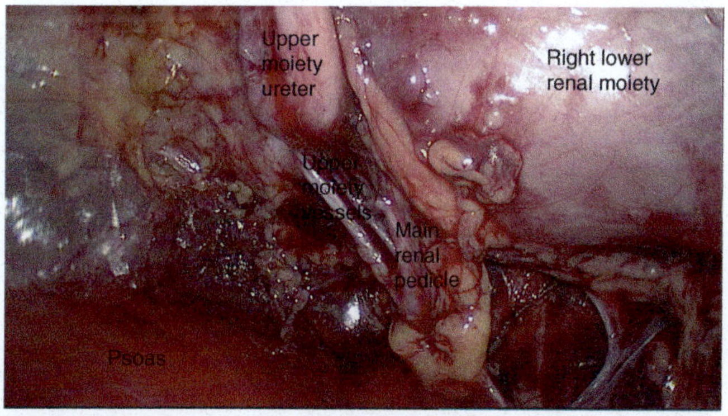

Fig. 36.15 Once the ureter is mobilized up to the upper moiety, it can be used for traction laterally to aid in identification of vessels to this moiety

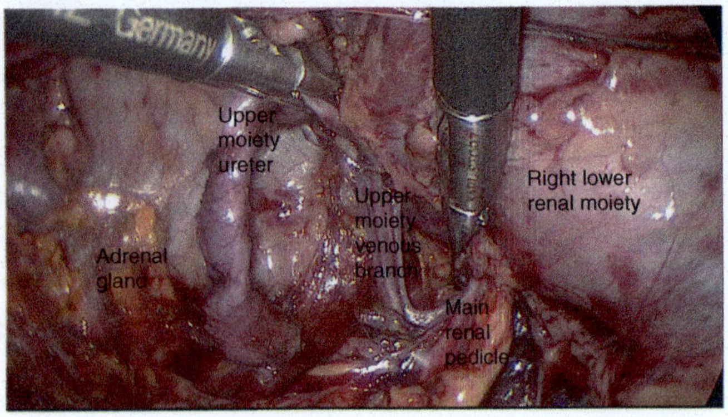

Fig. 36.16 The vessels are then individually isolated using the Maryland or right-angled forceps before being sealed and divided using hook monopolar diathermy if small or a bipolar or ultrasonic forceps if large. The arterial branches as shown in Fig. 36.15 should ideally be divided first before similarly sealing and dividing the vein(s), which should then become less engorged as shown above once there is no blood flow through the upper moiety

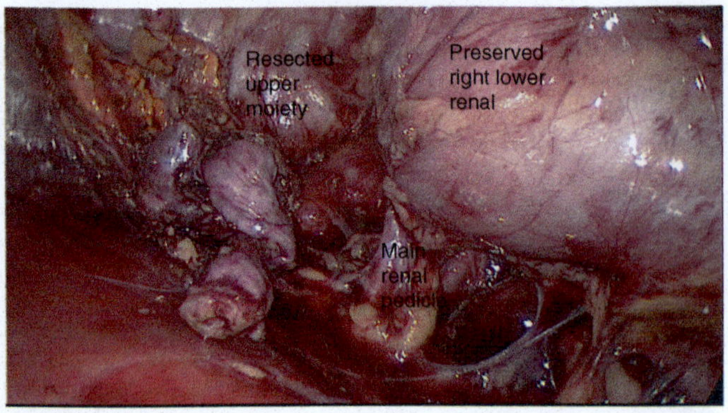

Fig. 36.19 The resected moiety should be checked to ensure that all dilated calyces are excised and hemostasis is achieved on the cut surface of the lower moiety. The distal ureter can be mobilized and resected if required if it is a refluxing unit, but care must be taken to preserve vascularity to the residual ureter

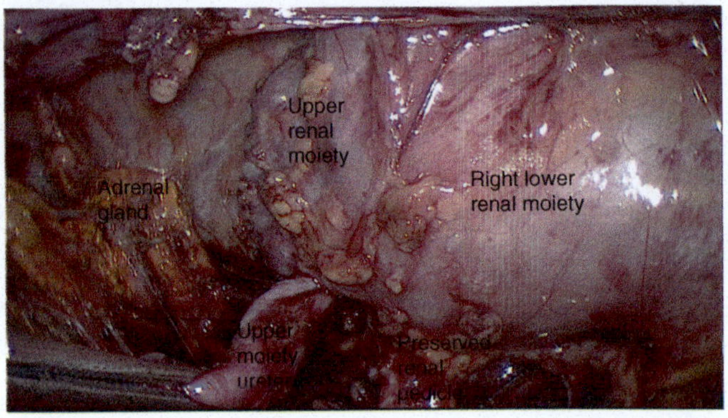

Fig. 36.17 There should be clear demarcation following successful devascularization of the upper renal moiety, which also appears dysplastic here as shown

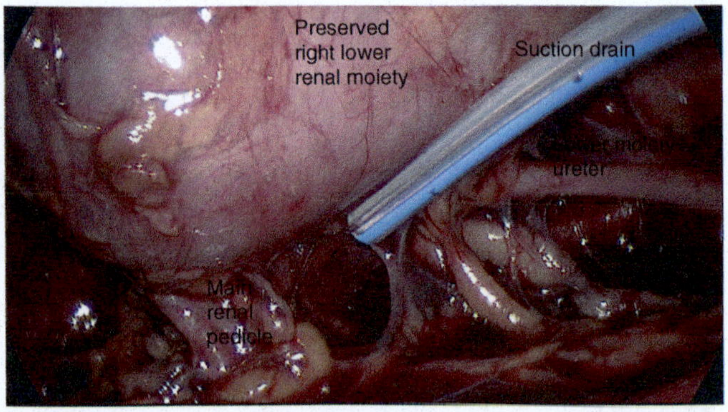

Fig. 36.20 A suction drain (10 Fr) can be inserted via the inferior port site down to the renal bed to minimize the risk of hematoma or urinoma formation. The resected specimen can then be retrieved via the primary port site, which seldom needs to be extended given the naturally small, dysplastic moiety

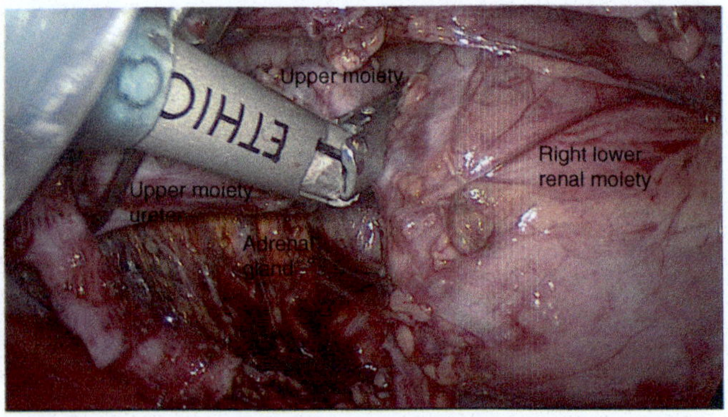

Fig. 36.18 The upper renal moiety is then freed from the adrenal gland and perinephric attachments with blunt dissection and hook diathermy. It is then resected from the lower renal moiety using either ultrasonic or bipolar forceps, aided with traction on the upper moiety ureter

36.6 Alternatives

The posterior prone approach can be used for retroperitoneoscopic renal surgery and has the advantage of gravity to aid the exposure of the renal pedicle, which may then only require two ports but is more limited for access to the lower ureter. Traditional laparoscopy can also be used for renal surgery and provides a more familiar anterior approach to the kidney but has the disadvantage and potential risks of transgression of the peritoneal cavity.

36.7 Highlights and Pitfalls

- Appropriate instruction to the camera assistant is essential to maintain correct orientation in the retroperitoneal space.
- Steps must be taken to avoid breaching the peritoneum in the initial dissection to create space for the primary port, ensuring adequate mobilization of the peritoneum before inserting working ports as well as care with instrument insertion and manipulation so as to avoid trauma to the peritoneum within the confined space.
- Careful dissection and identification of the vessels to both renal moieties is required in heminephrectomy to avoid vascular injury to the remaining moiety because the vascular anatomy is variable and branching may occur at a very short distance from the renal parenchyma.
- For a grossly dilated ureter in heminephrectomy, it may be useful to insert a ureteric stent into the normal ureter cystoscopically at the start of the procedure to aid identification.

Suggested Reading

Borzi PA. A comparison of the lateral and posterior retroperitoneoscopic approach for complete and partial nephroureterectomy in children. BJU Int. 2001;87:517–20.

Leclair MD, Vidal I, Suply E, Podevin G, Héloury Y. Retroperitoneal laparoscopic heminephrectomy in duplex kidney in infants and children: a 15-year experience. Eur Urol. 2009;56:385–91.

Valla JS. Retroperitoneoscopic surgery in children. Semin Pediatr Surg. 2007;16:270–7.

37

Pyeloplasty

Harish Chandran

Abstract

Pyeloplasty treats the anatomical and functional obstruction at the pelvi-ureteric junction (PUJ). This chapter discusses the endoscopic performance of this operation. Embryology, antenatal findings, postnatal investigations, and treatment options (including follow up and indications for surgical intervention) are not discussed, as these can be found in other textbooks addressing these issues.

Keywords

Antenatal hydronephrosis • Ultrasound • Anderson-Hynes • Pyeloplasty • Retroperitoneal • Transperitoneal • Laparoscopic

37.1 General Information

Pyeloplasty treats the anatomical and functional obstruction at the pelvi-ureteric junction (PUJ). This chapter discusses the endoscopic performance of this operation. Embryology, antenatal findings, postnatal investigations, and treatment options (including follow up and indications for surgical intervention) are not discussed, as these can be found in other textbooks addressing these issues.

Pyeloplasty can be performed in a variety of ways and by a number of different approaches. The technique described in detail here is endoscopic retroperitoneal pyeloplasty. A pyeloplasty aims to produce better drainage from the affected kidney. This goal is achieved by creating dependent drainage from the pelvis, by fashioning the pelvis and ureter to resemble a funnel. The original description of this procedure by Anderson and Hynes incorporates the excision of the narrowed segment of the PUJ, and the anastomosis of the spatulated ureter to the trimmed pelvis (Fig. 37.1).

In recent years, laparoscopic pyeloplasties have been performed with increasing frequency and in ever-younger patients. As surgeons' experience increases and instrumentation improves, this method of pyeloplasty has become the method of choice in paediatric hospitals. Access to the kidney can be via either the retroperitoneal route or via a transperitoneal approach.

I prefer retroperitoneoscopic pyeloplasty, which is described in further detail here. This approach avoids the need to mobilise other organs (colon) or unnecessarily open anatomical planes. It also lessens the risk of injury to other structures and avoids the sequelae of an intraperitoneal leak of urine, which irritates the peritoneum.

H. Chandran
Department of Paediatric Surgery and Urology, Birmingham Children's Hospital, Birmingham, UK

M. McHoney et al. (eds.), *Color Atlas of Pediatric Anatomy, Laparoscopy, and Thoracoscopy*,
DOI 10.1007/978-3-662-53085-6_37

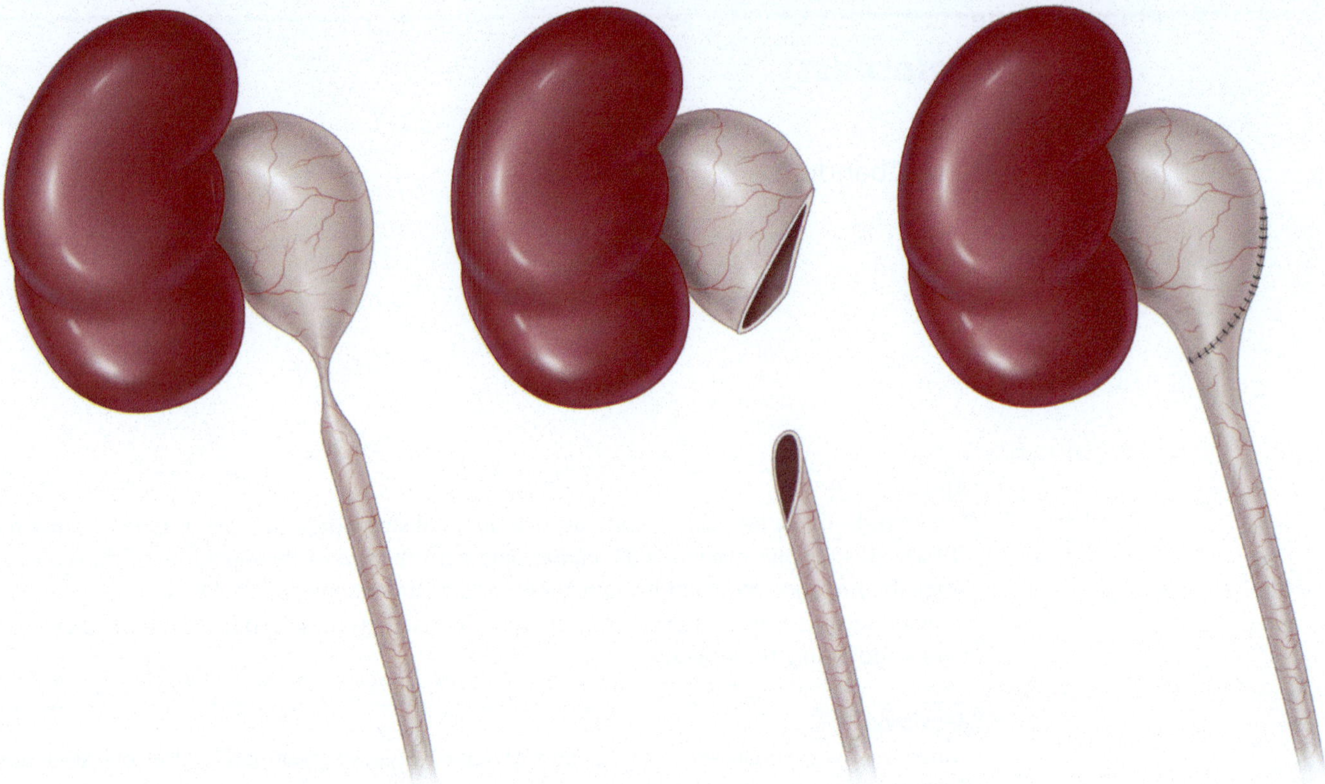

Fig. 37.1 The pyeloplasty: the excision of the narrowed segment of the PUJ and the ureter, and the anastomosis of the spatulated ureter to the trimmed pelvis

37.2 Working Instruments

- Three 5mm ports.
- 5mm, 0 degree telescope.
- Dissectors (Kellys) x 2
- Scissors.
- Laparoscopic needle holders x 2. These can be 3mm (with reducer) or 5mm.

37.3 Positioning, Port Siting, and Ergonomic Considerations

The patient is placed close to the edge of the operating table and in the kidney position (Fig. 37.2).

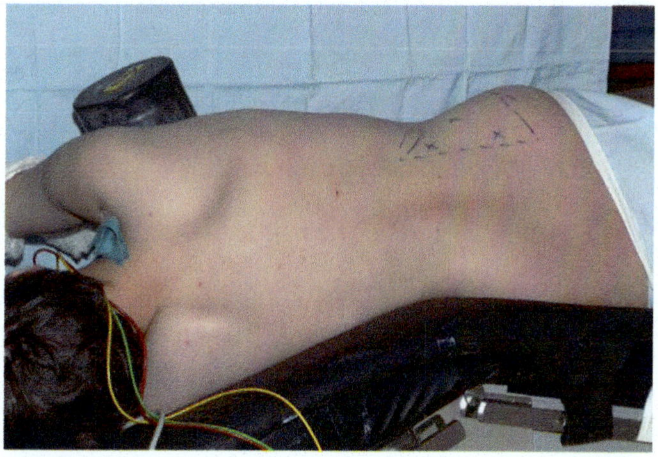

Fig. 37.2 The patient is placed close to the edge of the operating table and in the kidney position

The surgeon and assistant stand behind the patient, with the monitor screen facing them. The nurse is positioned opposite the surgeons, and can observe the operation on another monitor (Fig. 37.3).

A small (5-mm) incision is made close to the angle between the lowest rib and the vertical paraspinal muscles. A pair of Metzenbaum scissors is used to dissect and spread the muscle fibres to reach the retroperitoneum (Fig. 37.4).

An inflation device is used to create a space posterior to the kidney, lifting it medially off the psoas muscle. A 5-mm port is inserted into this space and secured (Fig. 37.5).

The central port is used for the telescope and the other two ports are used for the operating instruments (Fig. 37.6).

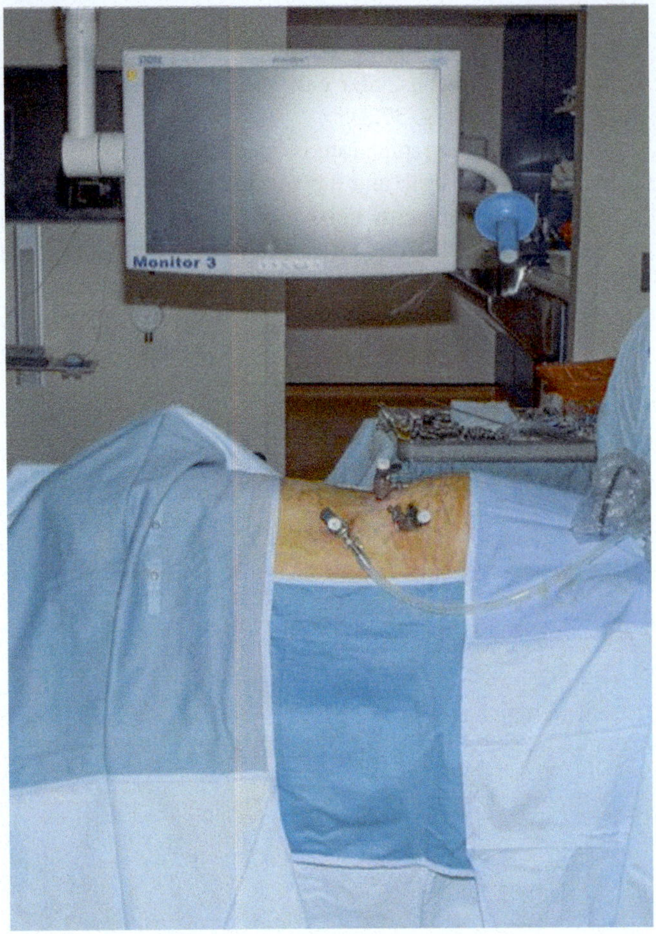

Fig. 37.3 The surgeon and assistant stand behind the patient, with the monitor screen facing them. The nurse is positioned opposite the surgeons, and can observe the operation on another monitor

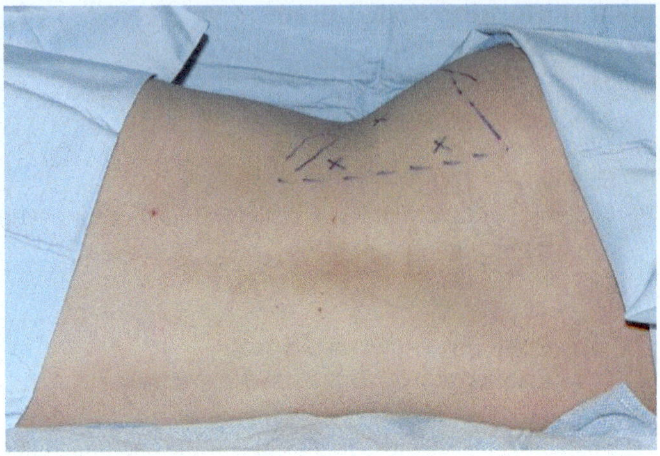

Fig. 37.4 A small (5-mm) incision is made close to the angle between the lowest rib and the vertical paraspinal muscles. A pair of Metzenbaum scissors is used to dissect and spread the muscle fibres to reach the retroperitoneum

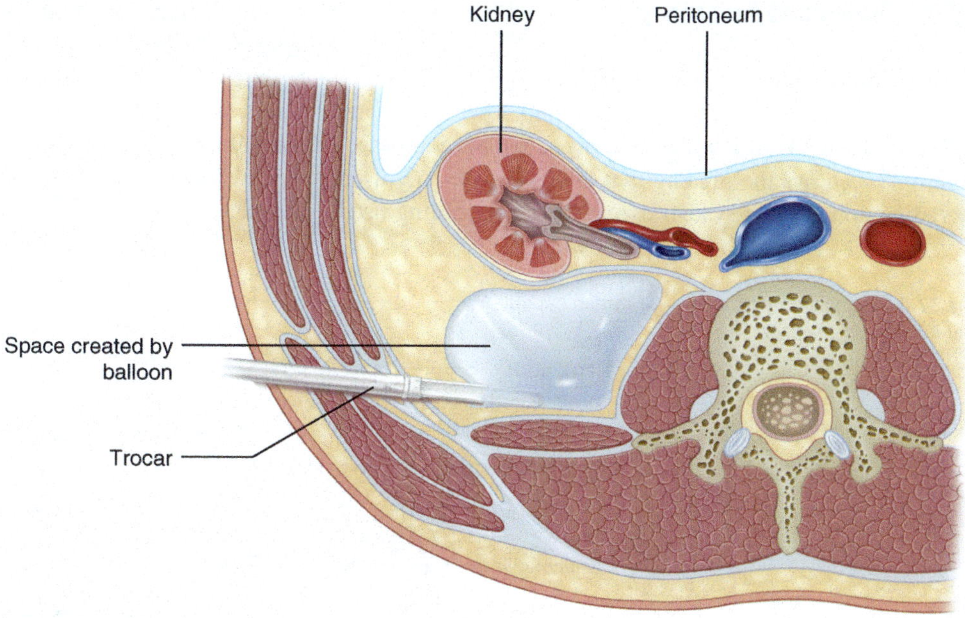

Fig. 37.5 An inflation device is used to create a space posterior to the kidney, lifting it medially off the psoas muscle. A 5-mm port with blunt trocar is inserted into this space and secured

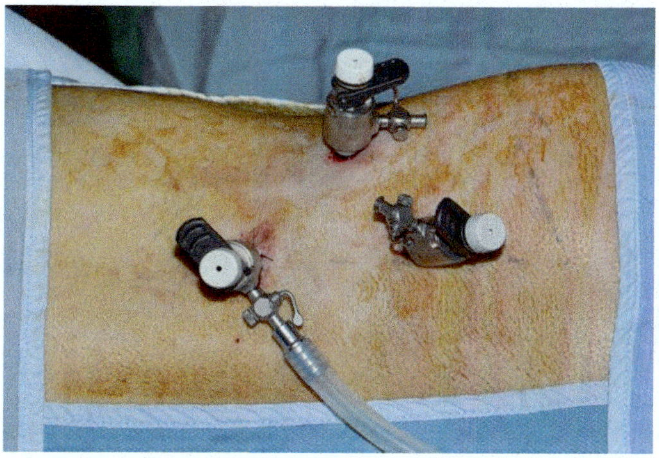

Fig. 37.6 The central port is used for the telescope and the other two ports are used for the operating instruments

37.4 Relevant Anatomy

Cross sectional anatomy of a normal kidney (Fig. 37.7).

Hydronephrosis is graded by ultrasound. There are many different ways in which the severity may be expressed. Figure 37.8 shows the grading that is most commonly used by radiographers; this system is related to the grading used by the Society for Fetal Urology (SFU). The SFU grading system (Table 37.1) is widely used to describe ultrasound findings in North America and in many countries in Europe. This system allows clinicians to communicate the severity of the hydronephrosis in a standardised manner and also informs the management of the child.

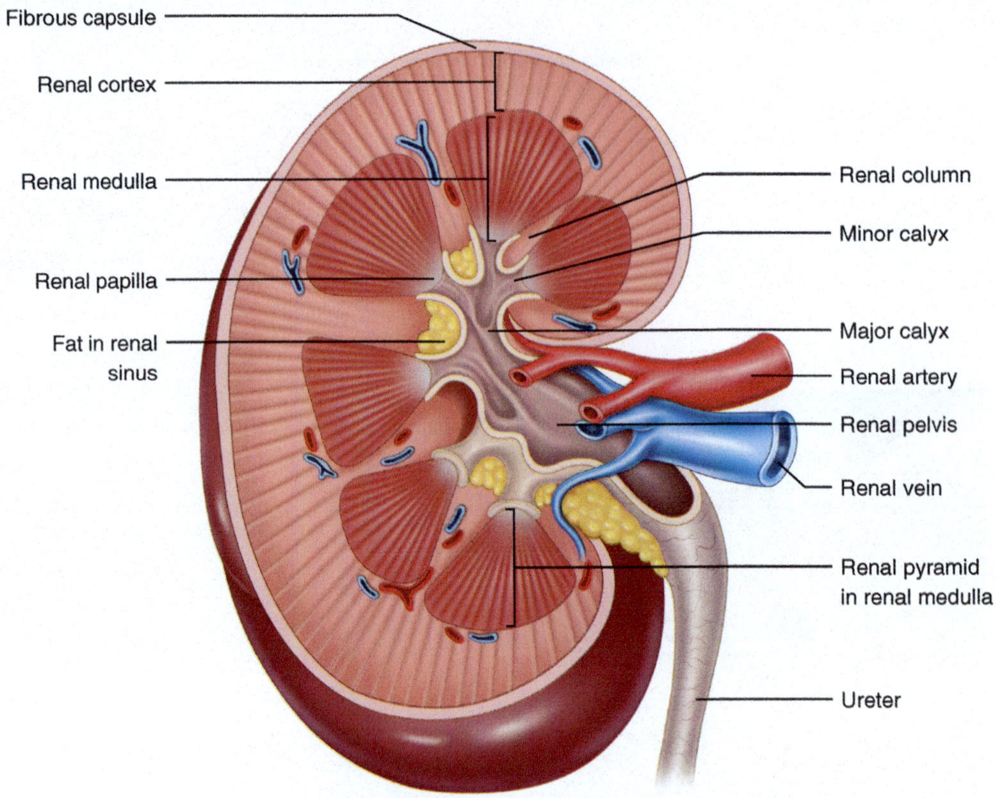

Fig. 37.7 The anatomy of a normal kidney

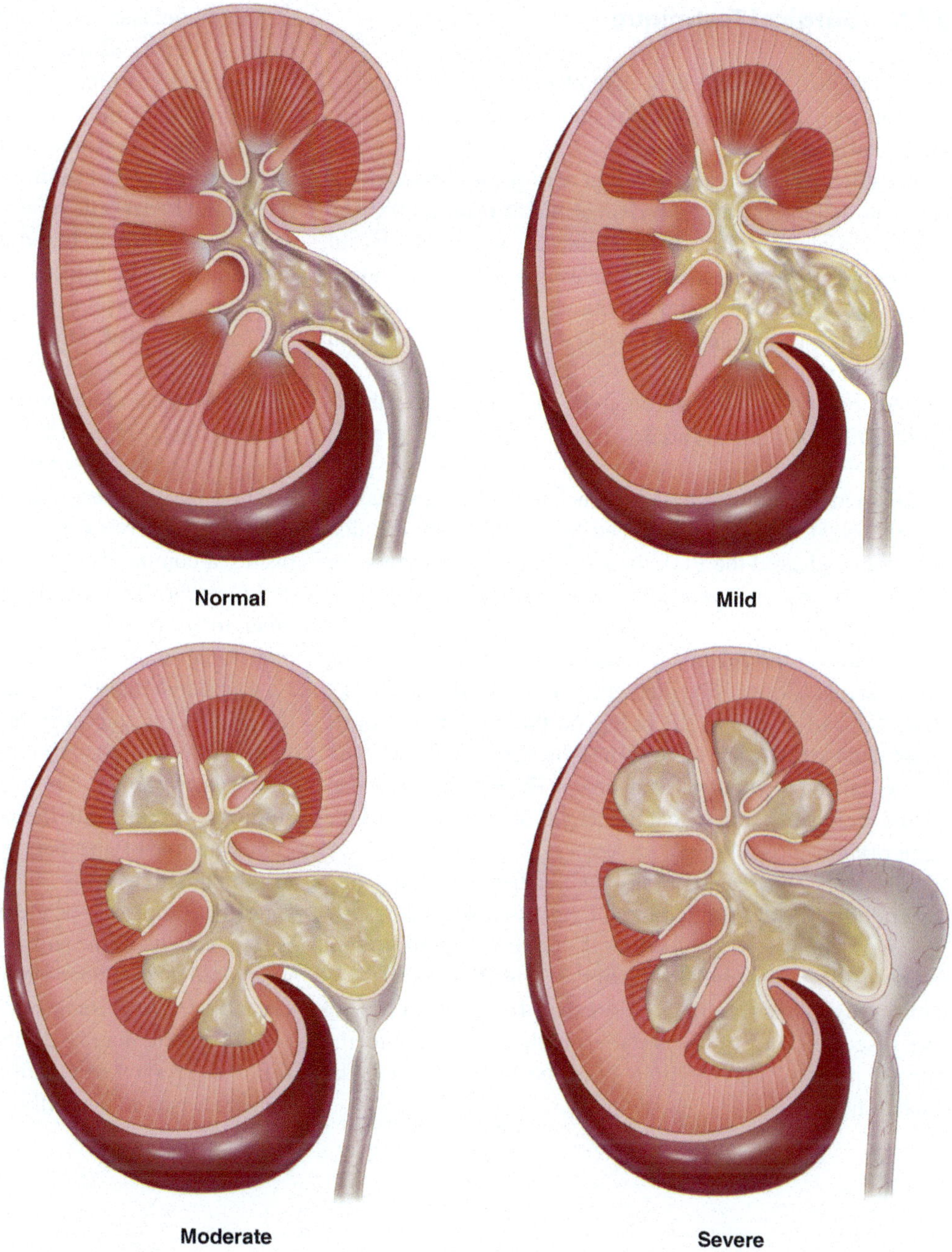

Fig. 37.8 Hydronephrosis is graded by ultrasound. The grading most commonly used by radiographers

Table 37.1 The Society for Fetal Urology (SFU) grading of hydronephrosis

Grade	Ultrasound appearance
0	No dilatation
1	Pelvis only dilated
2	Pelvis dilated and a few calyces visible
3	Pelvis and many calyces dilated
4	Pelvis and calyces dilated and parenchyma thinned

37.5 Surgical Technique

The anaesthetised patient is first placed in the supine or lithotomy position. A cystoscopy and retrograde pyelogram are performed.

This delineates the anatomy and may show the narrowed segment or kinking of the proximal ureter (Fig. 37.9). The pyelogram also confirms the hydronephrosis, with calyceal dilatation. When the ureteric catheter is withdrawn, waiting for 4–5 min will demonstrate minimal or delayed drainage, confirming the diagnosis. A double J stent is then inserted and left in situ.

The patient is re-positioned in the lateral decubitus ('kidney') position, secured (Fig. 37.2), and prepared as above.

Dissection commences by incising Gerota's fascia widely. Next, the ureter is traced to the PUJ. Having dissected and cleared the PUJ, the pathology (i.e., extrinsic or intrinsic obstruction) is usually identified by the completion of this stage.

The area of the pelvis and ureter to be resected is marked with a series of dots using diathermy (Figs. 37.10 and 37.11). An incision is made in the pelvis (at the point that will be most dependent when the patient is upright), and this incision is carried downwards at an angle. The length of this part of the incision should match the length of spatulation of the ureter (about 1.5–2.0 cm).

The incision is then angled further cranially and extended to resect the redundant pelvis (Fig. 37.12). The ureter is spatulated on its lateral surface. A narrow area of both the pelvis and the ureter, at the PUJ, are left in continuity. This area may be grasped for traction during the resection and to anchor the pelvis and ureter to the psoas muscle, to facilitate the performance of the anastomosis.

The pelvis is resected, using a pair of scissors, along the previously marked line. Care is taken not to cut the stent! The ureter is spatulated on its lateral aspect.

The first 5/0 monofilament polyglactin suture is placed between the pelvis and ureter and they are approximated, using intracorporeal suturing and a sliding knot (Fig. 37.13). If crossing lower pole blood vessels are present, the suture is passed anterior to the blood vessels. Other techniques achieve the same result by using a manufactured ready-made locking suture or extracorporeal knots and a knot-pusher.

Once the 'heel' stitch is in place, it is prudent to use another suture to secure this part of the anastomosis, prior to using two continuous sutures for the anterior and posterior anastomoses (Fig. 37.14).

A third suture (5/0 or 6/0 polyglactin) is placed anterior to the first two sutures, tied, and the needle brought into the inside of the pelvis (Fig. 37.15). With the needle now on the inside, a continuous suture is used to complete the anterior anastomosis. Care is taken to place sutures close to each other to prevent any gaps. Once the anastomosis is completed, the suture end is left long.

Another suture (5/0 or 6/0) is similarly used to complete the posterior anastomosis and the ends of the two sutures are tied (Figs. 37.16 and 37.17). This completes the uretero-pelvic anastomosis. The pelvis is now closed using a new suture to approximate the edges of the resected pelvis. The configuration of the pelvis leading into the ureter should resemble a funnel (Fig. 37.17).

After completion of the anastomosis, the retroperitoneum is lavaged with saline and all ports are removed. The port sites are closed using a deep suture (Vicryl 4/0) to muscle and fascia and 6/0 Vicryl Rapide or skin glue for the skin incisions.

The patient may be discharged the following day, if comfortable. Prophylactic antibiotics are given until the stent is removed, which is usually in 6 weeks.

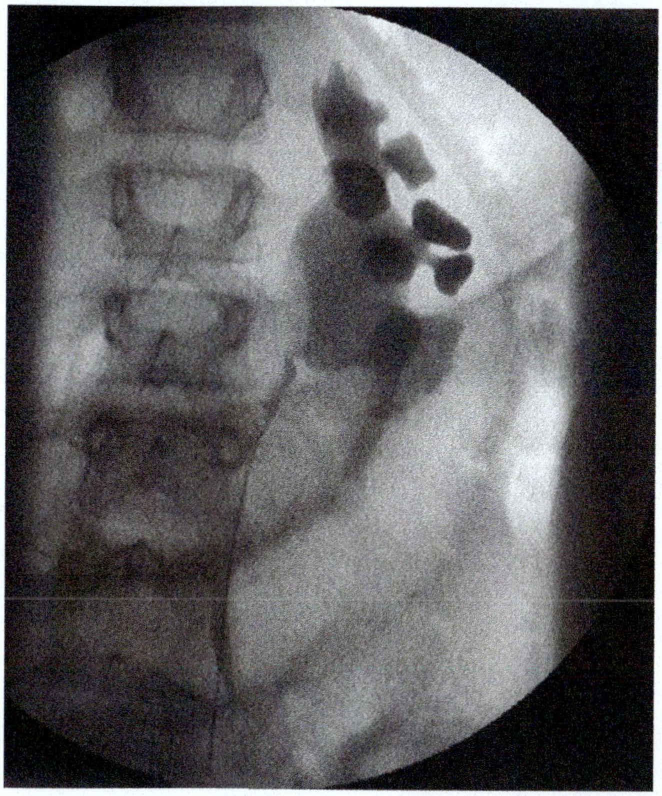

Fig. 37.9 Retrograde pyelogram: narrowed segment or kinking of the proximal ureter. The pyelogram also confirms the hydronephrosis, with calyceal dilatation

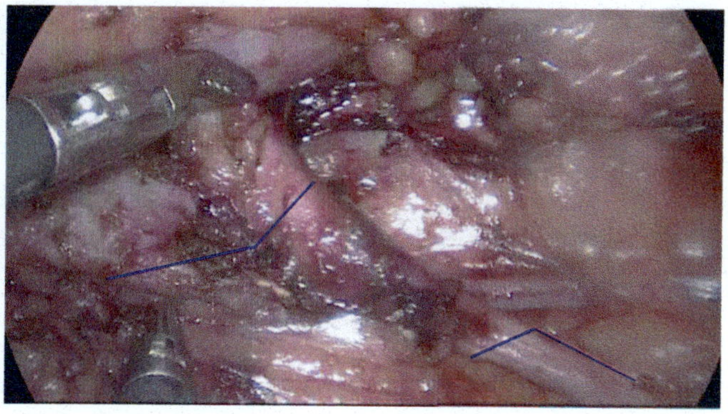

Fig. 37.10 The area of the pelvis and ureter to be resected is marked with a series of dots using diathermy

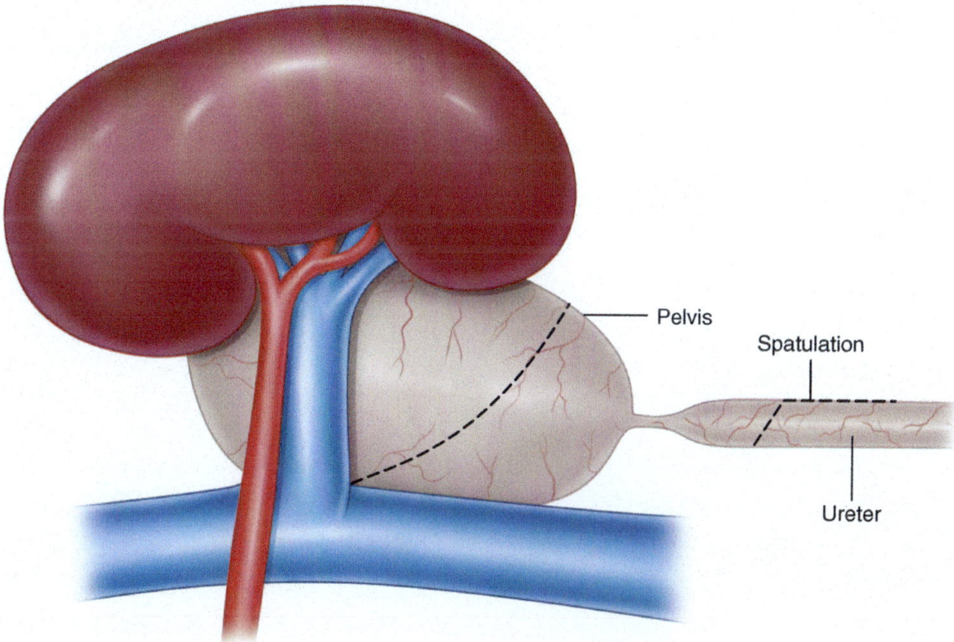

Fig. 37.11 An incision is made in the pelvis (at the point that will be most dependent when the patient is upright), and this incision is carried downwards at an angle. The length of this part of the incision should match the length of spatulation of the ureter (about 1.5–2.0 cm)

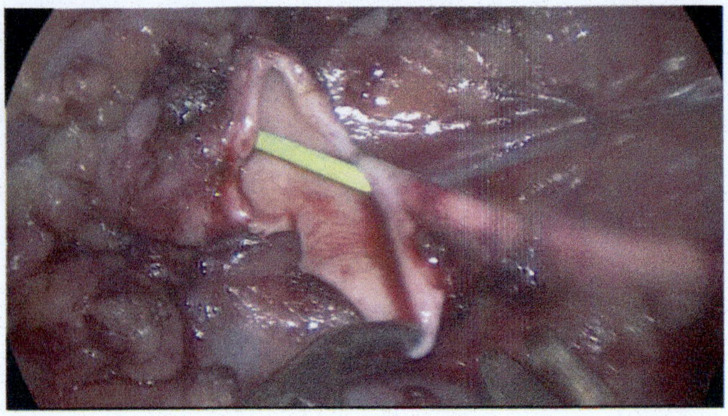

Fig. 37.12 The incision is then angled further cranially and extended to resect the redundant pelvis

Fig. 37.13 The first 5/0 monofilament polyglactin suture is placed between the pelvis and ureter and they are approximated

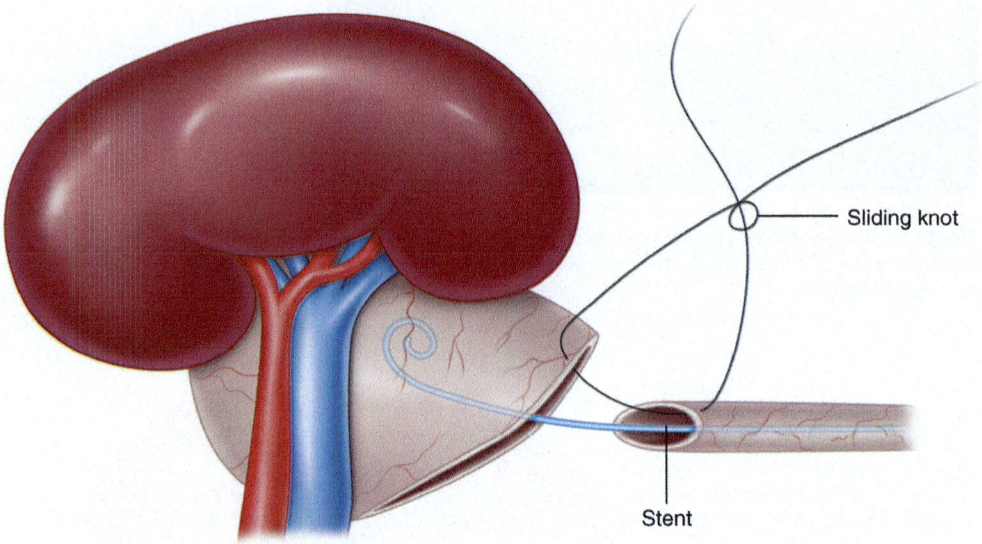

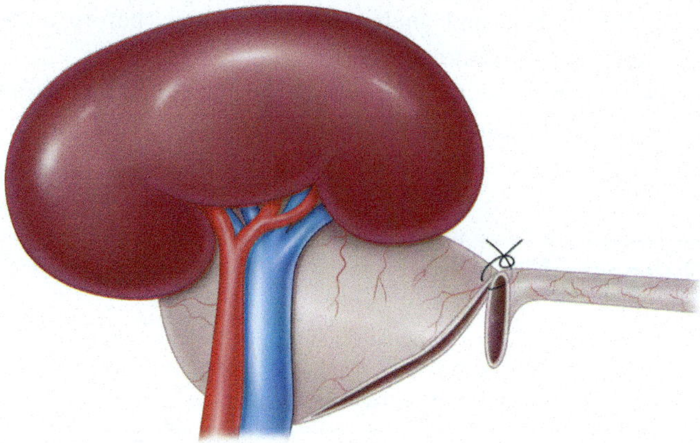

Fig. 37.14 Once the 'heel' stitch is in place, it is prudent to use another suture to secure this part of the anastomosis, prior to using two continuous sutures for the anterior and posterior anastomoses

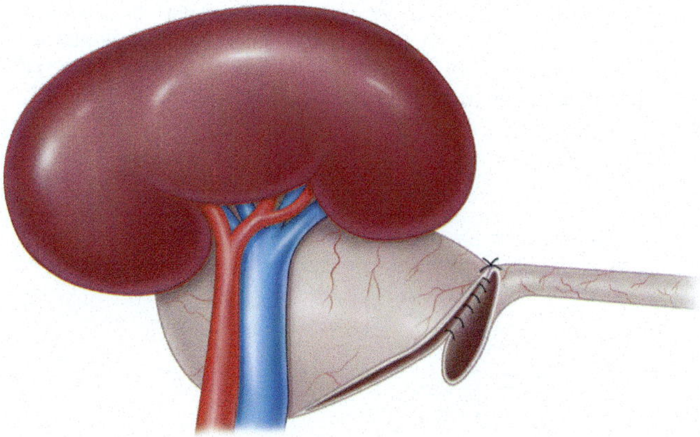

Fig. 37.15 A third suture (5/0 or 6/0 polyglactin) is placed anterior to the first two sutures, tied, and the needle brought into the inside of the pelvis to faciliate a continuous suture

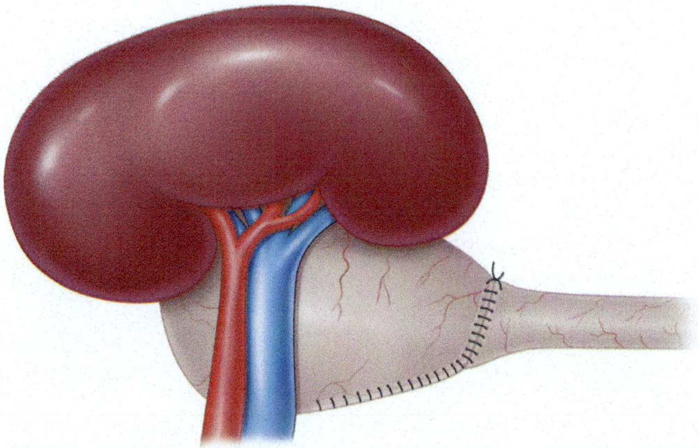

Fig. 37.16 Another suture (5/0 or 6/0) is similarly used to complete the posterior anastomosis and the ends of the two sutures are tied. The pelvis is now closed using a new suture to approximate the edges of the resected pelvis

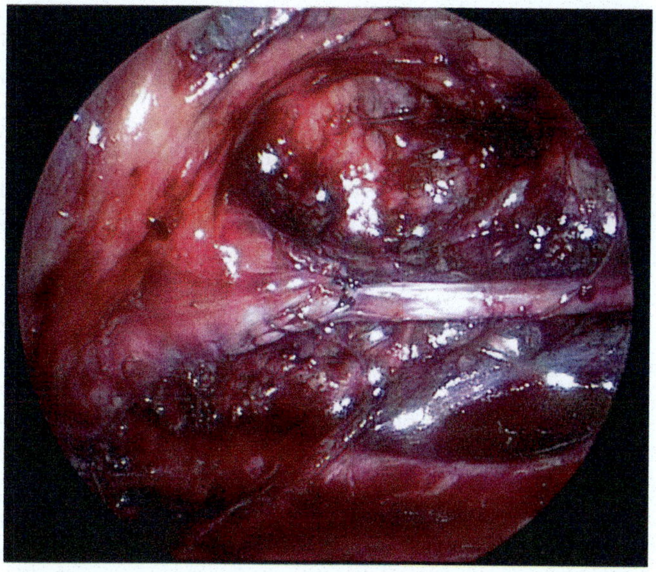

Fig. 37.17 The configuration of the pelvis leading into the ureter should resemble a funnel (when upright)

37.6 Alternatives

Transperitoneal laparoscopic pyeloplasty can be performed in a similar manner; the patient is placed semi-supine, with the affected side raised. Three 5-mm ports are inserted, the colon is mobilised, or a "window" is made in the meso-colon to access the kidney and PUJ. Resection of the PUJ and anastomosis are performed as described above.

37.7 Highlights and Pitfalls

- It is important to place the two operating ports horizontally in line with each other to facilitate ergonomic suturing.
- Do not dismember completely, leave a small portion of pelvis attached; placing a stay suture from the PUJ to the psoas muscle, helps to display the pelvis and ureter for suturing. After the anastomosis is almost complete, this area can be divided and the specimen extracted.
- When there are lower polar crossing vessels, the pelvis and ureter are dismembered, the pelvis is brought anterior to the vessels and anchored to the psoas, before commencing the anastomosis.
- Do not mobilise the ureter more than required and avoid grasping it during the anastomosis.

Suggested Reading

Canon SJ, Jayanthi VR, Lowe GJ. Which is better—retroperitoneoscopic or laparoscopic dismembered pyeloplasty in children? J Urol. 2007;178:1791–5; discussion 1795.

Davenport K, Minervini A, Timoney AG, Keeley Jr FX. Our experience with retroperitoneal and transperitoneal laparoscopic pyeloplasty for pelvi-ureteric junction obstruction. Eur Urol. 2005;48:973–7.

Diamond DA, Peters CA. Perinatal urology. In: Barratt TM, Avner ED, Harmon WE, editors. Pediatric nephrology. 4th ed. Baltimore: Lippincott Williams & Wilkins; 1999. p. 897–912.

Inagaki T, Rha KH, Ong AM, Kavoussi LR, Jarrett TW. Laparoscopic pyeloplasty: current status. BJU Int. 2005;95:102–5.

Singh V, Sinha RJ, Gupta DK, Kumar V, Pandey M, Akhtar A. Prospective randomized comparison between transperitoneal laparoscopic pyeloplasty and retroperitoneoscopic pyeloplasty for primary ureteropelvic junction obstruction. JSLS. 2014;18(3): e2014.00366. doi:10.4293/JSLS.2014.00366.

Index

A

Aortopexy, minimally invasive. *See* Minimally invasive aortopexy
Appendicectomy
- appendix
  - anatomical position, 150
  - dividing mesoappendix, 151, 153
  - inspection, 150
  - loop position, 154, 156–157
  - mobilising, 150–152
  - removal, 154, 155, 158
  - resection, 154, 157
- Endoloop® ligation technique, 151, 154
- GIA™ stapler, 151
- instrument port placement, 150, 151
- intravenous antibiotics, 155
- patient positioning, 148
- port positions, 148
- postoperative phase, 155
- preoperative setup, 148
- single-incision laparoscopic surgery, 159
- single-port/multiport, 147
- umbilical port insertion, 150
- working instruments, 149

Autoimmune myasthenia gravis, thymectomy for, 49. *See also* Thymectomy

B

Bilateral synchronous sympathectomy, 47
Bilateral thoracoscopy, 64
Bronchogenic cysts, 55–57
Button gastrostomy, primary
- advantages, laparoscopic approach, 103
- atraumatic grasper, 104
- guide wire placement, 104, 105, 107
- Hassan cut down technique, 104
- infraumbilical port insertion, 104
- MIC-KEY button, 104, 106
- patient positioning, 104
- site for, 104
- stay sutures, 104–107
- upper gastrointestinal endoscopy, 107
- working instruments, 103

C

Calot's triangle, 168–170
Carbon dioxide insufflation, 3
Cholecystectomy, laparoscopic. *See* Laparoscopic cholecystectomy (LC)
Choledochal cyst, laparoscopic surgery for
- cystectomy, 185, 186
- ductoplasty, 187
- hepatic ducts, transection of, 187
- hepaticoduodenostomy, 187
- hepaticojejunostomy, 185
- intraoperative complications, 187
- pancreatic duct injury, 187
- patient positioning, 184
- portal vein injury, 187
- single-port laparoscopic cystectomy, 187
- trocar placement, 184
- working instruments, 183

Cholelithiasis, 165. *See also* Laparoscopic cholecystectomy (LC)
Chronic immune thrombocytopenia, 173
Congenital pulmonary airway malformations (CPAMs), lobectomy for, 81
- bronchial and pulmonary artery anatomy, 82, 83
- clipping systemic artery, 88
- extralobar sequestration, 86–88
- LigaSure coagulation apical branch, 83, 85
- LigaSure division, of lung parenchyma, 83, 84
- linear stapler, 88
- lower lobe pulmonary artery branches, 83–85
- oblique fissure, 83, 84
- patient positioning, 82
- port siting, 82
- sealing device, 88
- single lung ventilation, 82
- underwater test, 86, 87
- working instruments, 82

Cystectomy, for choledochal cyst, 185, 186
Cystohepatic triangle. *See* Calot's triangle

D

Diaphragmatic hernia, 173
Diaphragm eventration, thoracoscopic placation for, 94
- advantage, 93
- head-up port position, 96

M. McHoney et al. (eds.), *Color Atlas of Pediatric Anatomy, Laparoscopy, and Thoracoscopy*,
DOI 10.1007/978-3-662-53085-6

Diaphragm eventration, thoracoscopic placation for (*cont.*)
  indications, 93
  laparoscopic repair, 96
  patient positioning, 93
  plication, 95
  port siting, 93
  right-sided eventration, 94
  suturing technique, 96
  working instruments, 93
Ductoplasty, 187
Duhamel procedure, 225
Duodenal atresia
  repair of (*see* Duodenoduodenostomy)
  types of, 134, 135
Duodenoduodenostomy, 141
  anastomotic leak rate, 133
  continuous sutures, 137, 140
  falciform ligament, 134, 137
  Hasson port, 136
  Kimura type anastomosis, 137, 139
  pancreas, 134
  patient positioning, 136
  proximal duodenum enterotomy, 137, 138
  right lobe of liver, 134
  suture placement, 137
  transverse colon, 134
  working instruments, 134

E

Empyema, thoracoscopic debridement of
  bleeding, 30
  debrided visceral pleura, 28–30
  ergonomic considerations, 28
  long cotton tip applicators, 30
  patient positioning, 28
  port placement, 28
  postoperative radiographic appearances, 30
  using forceps and suction, 28, 29
  working instruments, 27
Endoscopic bullectomy, 36
Endoscopic retrograde cholangio-pancreaticography
  (ERCP), 166
Endoscopic retroperitoneal pyeloplasty
  cystoscopy and retrograde pyelogram, 286, 287
  description, 279, 280
  diathermy, 286, 287
  heel stitch, 286, 289
  hydronephrosis, 284, 285
  incisions, 286–288
  inflation device, 281, 283
  kidney, cross sectional anatomy of, 284
  5/0 monofilament polyglactin suture, 286, 288
  patient positioning, 281
  port siting, 281, 283
  sutures, 286, 289
  working instruments, 281
Endoscopic stapling, 36
Esophageal atresia, thoracoscopic repair of
  anterior wall anastomosis., 68, 72
  azygos vein
    dissection, 68
    division, 68, 73
    identification, 68
  ergonomic considerations, 66
  lower pouch
    dissection, 68, 69
    identification, 68
    transanastonomic tube, 68, 71
    transfixion, 68, 69
  patient positioning, 66
  port siting, 66
  posterior wall anastomosis, 68, 71
  right posterior mediastinum, anatomy of, 67
  upper pouch
    alternative mobilization, 68, 70
    diathermy mobilization, 68, 70
    dissection, 73
    identification, 68, 69
    ligation, 68, 69
    opening, 68, 70
  working instruments, 65

F

First-stage Fowler-Stephens procedure, 245, 246
Fowler-Stephens procedure (FSP)
  first-stage, 245, 246
  second-stage, 247–251

G

Gerota's fascia, 262, 264

H

Heminephrectomy. *See* Retroperitoneoscopic nephrectomy
Hepaticoduodenostomy, for choledochal cyst, 187
Hepaticojejunostomy, for choledochal cyst, 185
Hepatobiliary triangle. *See* Calot's triangle
Hereditary spherocytosis, laparoscopic splenectomy for, 173
Hirschsprung disease, surgical correction of
  laparoscopic-assisted Soave pullthrough procedure
    (*see* Soave pullthrough procedure)
  laparoscopic Duhamels pull-through for
    (*see* Laparoscopic Duhamels pull-through method)
  Swenson-like transanal pullthrough, 225–232
Horner syndrome, 47
Hydronephrosis, 284, 285
Hyperhidrosis, 45

I

Impalpable testicle (IPT), 241
  laparoscopic assessment of, 243, 244
  port placement for, 242
Inguinal herniotomy
  avoiding vas and vessels, 238
  closed right inguinal canal, 237
  closure of ring, 239
  ergonomic considerations, 236
  herniated contents, reduction of, 238
  in infants, 236
  intracorporeal knot, 239
  nonabsorbable braided suture, 240
  open left inguinal canal, 237
  patent processus vaginalis, 237
  port placement, 236
  reduced ovary, 237
  sutures, 238–239
  working instruments, 235
Intra-abdominal testis, 242–248, 252

Intussusception
incidence, 161
laparoscopy for
ergonomic considerations, 162
intussusceptum, 164
patient positioning, 162
port siting, 162, 163
surgical technique, 164
working instruments, 161
IPT. *See* Impalpable testicle (IPT)

L
Laparoscopic adrenalectomy, 259
Laparoscopic cholecystectomy (LC)
American (Reddick Olsen) operation setup, 166
Calot's triangle, 168–170
cholangiography catheter, 170, 171
expert hepato-biliary centre, 172
French (Dubois) operation setup, 166
gallbladder retrieval, 170, 171
Gelpoint® platform, 171
iatrogenic injury, 171
intraoperative cholangiography, 170
Liga clips, 171
Mascagni's/Lund's node, 168, 169
Mirizzi syndrome, 172
operating room set-up, 166, 167
patient positioning, 166
patient selection and pre-operative evaluation, 171
peritoneal dissection, 170
port position, 166, 168
post-cholecystectomy syndrome, 172
post-operative bile leak, 171
reduced port and 3-mm cholecystectomy, 171
safety, 168
single-port laparoscopic cholecystectomy, 171
vascular and ductal anatomy, variations in, 168
working instruments, 166
Laparoscopic colectomy, 189, 199
descending colon, 193
distal transverse colon and splenic flexure, 194
hepatic flexure and proximal transverse colon, 196
ileostomy formation, 198
insufflation pressures, 191
patient positioning, 189–190
port sites, 190
rectum removal, 197
sigmoid colon, 192, 193
splenic flexure, 193, 194
surgeon's position, 191
terminal ileum, appendix, and ascending colon, 195
working instruments, 189
Laparoscopic Duhamels pull-through method, 201, 209
aganglionic bowel and coloanal anastomosis, 207
biopsy site, 205
ergonomic considerations, 202
patient positioning, 202
pelvis, 203
port siting, 202
posterior incision, in anal canal, 207
pull-though and anastomosis, 208
rectosigmoid colon, 205–206
rectum and colon, stapling of, 207
retrorectal dissection, 205
seromuscular biopsies, 204
working instruments, 202
Laparoscopic extraperitoneal repair, 240
Laparoscopic (transperitoneal) nephrectomy, 265, 270
bowel loops distention, 270
ergonomic considerations, 266
Ligaclips, 269
LigaSure, 268, 269
monopolar diathermy, 268
port siting, 266
renal vessels, 269
transmesocolic approach, 269
working instruments, 265
Laparoscopic repair, of Morgagni hernia. *See* Morgagni (anterior) diaphragmatic hernia
Laparoscopic Thal fundoplication. *See* Thal fundoplication
Lobectomy, for congenital pulmonary airway malformations, 81–88

M
Magnetic resonance cholangiopancreatography (MRCP), 165
Meckel diverticulum, laparoscopy for
Alexis wound retractor, 145
Endo GIA stapler, 145
Endoloop suture, 145
end-to-end small bowel anastomosis, 144, 145
ergonomic considerations, 144
intracorporeal anastomosis, 145
patient positioning, 144
port siting, 144
working instruments, 143
Mediastinal cysts
aspiration, 57
differential diagnosis, 55
ergonomic considerations, 55
experienced surgical team, 58
haemostasis, 57, 58
left-sided lesions, 56
mediastinial pleura, 57
monopolar hook diathermy, 58
oesophageal duplication cysts, 57
patient positioning, 55, 58
port siting, 55
right-sided lesions, 56
working instruments, 55
Minimal access surgery (MAS)
considerations in infants, 21–22
disposable instruments, 15
electrosurgical devices, 17
ergonomics
adjustable monitor positions, 25
cognitive, 18
definition, 18
first order paradoxical movement, 21, 24
integrated theaters, 25
intracorporeal to extracorporeal shaft ratio, 25
mechanical constraints, 19, 20
off axis endoscopic viewing, 21, 22
optical axis to instrument plane angle, 21, 23
optical axis to target view angle, 21, 22
organizational, 18
physical, 18
SCOPe position, 25
second order paradoxical movement, 21, 24
table height, 25
visual constraints, 21

Minimal access surgery (MAS) (*cont.*)
Hassan port system, 15, 16
internal valve mechanisms, 15, 16
interrelated performance enhancing elements, 10
knot pushers, 17
monopolar/bipolar diathermy, 17
Nathanson retractor, 17
operative field visualization, 12
charge coupled device, 14
high definition camera systems, 14
light sources, 13
white balancing, 14
operative workspace creation, 11
pretied surgical loops, 17
pyloromyotomy, 17
reusable instruments, 15
specimen retrieval bags, 17
stapling devices, 17
suction-irrigation devices, 17
titanium clips, 17
trocar tip types, 15
ultrasonic scalpels, 17
Veres needles, 17
vessel-sealing technology, 17
working instruments, 11
Minimally invasive aortopexy
ergonomic considerations, 76
patient positioning, 76
pericardium identification, 77
port siting, 76
Prolene sutures, 78–80
simultaneous bronchoscopy, 80
superior mediastinum, 77
thoracoscopic port sites, 78, 79
working instruments, 75
Minimally invasive pectus excavatum repair. *See* Nuss procedure, for pectus excavatum
Minimally invasive surgery (MIS)
active bleeding, 4
benefits, 1–2
carbon dioxide absorption from surgical cavity, 2–3
carbon dioxide insufflation, 3
clinical status, 4
continued professional development, 5
contraindications, 4
cosmetic advantages, 2
energy metabolism, 2
hazards, 2–3
indications, 4
intracorporeal suturing, 3
laparoscopic training sets, 7
magnification, 2
mentorship, 5
metabolic response, blunting of, 2
and open surgery, 7
operating rooms, 6
pediatric anaesthetists, 6
port insertions, 7
postoperative pain and recovery, 1–2
primary and secondary ports, 7
research, 6
specific operations, 4
thermoregulation, 2
training and competence, 5
visualisation, 2
Mirizzi syndrome, 172
MIS. *See* Minimally invasive surgery (MIS)
Morgagni (anterior) diaphragmatic hernia
ergonomic considerations, 97
Gore-Tex sheet, 99, 101
laparoscopic needle holder, 99
monopolar hook diathermy/scissor, 99
2/0 nonabsorbable suture, 99
patient positioning, 97
port siting, 97
stab incision, 99
sutures, 99–100
transverse colon, 98
working instruments, 97

N

Nephrectomy
laparoscopic (transperitoneal), 265–270
retroperitoneoscopic nephrectomy, 271–277
Neuroblastomas. *See* Thoracic neuroblastomas
Nissen fundoplication
advantages, 109
closed oesophageal hiatus, 115
completed posterior window and crural exposure, 113
ergonomic considerations, 109
final fundal suture, 116
Hook diathermy dissection, 111
lower oesophagus and GOJ, 112
mobilisation
of fundus, 114
of oesophagus, 112, 117
Nathenson's retractor, 117
non absorbable suture, 114
oesophageal hiatus, anatomy of, 110, 111
patient positioning, 109
port siting, 109
ski needle, 114
suturing of wrap, 116
working instruments, 109
Nuss procedure, for pectus excavatum
bilateral thoracoscopy, 64
fibrous pericardium, 61
Nuss bars, 62–63
patient positioning, 60
pectus deformity, 61
phrenic nerve and pericardiophrenic artery, 61
port siting, 60
right hemidiaphragm, 61
wire tethering sutures, 64
working instruments, 59

O

Open Hasson technique, 245, 247, 254

P

Palmar hyperhidrosis, 45
Palomo technique, 254–256
Pelvi-ureteric junction (PUJ), anatomical/functional obstruction. *See* Pyeloplasty
Pleurectomy, 36, 37
Post-cholecystectomy syndrome, 172
Posterior prone approach, 277
Postpneumonic empyema, 27. *See also* Empyema, thoracoscopic debridement of

Primary laparoscopic Duhamel pull-through method, 201
Pyeloplasty, 279, 280. *See also* Endoscopic retroperitoneal pyeloplasty
Pyloric stenosis, treatment for. *See* Pyloromyotomy, laparoscopic approach for
Pyloromyotomy, laparoscopic approach for
  advantages, 125
  blade position, 128
  bowel-holding (Johan) forceps, 130
  cosmetic outcome, 130
  ergonomic considerations, 126
  grasper insertion, 128
  hook diathermy, 130
  hypertrophied pyloric muscle, 128
  leaks, 129
  myotomy incision, 130
  patient positioning, 126
  port siting, 126
  pyloric tumour, anatomy of, 127
  stab incision, 128
  supraumbilical incision, 126
  working instruments, 125
  wound closure, 129

R
Renal hilum, 268
Renal surgery, retroperitoneoscopic approach to, 271. *See also* Retroperitoneoscopic nephrectomy
Rest energy expenditure (REE), 2
Retrograde pyelogram, 286, 287
Retroperitoneoscopic adrenalectomy, 259, 264
  Endopouch® device, 262
  ergonomic considerations, 260, 261
  Gerota's fascia, 262, 264
  LigaSure™ instrument, 262
  patient positioning, 260, 261
  port siting, 260, 261
  working instruments, 260
Retroperitoneoscopic nephrectomy, 271, 277
  ergonomic considerations, 272
  heminephrectomy, 275
  nephrectomy, 274
  patient positioning, 272
  peritoneal cavity, 271, 277
  port siting, 272
  posterior prone approach, 277
  retroperitoneal space, 273, 274
  traditional laparoscopy, 277
  working instruments, 271
Retroperitoneoscopic pyeloplasty, 272

S
Scrotal varicoceles, 253
SEAL technique. *See* Subcutaneous endoscopically assisted ligation (SEAL) technique
Second-stage Fowler-Stephens procedure, 247–251
Severe tracheomalacia, treatment option for, 75. *See also* Minimally invasive aortopexy
Single-incision laparoscopic surgery (SILS), 159
Single-port laparoscopic cholecystectomy, 171
Single-port laparoscopic cystectomy, 187
Soave pullthrough procedure
  absorbable sutures, 216, 222, 224
  anastomosis, 216, 222–223
  breakthrough and prolapse, 216, 221, 222
  colonic mobilisation, 216, 218–219
  mini lap dissectors, 212
  mucosal tube development, 216, 221
  needle-tipped diathermy, 216, 220
  on-table washout, 224
  patient positioning, 212
  patient preparation, 212, 213
  patient repositioning, 216, 219
  patient's skin protection, 224
  point diathermy, 216, 220
  port positions, 212, 214
  proximal dilated bowel, 215
  rolled circular muscle cuff, 216, 221
  serosubmucosal biopsies, 216, 217
  submucosal dissection, 216, 219
  surgeon and assistants position, 212, 214
  working instruments, 211
Society for Fetal Urology (SFU) grading system, 284, 285
Splenectomy, laparoscopic
  chronic immune thrombocytopenia, 173
  diaphragmatic hernia, 173
  ergonomic considerations, 174
  indications, 173
  partial splenectomy, 181
  patient positioning, 174, 175
  pfannensteil incision, 181
  port siting, 174
  preoperative preparation, 173–174
  preoperative splenic artery embolization, 181
  preoperative ultrasound, 173
  risks, 173
  sickle cell disease, 173
  single incision splenectomy, 181
  specimen retrieval and morcellation, 177, 180
  spleen
    blood supply, 176
    detachment, 177, 179
    lower pole mobilization, 177, 178
    sealing of artery, 177
    splenic hilum exposure, 177
    splenic vein division, 177, 178
  suprapubic port, 181
  thalassemia, 173
  working instruments, 174
Spontaneous pneumothorax
  apical blebs, Endoloop ligation of, 34
  apical subpleural blebs, 31
  dissection of pleura, 34, 35
  Endoloop placement, 34
  endoscopic bullectomy, 36
  endoscopic stapling, 36
  ergonomic considerations, 33
  incision onto rib bed, 34
  left mediastinum, anatomy of, 32, 33
  lung reinflation, 34, 36
  medial chest wall dissection, 34, 35
  patient positioning, 33
  port siting, 33
  ruptured bleb, 32
  stripped right chest wall, anatomy of, 32, 33
  visualisation of apical blebs, 34
  working instruments, 32
Subcutaneous endoscopically assisted ligation (SEAL) technique, 240
Swenson-like transanal pullthrough procedure
  advantages, 225
  colonic mobilization, 227, 229

Swenson-like transanal pullthrough procedure (*cont.*)
ergonomic considerations, 226
full-thickness colonic biopsy, 227
hook monopolar diathermy, 227
Lone Star retractor, 227, 230
patient positioning, 226
port siting, 226
seromuscular colonic biopsy, 227, 228
transanal approach, 227, 230–231
two-layer coloanal anastomosis, 227, 232
working instruments, 226
Sympathectomy, 45–48

T
Thal fundoplication
ergonomic considerations, 120
liver retraction, 121, 124
minimum crus dissection, 121, 122
Nathanson retractor, 121
operative field exposure, 121
patient positioning, 120
phreno-oesophageal ligament, dissection of, 121, 122
port siting, 120
success rate, 119
suture placement, 122, 123
umbilical/supraumbilical Hasson entry, 121
working instruments, 120
Thoracic neuroblastomas
electrosurgical devices, 92
monopolar hook diathermy, 91
patient positioning, 89
port siting, 89
posterior thoracic wall, 90
working instruments, 89
Thoracic sympathectomy
anatomy, 46
ergonomic considerations, 45
indication, 45
patient positioning, 45
port siting, 45
spatulated diathermy blade, 48
sympathetic chain, 46, 47
working instruments, 45
Thoracoscopic lung biopsy and segmentectomy
CT scan, 41, 42
ergonomic considerations, 39
patient placement, 39
port positions, 39, 40
preoperative workup, 39
segmental lesion, 42, 43
using Endoloop, 42
using stapling device, 42, 43
working instruments, 39
Thymectomy
advantages, 54
antibiotics, 51
dissection, 51, 52
port positioning, 51
preoperative CT scan, with intravenous contrast, 50
short thymic veins, 51, 52
thymus removal through anterior port, 51, 53
working instruments, 50
Tracheoesophageal fistula (TOF), thoracoscopic repair of
division, 68, 70
ergonomic considerations, 66
identification, 67, 68
ligation, 68, 69
patient positioning, 66
port siting, 66
transfixion, 68, 69
working instruments, 65
Transperitoneal laparoscopic pyeloplasty, 290

U
Undescended testis, Fowler-Stephens orchidopexy for, 241–242, 252
atrophic intra-abdominal testicle/nubbin, 243, 244
ergonomic considerations, 242
first-stage Fowler-Stephens procedure, 245, 246
laparoscopic assessment, 243, 244
patient positioning, 242
port siting, 242
second-stage Fowler-Stephens procedure, 247–251
working instruments, 242
Upper gastrointestinal (UGI) endoscopy, 107

V
Varicocelectomy, laparoscopic, 253
laparoscopic Palomo technique, 254–256
patients evaluation, 253–254
postoperative follow-up, 257
treatment indications, 254
Varicoceles, 253. *See also* Palomo technique
Video-assisted thoracoscopic surgery (VATS), 22, 30

W
Windsock deformity, 134, 135

The manufacturer's authorised representative in the EU is Springer
Nature Customer Service Centre GmbH, Europaplatz 3, 69115 Heidelberg,
Germany. If you have any concerns regarding our products, please
contact ProductSafety@springernature.com

Printed and bound by CPI Group (UK) Ltd, Croydon, CR0 4YY
26/09/2026
02222871-0002